Studies in Computational Intelligence

Volume 1035

The series “Studies in Computational Intelligence” (SCI) publishes new developments and advances in the various areas of computational intelligence—quickly and with a high quality. The intent is to cover the theory, applications, and design methods of computational intelligence, as embedded in the fields of engineering, computer science, physics and life sciences, as well as the methodologies behind them. The series contains monographs, lecture notes and edited volumes in computational intelligence spanning the areas of neural networks, connectionist systems, genetic algorithms, evolutionary computation, artificial intelligence, cellular automata, self-organizing systems, soft computing, fuzzy systems, and hybrid intelligent systems. Of particular value to both the contributors and the readership are the short publication timeframe and the world-wide distribution, which enable both wide and rapid dissemination of research output.

Indexed by SCOPUS, DBLP, WTI Frankfurt eG, zbMATH, SCImago.

All books published in the series are submitted for consideration in Web of Science.

More information about this series at https://link.springer.com/bookseries/7092

Pedro Y. Piñero Pérez · Rafael E. Bello Pérez ·
Janusz Kacprzyk
Editors

Artificial Intelligence in Project Management and Making Decisions

Springer

Editors
Pedro Y. Piñero Pérez
Departamento de Investigaciones en Gestión de Proyectos
Universidad de las Ciencias Informáticas
La Habana, Cuba

Rafael E. Bello Pérez
Centro de Investigaciones de la Informatica
Cuban Sciences Academy
Universidad Central "Marta Abreu" de las Villas
Santa Clara, Cuba

Janusz Kacprzyk
Systems Research Institute
Polish Academy of Sciences
Warsaw, Poland

ISSN 1860-949X ISSN 1860-9503 (electronic)
Studies in Computational Intelligence
ISBN 978-3-030-97268-4 ISBN 978-3-030-97269-1 (eBook)
https://doi.org/10.1007/978-3-030-97269-1

This Springer imprint is published by the registered company Springer Nature Switzerland AG
The registered company address is: Gewerbestrasse 11, 6330 Cham, Switzerland

Preface

Project management, like other branches of knowledge, has changed since the development of computer science and informatics. Project-oriented organizations are advancing digital transformation and adopting enterprise architectures. In this way, these organizations improve processes and information systems simultaneously.

For example, in the construction sector, BIM technologies allow simulation in virtual environments of the complete execution of projects. At the current moment of development, BIM technologies have fundamentally advanced in the integration and parameterization of the designs or plans of the works with the planning activities. But, the efforts to integrate these technologies with artificial intelligence and other intelligent data analysis techniques are still incipient. Various sources identify that around 34% of the resources invested in a project are wasted. In addition, they point out that approximately 95% of the information that is generated later is underused and is not used as lessons learned in new projects. The same happens in the development of software projects. In this area of knowledge, several sources identify that, in medium- and large-sized projects, approximately 26% need to be renegotiated and 31% are cancelled, one of the main factors of failure being difficulties in control and decision-making. But, the efforts to integrate these technologies with artificial intelligence and other intelligent data analysis techniques are still incipient.

In these different application scenarios, project management is presented as an interdisciplinary area where elements of leadership, management techniques, economic management, logistics, mathematical and computer science converge with applications in numerous sceneries. In the different scenarios, the introduction of information systems has allowed the collection of large volumes of useful data to aid decision-making involving the different areas of knowledge of project management. But, the effective introduction of artificial intelligence techniques and other data analysis techniques to aid decision-making is still insufficient.

This book presents a selection of papers presented at the IX International Workshop on Project Management and Decision-Making. The papers collect different experiences in the application of artificial intelligence techniques to help decision-making in project management. In this context, we describe artificial intelligence as the set of software and hardware techniques that simulate different human reasoning methods and the collective intelligence of natural ecosystems to solve complex problems where the algorithm for its solution has non-polynomial complexity, or there are no algorithms for its solution.

The book is structured in four parts. Part I groups together a set of works that apply linguistic data summarization techniques to aid decision-making in project management. In Part II, different artificial intelligence techniques are combined in planning and sustainable management of project-oriented organizations. Part III presents experiences in knowledge management and human resource management assisted by artificial intelligence. Finally, Part IV presents different trends in BIM technologies and other emerging technologies in project management.

La Habana, Cuba — Pedro Y. Piñero Pérez
Santa Clara, Cuba — Rafael E. Bello Pérez
Warsaw, Poland — Janusz Kacprzyk

Acknowledgments

This book collects a group of experiences in the application of artificial intelligence techniques and other data analysis techniques in project management. Papers are presented that propose new algorithms and techniques to help decision-making in different project management processes, among which the following stands out: scope and requirement management, human resource management, risk management, time management, quality assurance, procurement management, monitoring and control.

We would like to thank all the engineers, professors and researchers without whose efforts this book could not have been written. We are indebted to Iliana Pérez Pupo for her comments that enable us to correct errors and advance satisfactorily in the development of this book.

And, last but not least, we wish to thank Dr. Tom Ditzinger and Mr. Holger Schaepe for their dedication and help to implement and finish this important publication project on time, while maintaining the highest publication standards. We also would like to thank Mr. Ramamoorthy Rajangam, from Springer, for their help and support.

Pedro Y. Piñero Pérez
Rafael E. Bello Pérez
Janusz Kacprzyk

Contents

Part I
Linguistic Data Summarization for Decision-Making in Project Management

Chapter 1
Linguistic Data Summarization: A Systematic Review

Iliana Pérez Pupo, Pedro Y. Piñero Pérez, Rafael E. Bello Pérez, Roberto García Vacacela, and Nicolas Villavicencio Bermúdez

Abstract This paper constitutes a systematic review of Linguistic Data Summarization theory and its trends. Authors analyzed the advantages and limitations of different techniques to generate linguistic summaries. In this investigation, authors discuss about the high variability of protoforms used by researches in several domains. This review exposes the strategy of validation of researches reported in bibliography. We identify the approaches in data summarization followed by the main authors. As conclusions, of this investigation, authors identify the following opportunities for improvement in this research area. One opportunity deal with the use of linguistic summaries not only for description, but also for prediction in decision-making problems. Other is about the necessity of new algorithms to facilitate the generation of linguistic summaries and its applications in several domains.

Keywords Artificial Intelligent · Linguistic data summarization · Linguistic summaries · Project management

---

I. Pérez Pupo (✉) · P. Y. Piñero Pérez
Grupo de Investigaciones en Gestión de Proyectos, Universidad de las Ciencias Informáticas, La Habana, 54830 Havana, CP, Cuba
e-mail: iperez@uci.cu

R. E. Bello Pérez · R. García Vacacela · N. Villavicencio Bermúdez
Centro de Investigación en Informática, Universidad Central Marta Abreu de las Villas, Santa Clara, Cuba
e-mail: rbellop@uclv.edu.cu

R. García Vacacela
e-mail: roberto.garcia@cu.ucsg.edu.ec

N. Villavicencio Bermúdez
e-mail: nicolas.villavicencio@cu.ucsg.edu.ec

I. Pérez Pupo
Facultad de Empresariales, Universidad Católica Santiago de Guayaquil, Guayaquil, Ecuador

P. Y. Piñero Pérez et al. (eds.), *Artificial Intelligence in Project Management and Making Decisions*, Studies in Computational Intelligence 1035,
https://doi.org/10.1007/978-3-030-97269-1_1

1 Introduction

One of the latent problems in decision-making processes is associated with the discovery of behavior patterns and their interpretability. This problem increases with the rapid growth of data generated by information systems. In general, data taken directly from information systems is not readable or understandable to the naked eye by decision-makers, which makes the decision-making process difficult. Thus, arises the so-called "rich data, poor information" dilemma present in decision-making [1]. In this sense, the existence of tools that allow the discovery of hidden non-trivial dependencies in the data is an object of interest for decision makers.

In general, knowledge discovery techniques from data have been classified into two classes: predictive techniques and descriptive techniques. The goal of predictive techniques is to predict the value of a specific attribute (known as a dependent variable) based on the values of other attributes. While the objective of descriptive techniques is to explore patterns (correlations, trends, cluster or outliers) that summarize the underlying relationships in the data [2].

In this context, linguistic data summarization (LDS) arises as one of the descriptive knowledge discovery techniques with an interesting and promising approach to produce summaries from a database using natural language [3]. Authors such as Yager and Zadeh were pioneers in the development of linguistic data summarization. His publications [4] and [5] set guidelines in the conceptualization and subsequent development of linguistic data summarization as a technique in the field of soft computing. Different authors have characterized this technique as shown below.

Kacprzyk in [6] states that linguistic summaries are in line with the new paradigms of word computing, being able to capture the essence of the data very well. Later, with Zadrożny he identified that the essence of this technique is that a set of data can be summarized linguistically with respect to a selected attribute or attributes, using linguistically quantified propositions [7].

Boran, Akay and Yager in [8] suggest that LDS is one of the powerful techniques for the discovery of descriptive knowledge from a large data set, capable of extracting potential, useful and summaries knowledge from both data numerical and categorical, whose objective is to discover the patterns that summarize the existing relationships between attributes in a large database with more abstract conceptual representations.

For her part, for Anna Wilbik, the objective of linguistic summaries is to provide a quick overview, since they facilitate the understanding of large amounts of data by describing the main properties of the data in a linguistic way [9]. While other authors identify linguistic summaries as an attempt to describe, through linguistic expressions, patterns that emerge in the data [10].

Few review papers associated with this topic are identified in the bibliography consulted. For example, in the work [8] the authors have concentrated on the methods of generation and evaluation of summaries and in [11] a collection of examples of summaries generated by different authors between from 2001 to 2016. While in [12] the authors concentrate on synthesizing the protoforms used, ranging from the article [4], to 2013 and in [13] the authors present a table listing the types of protoforms most used between 2006 and 2016. These reviews found focus only on specific aspects of the theory, they do not apply systematic or meta-analysis techniques and fail to fully characterize each of the works they reference.

Based on this problematic situation, the objective of this work is: to build a theoretical referential framework through a tertiary review that allows to analyze the evolution and trends in the development of the linguistic summarization of data and its applications. The work is organized as follows: the second Methodology section explains the experimental design used, the Development section shows the results of the analysis and finally the Conclusions are presented.

2 Methodology

The study that is presented constitutes an exploratory investigation [14] in it that develops the following design of experiments.

Step 1. Design of a systematic review protocol that allows building a theoretical framework associated with the topic.
Step 2. In-depth analysis of the bibliography consulted by each of the established categories.
Step 3. Application of descriptive statistics and meta-analysis techniques for the presentation of the research results.
Step 4. Identify open lines of investigation.

In line with the silver, a systematic review protocol is designed [15] and the historical evolution of linguistic summarization in different stages is analyzed: publications before 2000, publications between 2000 and 2015, and from 2015 to 2020.

Systematic review protocol applied:

1. Definition of the object of study, field of action and objective of the research.
    a. Research object: linguistic summarization of data.
    b. Course objective: Build a theoretical framework associated with the methods related to the linguistic summarization of data, trends and applications in project management.
    c. Research field: structures, methods and applications for generating linguistic summaries of data.
2. Definition of a bibliographic manager.
3. Define sources of academic information for the development of the review: "Semantic Scholar", "Google Scholar" and "Scopus" and other meta-search engines based on open science.
4. Definition of the following set of key phrases to carry out the searches: "linguistic data summarization", "linguistic data summaries", "linguistic summarization", "linguistic summaries".

5. Define the goals of the bibliometric analysis in the form of research questions and inclusion—exclusion criteria:
   i. How has been the trend of publications per year?
   ii. Who are the main authors?
   iii. What are the affiliations and countries of the main authors?
   iv. How are the publications distributed considering the types of documents in: articles, books, theses and proceedings of conferences or congresses?
   v. Exclusion of works published in spaces with a low level of arbitration.
   vi. Exclusion of text mining works not associated with the use of linguistic data summarization techniques.
6. Sort and filter posts in the following set of categories:
   - Classic: refers to pioneering publications in linguistic data summarization, where the fundamental principles that mark the theory are exposed.
   - Extensions to theories: refers to publications that extend the theory proposed in the publications understood as classical, do not set guidelines that significantly change the previously proposed methods, although they do develop contributions to knowledge.
   - Application results: publications that focus on the use of existing theory in concrete practical scenarios.
   - Tertiary reviews: refers to review articles on trends and developments in the subject in question.
7. Synthesize the main trends.

As a second step in the methodology used, a detailed analysis of each of the works is carried out and they are characterized with respect to the following elements:

- Structure of the summaries that are generated (protoforms).
- Methods or techniques for the generation of linguistic data summaries.
- Main validation techniques and methods used in the investigations.
- Areas of application of the proposal.

Then, descriptive statistics techniques and meta-analysis are applied for the presentation of the research results [15]. Finally, open lines are analyzed and proposed in the object of study analyzed.

3 Linguistic Data Summarization Review

From the study carried out, it is identified that European and North American authors were pioneers in the development of the theory in the 1980s, while at present there is a greater dispersion of the works with respect to geographical areas, see Fig. 1. A characterization of the bibliography consulted in this work is shown below:

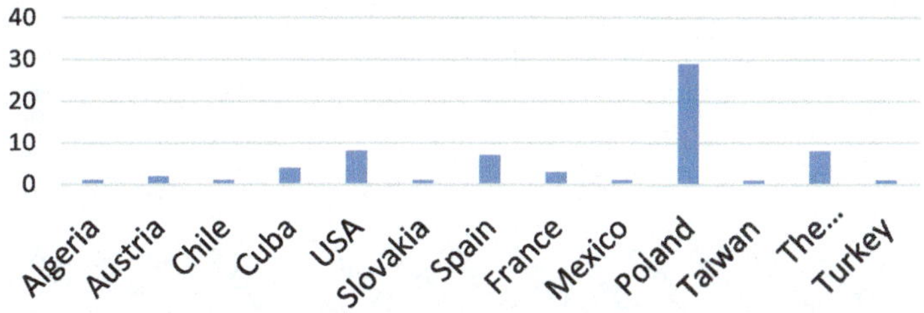

Fig. 1 Publications by country of the consulted bibliography

- Most of the researches consulted are articles in journals, it is identified that in this thematic area there are few published doctoral or other theses, see Fig. 2.
- Most of consulted publications are indexed in Scopus and Web Of Science, Fig. 3.

The main sources of information and the main publishers for the researches consulted were IEEE, Elsevier and Springer respectively, see Fig. 4.

The consulted bibliography identifies a set of elements that characterize or form part of the notation established in this thematic area [1, 4, 16].

- *D*: Database, for example, of workers, or projects.
- $Y = \{y_1, \ldots, y_n\}$, set of objects (records) in the database.
- $A = \{A_1, \ldots, A_m\}$, set of attributes (fuzzy variable) that characterize the objects Y, for example "salary" and "age".

Fig. 2 Bibliography by type of publication

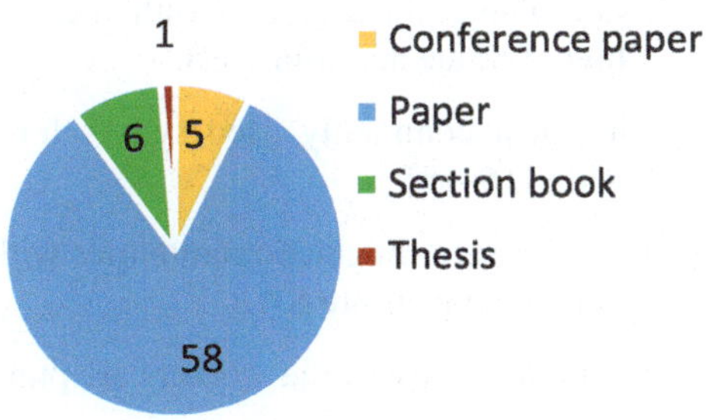

Fig. 3 Indexing of publications in magazines and conferences analyzed

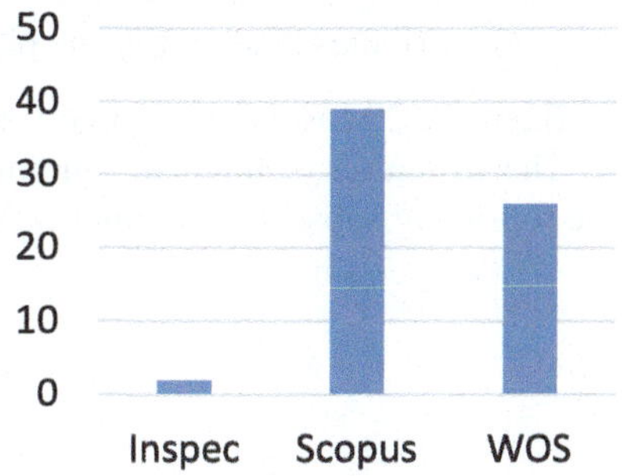

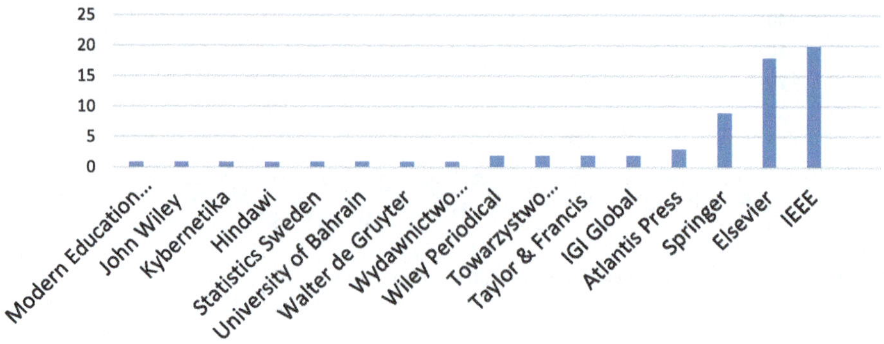

Fig. 4 Main sources and editorials of the bibliography analyzed in depth

- $A_j(y_i)$: Denotes the value of the A_j attribute for the y_i object, for example, "Low salary" for the "salary" attribute.
- Q: quantifier, it is a fuzzy set with the universe of discourse in the interval [0, 1] expressing a quantity, for example, "Most", "60%" or "More than half".
- A: Qualifier or filter, it is another attribute with a linguistic value (fuzzy predicate) that determines a fuzzy subset of the y_i object, for example "young" for the age attribute.
- S: Summarizer, is an attribute with a linguistic value (fuzzy predicate) defined in the domain of the A_j attribute, for example "Low salary" for the "salary" attribute.
- T: Degree of truth (validity) of the summary, it is a number in the interval [0, 1] that evaluates the degree of truth of the summary; generally, only summaries with a high T value are of interest.

The most commonly used syntax for the construction of summaries are the following [16–18]:

- *Qy's are S*: for linguistic summaries without filter, in which the statement has the form: "Q objects are S".

  Example: "Most of the workers are punctual".

- *QRy's are S*: for linguistic summaries with a filter, in which the declaration has the form: "*Q R* objects are *S*", where *R* is a filter that can be made up of one or more attributes that qualify the object.

  Example: "Most young workers are late".

The structures of linguistic summaries have been conceptualized in protoforms; the following subsection summarizes the trends in this regard.

3.1 *Evolution and Trends in Protoforms for the Construction of Summaries*

The protoforms that have been used the most in the bibliography for the construction of linguistic summaries follow the structure proposed by Zadeh in [19] and later extended by Kacprzyk and Zadrożny in [20, 21]. Table 1 synthesizes the main elements of these protoforms.

In these protoforms, hierarchies are established where the levels are passed from highest to lowest abstraction and are characterized by the following elements:

- The term $S^{structure}$ means that in the summary it is known which variables make up it; while S^{value} denotes that the value of the summarizer is unknown.
- Protoform 0: it responds to the "*QRy's are S*" structure. It is the one with the lowest level of abstraction, since it is assumed that all the elements that make up the summary are known. What you want to know is its degree of truth *T* [5, 20].
- Protoform 1: "*Qy's are S*" structure, the summarizer (*S*) is known and we want to discover the quantifier (*Q*). In the case of the application of protoforms 1 and 2, they can generate summaries from SQL statements and their extensions based on FQUERY [20].
- Protoform 2: "*QRy's are S*" structure, the summarizer (*S*) and the filter (*R*) are known, it is desired to discover the quantifier (*Q*). As in protoforms 0 and 1, these summaries can be obtained through queries to the database, although a little more effort is required to be compared to protoform 1.
- Protoform 3: "*Qy's are S*" structure, the quantifier (*Q*) and the summary structure are known, we want to discover the summarizer (*S*).
- Protoform 4: "*QRy's are S*" structure, the quantifier (*Q*) and the summary structure are known, we want to discover the summarizer (*S*) and the filter (*R*).
- Protoform 5: "*QRy's are S*" structure. It is the one with the highest level of abstraction, since no element is known and therefore it is about discovering everything. Summarization in this case can be time-consuming due to the size of the search space; but the summaries could be more interesting. These are the most complex summaries and are the main objective of this work.

Table 1 Classification of linguistic summarization

Type	Protoform	Known	Doubt	Comments
0	*QRy's are S*	All	T	Conditional summaries through ad-hoc queries
1	*Qy's are S*	S	Q	Simple summaries through ad-hoc queries
2	*QRy's are S*	S R	Q	Conditional summaries through ad-hoc queries
3	*Qy's are S*	Q $S^{structure}$	S^{value}	Simple value-oriented summaries
4	*QRy's are S*	Q $S^{structure}$, R	S^{value}	Summaries oriented to conditional values
5	*QRy's are S*	Nothing		General fuzzy rules

In the linguistic summaries, the most used types of protoforms respond to: classical protoforms, time series protoforms and temporal protoforms [12]. The classical protoforms summarize the attributes of the entire data set or the relationships between them [4, 20]. These summaries are "*Qy's are S*" and "*QRy's are S*" structures, as mentioned above. A Summary example with this type of protoform is: "Most houses have high gas consumption" and "Most old houses have high gas consumption", respectively. These protoforms have been used in dissimilar application scenarios, an example of this is presented in [22] with applications in glucose management in the Intensive Care Unit in the Netherlands, an example of a summary obtained is "All insulin boluses administered are higher than the protocol".

In the consulted bibliography there are investigations that particularize the protoforms for specific problems [12]. For example, in some investigations related to time series protoforms are introduced with the structure: "*Q Bs are A* Q_T *time*", where Q_T is a quantifier applied to the time attribute. Examples of summaries that follow these protoforms are: "most of the trends in topic *B* are low variability" and "about half of small businesses have short response times most of the time". There are some extensions associated with the sequence of events or activities, such as the case of [9] where we can find summaries such as "When the processing time was long, the case contained a sequence such as '*abcdefgh*'". Other examples by the same authors are presented in [18] in the analysis of logs associated to records of appeals processes of the municipality of Dutch, where summaries such as "Many cases have short operation time" are obtained. Other extensions to protoforms are presented in application scenarios where you want to incorporate epochs or temporal phenomena, where the consulted bibliography uses protoforms that do not use linguistic quantifiers. This type of summaries has the structure *P*, data are *A*, where *P* is a temporary term and *A* is a fuzzy linguistic term. For example, "regularly in autumn, with high rainfall".

Also, we can find hybrid models, as for example in [23] simple temporary protoforms such as "T_t between all segments, *Q* is/are *P*" are introduced, for the analysis of logs in web servers. In this case summaries are generated such as "Recently, among all segments, most are slowly increasing"; and extended temporal protoforms "T_t among all *R* segments, *Q* is/are *P*", generating summaries such as "Initially, among all short segments, most are slowly increasing" with T_t being a temporal term.

There are other variants of protoforms related to climate data [13], where the authors establish protoforms with the structure: "*Q Xs are TEMP*" and "*Q Ts are TEMP*". In this research, the authors establish that *Q* is the quantifier, *Xs* is the referential (maximum daily temperature, in a given month of a given location), *TEMP* is one of the linguistic terms of the linguistic variable "Temperature" and *Ts* refers to one of the weekly subdivisions of the month. The authors Ramos-Soto and Martin-Rodillab combine these two types of protoforms with the terms, "but" and "specially" to establish contrast and emphasis sentences respectively. Below are different examples of both cases:

Contrast protoform examples:

"Many values in Ancares are normal", **but**
"Most of the values of the third week in Ancares are hot"

"Most of the values of the fourth week in Ancares are cold"

Example of an emphasis protoform:

"Many values in Pontevedra-Campolongo are hot", **especially**
"Many values from the second week in Pontevedra-Campolongo are very hot"

Other extensions to the protoforms that they propose in the bibliography focus on the use of multiple summarizers where summaries are generated such as "Many cases have activities "confirm reception", "register appeal" and "person A involved" [18].

In the researches consulted, the authors generally construct the summaries for specific languages. In this sense, the possibility of representation of the same summary with different controlled natural languages is identified as an open line to research [24], facilitating the internationalization of the results.

3.2 Methods or Techniques for the Generation of Linguistic Data Summaries

In the consulted bibliography, different trends are identified for the construction of linguistic summaries, among which the following stand out:

- Construction of summaries from descriptive statistics techniques [8, 25].
- Generation of linguistic summaries from query language, initially developed by Kacprzyk et al. [26].
- Generation of linguistic summaries from association rules [27] and production rules [28].
- Generation of linguistic summaries from meta-heuristics, where the first works are generated by George and Srikant [29].
- Generation of summaries from clustering techniques [9], and combinations of these with problems of summaries generation from outliers [30].
- Generation of summaries from rough sets [31].

3.2.1 Statistical Approach and Query-Based Approach

Descriptive statistics techniques are generally used for the construction of summaries based on protoforms 1 and 2 [8]. These techniques are generally characterized by generating quantitative summaries and with a limited ability to discover new knowledge [25].

Another example of the approach based on descriptive statistics is [32], where preliminary results are presented on linguistic summaries applied to the average daily duration of outgoing calls based on smartphones. In this work the algorithms were applied to patients with bipolar disorder. The study aimed to develop a diagnostic tool based on the use of smartphones for the detection and prediction of the first symptoms of a phase change. The linguistic summaries obtained in this research are personalized, based on the short protoform that takes the form "Among all the objects that belong to patient A, Q are P", where Q is the quantifier and P is the summarizer.

As an advantage, the processing of personalized linguistic summaries instead of real numbers is identified, thus avoiding the storage of confidential data on the historical records of the patient that are processed using statistics.

In the generation of linguistic summaries from query language, the FQUERY Package developed by Kacprzyk and Zadrożny in [2] and [6] stands out. These authors use fuzzy queries that extend to SQL for the construction of summaries with the protoforms described in Table 1, it is a package integrated to the Microsoft Access manager [23, 33]. Other examples are proposed in [34] and [35] where linguistic summaries are generated from a query language called SummarySQL. This approach has the advantage of the flexibility and versatility of query languages such as those based on SQL, some representative works of this approach are listed below.

Both the statistical and the query-based approaches generally work with numerical data, they do not consider the correlation between the variables, they are simply based on an observation of the objects, they identify things that are explicit, but not implicit, so they can present difficulties to identify outliers.

3.2.2 Rules-Based Approach

The rule-based summarization approach is based on production rule construction techniques such as decision trees and others. In this sense, authors such as Dubois and Prade [28], Wilbik and collaborators [18, 27], Wu et al. [36, 37].

In [28] gradual inference rules are generated in the form "The more X is F, the more Y is G", which express a progressive change in the degree to which the entity Y satisfies the gradual property G when it is modified the degree to which the entity X satisfies the gradual property F. Then the rules are transformed using fuzzy sets until they become linguistic summaries.

Kacprzyk and Zadrożny in [27] and [36] propose the generation of linguistic summaries from association rules and production rules respectively. In these works, it is possible to find summaries in the most complex protoforms.

In [18], A. Wilbik, U. Kaymak and R. M. Dijkman propose a method for the generation of linguistic summaries inspired by the apriori algorithm. They assume that the possible linguistic quantifiers and labels are given for the attributes of the data set to be analyzed. They apply the pruning of the summaries and in the case of the extended summaries, they update the filter.

In [38], I. Pérez and collaborators present a proposal based on the generation of association rules that are then refined, based on the support and trust of the rules, transforming them later in linguistic summaries. This proposal is presented in combination with other approaches that consider the statistical behavior of the data.

In [39] they combine different strategies and take advantage of the statistical knowledge associated with the distribution of the data. Then they calculate the relative cardinality in the case of a numeric attribute and the mode in the case of categorical values to finally apply an algorithm based on the principles of the apriori algorithm. They focus on generating a linguistic summary for which its cardinality is calculated and subsequently identifying the quantifier that best describes it.

As the main limitation of rule-based approaches, it is pointed out that they are often based on the construction of a candidate rule base that is then transformed into linguistic summaries, generating many summaries that are eliminated in a second stage. This approach generally does not sufficiently exploit the information associated with the correlation of the data in the problem in question.

3.2.3 Meta-Heuristic-Based Approach

Another approach in the construction of linguistic summaries has been based on the use of meta-heuristics, where authors such as George and Srikant [29] stand out, followed by Kacprzyk and Strykowski [40], Donis-Diaz et al. [41]. Under this approach, genetic algorithms have been the most widely used technique.

In [41] a hybrid Genetic Algorithm (GA) model with a hill climber is proposed for the identification of linguistic summaries. These same authors also experiment with different meta-heuristics of collective intelligence such as ant colonies (ACO) [16]. As a weakness of this approach, it is pointed out that it does not take advantage of the information associated with the correlations between the data. Furthermore, in both works the authors establish an algorithm for generating summaries with a high computational cost that can be seen from the number of evaluations of linguistic summaries that this algorithm must perform.

3.2.4 Cluster-Based Approach and Outlier Discovery

In the approach based on clustering, works such as [9] stand out, where they build partition matrices and use the Single Linkage (SL) and Non-Euclidean Relational Fuzzy CMeans (NERFCM) algorithms. Other works such as [42] are based on the construction of a tree from considering all the combinations of filters and summarizers and then pruning the branches represented by a combination with low support in the database. The nature of these works implies a high computational cost during the search process.

The methods based on this approach generally build the clusters and then for each cluster they generate a linguistic summary that explains its behavior. Among the main limitations of this approach, it is identified that they generally depend on parameters that limit the number of clusters to be built. In other cases, distance-based strategies tend to form hyper-spheres and fail to identify clusters that describe other topologies and focus the analysis on numerical values that limit work with heterogeneous data.

The papers [43] and [44] are examples of generation of summaries from anomalous data (outliers), in these cases, for the process of construction of the summaries Linguistics, Duraj et al. allude to the method based on exhaustive construction of all

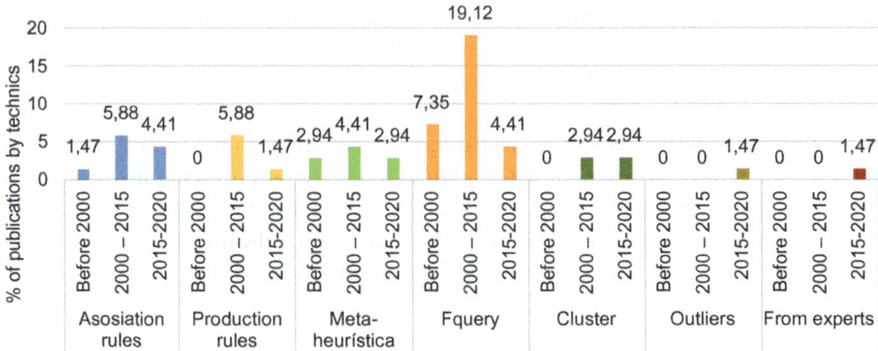

Fig. 5 Trend in methods or techniques for generating linguistic summaries of data

possible combinations of predefined quantifiers and summarizers; therefore, these proposals do not use heuristics or other methods that simplify the complexity of the search in the construction of the summaries.

Other authors such as Iliana Pérez and collaborators have published different works associated with the discovery of outliers combining different techniques, among which the linguistic summaries stand out [30, 45, 46].

As a summary of this analysis, it is identified that before 2015 the most used methods are those based on fuzzy queries, one of the advantages of this approach is the facilities provided by the use of query languages. While in the 2015–2020 period there is a greater diversity of methods without significant differences in the frequency of their use. This behavior shows the inclusion of new methods based on other theories and that provide promising results with potential for the use of summaries in other problems associated with prediction problems, among others.

Figure 5 shows a summary of the main trends in generation methods and their evolution over time.

3.3 Main Validation Techniques and Methods Used in the Investigations

In the studied bibliography, the main validation strategies used in research on linguistic data summarization were also analyzed, see Fig. 6.

From this analysis it is identified that before 2015, most of the investigations were only validated from observation and simple qualitative analysis in case studies. While, in the last five years, there is a wide use of basic descriptive statistics techniques applied to the quality indicators of the summaries [32, 47, 48]. In this subject, very few studies apply rigorous statistical techniques based on parametric and non-parametric tests [41] and [16].

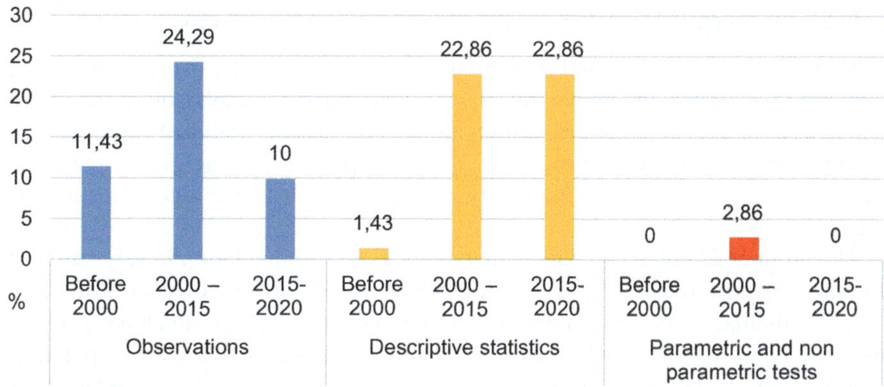

Fig. 6 Validation strategies used in the studied bibliography

Regarding external validity, an analysis is carried out observing how each of the researches consulted makes use of one of the following validation strategies: data triangulation, method triangulation, expert triangulation and case studies. The result shown in Fig. 7 was obtained.

Regarding the triangulation of data, only in recent years have there been investigations that apply it, some of these investigations are shown below:

- In [13] they use data from 15 meteorological stations in Galicia, in [48] the research is applied to the Pima Indian Diabetes Dataset and Wisconsin Breast Cancer Dataset (WBCD) both from the UCI Repository of Machine Learning [49].
- Kacprzyk and Zadrożny in [17] in the analysis of mutual fund quotes on the Warsaw, Zagreb and Moscow Stock Exchanges and to web server log records.
- In [50] summaries are generated from the level of pollution of different districts of a city.
- In [39] it is applied to data on commercial flights in the United States in the period 1987–2008 and to descriptions of second-hand cars considering the number of doors, price, mileage, year, horsepower and new starting price.

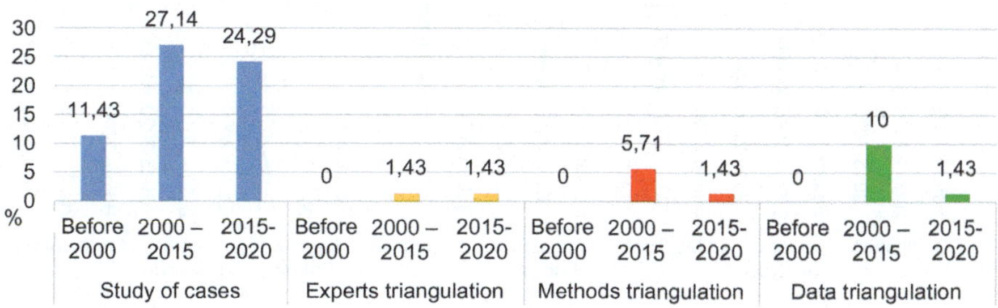

Fig. 7 Scope of the validation used in the studied bibliography

- In [25] it is applied to data on real flights and data on smart city.

Some authors have applied triangulation of methods in their research, among them are Donis-Díaz and collaborators [41], who make a comparison of the results using three variants of genetic algorithm models: the classic, modified classic and the hybrid proposed in the same article. Another example is [25], where a comparison is made regarding the execution time between two summarization algorithms applied to two databases.

The number of published papers that apply expert triangulation is very small. In this case, one could cite [50], where they apply surveys to validate the results.

In this thematic area, the validation techniques that have predominated in research are those based on case studies. It is an opportunity to improve the development of research that compares in depth the different methods of generating summaries.

3.4 Areas of Application of Linguistic Summaries

The linguistic summarization has been applied in numerous domains and in different types of application, whether it is descriptive application, such as for prediction or classification. Figure 8 shows the analysis carried out in this regard.

From the analysis carried out, it is identified that the area with the greatest number of applications in the period 2000–2015 was that associated with the financial and economic sectors. However, the trend in the last five years is concentrated in medicine, smart devices and sensor networks.

The consulted bibliography identifies different applications of linguistic summaries in decision-making problems as shown below:

- Economy and sales [51].
- Log records [9, 17, 23, 42], Kaymak [18, 52].
- Device and sensor networks [25, 32, 47, 53–56].

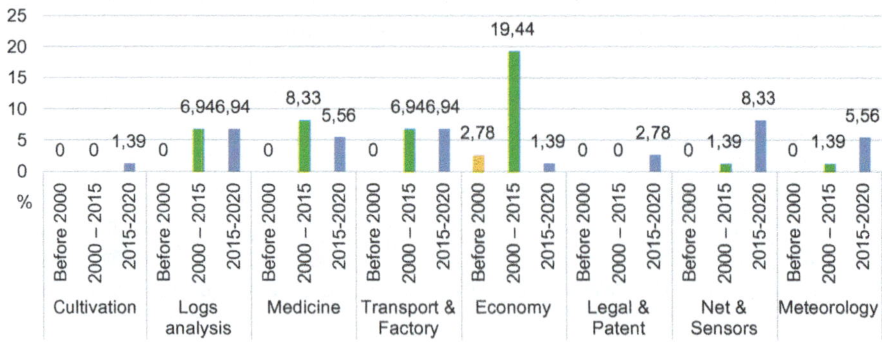

Fig. 8 Analysis of publications regarding application environments

- Health sector, in glucose management [22] heart diseases [47], physical activity [57], diabetes and cancer [48], among others.
- Meteorology [13, 25, 58].
- Agriculture [12, 50] and industry [41].
- Traffic management and urban mobility [59, 60].
- Project management [30, 31, 61, 62].
- Legal issues [18, 42], patent management [63] and consensus management [64, 65].

To a lesser extent, summaries have been used in prediction or classification problems as shown in [48].

Based on the methodology applied as part of the design of experiments and the detailed analysis of the different bibliographic sources, it is established in this research that the linguistic summarization of data is an "Area of interdisciplinary knowledge, which is part of soft computing and involves a set of knowledge areas among which fuzzy logic, machine learning, statistics, databases and others stand out. Its objective is to help decision-making from the automatic construction of summaries in natural language that reflect the implicit knowledge and the intrinsic relationships that exist between the variables in heterogeneous databases".

4 Conclusions

In this work, a review of the state of the art of linguistic data summarization has been carried out from the following perspectives: the notation and structure of summaries, generation techniques, research validation strategies and application areas. It is identified that there is consensus in the elements that make up the summaries and the notation used in the investigations.

Regarding the structure of the summaries, different protoforms have been defined, the most general ones being those proposed by J. Kacprzyk and S. Zadrożny. However, for each particular scenario it has its characteristics and this situation has motivated researchers to develop different extensions. The development of extensions that allow the integration of multiple summaries in a single analysis is particularly interesting, facilitating decision-making. An open line to research in this field is located in the possibility of representing the same linguistic summary using different controlled natural languages, an element that would facilitate the internationalization of the results.

Summaries generation techniques are diverse, the use of query-based methods that exploit the facilities of database query languages by combining these with other techniques stands out before 2015. In recent years there is a greater variety in the methods of generating linguistic summaries, the use of techniques associated with machine learning and data mining has increased. These new trends incorporate greater robustness into the algorithms, but it is identified that there are still lines open to research based on the possibility of hybridization of different techniques and a better exploitation of information associated with the data such as the correlation between the variables or specific aspects' application domains.

Regarding the validation of published research, it is identified that there are few studies that apply the techniques of data triangulation, method triangulation or others that help to verify the generalizability of the research. It is an opportunity to improve the development of research that compares in depth the different methods of generating summaries and allows identifying their potentialities for different scenarios.

Linguistic data summaries have generally been used descriptively, with the most exploited application areas being decision-making in economic management and sales from the early stages of theory development. However, as of 2016 there is a significant increase in applications in medicine, the analysis of data from smart devices and the analysis of sensor networks as part of the internet of things. The use, to a greater extent, of linguistic data summarization techniques in solving prediction problems and in the development of conversational systems, as part of decision support processes, is identified as an open line for research.

References

1. Wu, D., Mendel, J.M.: Linguistic summarization using IF–THEN rules and interval type-2 fuzzy sets. IEEE Trans. Fuzzy Syst. **19**(1), 136–151 (2010). https://doi.org/10.1109/TFUZZ.2010.2088128
2. Kacprzyk, J., Zadrożny, S.: On a fuzzy querying and data mining interface. Kybernetika **36**(6), 657–670 (2000)
3. Yager, R.R.: On Linguistic summaries of data. Knowledge Discovery in Databases, pp. 378–389 (1991)
4. Yager, R.R.: A new approach to the summarization of data. Inf. Sci. **28**(1), 69–86 (1982). https://doi.org/10.1016/0020-0255(82)90033-0
5. Zadeh, L.A.: A computational approach to fuzzy quantifiers in natural languages. Comput. Math. Appl. **9**(1), 149–184 (1983)
6. Kacprzyk, J.: Fuzzy logic for Linguistic summarization of databases. FUZZ-IEEE'99 International Fuzzy Systems, Conference Proceedings, vol. 2, pp. 813–818 (1999). https://doi.org/10.1109/FUZZY.1999.793053
7. Kacprzyk, J., Zadrożny, S.: Fuzzy Linguistic data summaries as a human consistent, user adaptable solution to data mining. Do Smart Adaptive Systems Exist? Best Practice for Selection and Combination of Intelligent Methods, pp. 321–340. Springer, Berlin (2005). https://doi.org/10.1007/3-540-32374-0_16
8. Boran, F.E., Akay, D., Yager, R.R.: An overview of methods for Linguistic summarization with fuzzy sets. Expert Syst. Appl. **61**, 356–377 (2016). https://doi.org/10.1016/j.eswa.2016.05.044
9. Wilbik, A., Dijkman, R.M.: On the generation of useful Linguistic summaries of sequences. IEEE International Conference on Fuzzy Systems (FUZZ-IEEE), pp. 555–562 (2016). https://doi.org/10.1109/FUZZ-IEEE.2016.7737736
10. Eciolaza, L., Pereira-Fariña, M., Trivino, G.: Automatic Linguistic reporting in driving simulation environments. Appl. Soft Comput. **13**(9), 3956–3967 (2013). https://doi.org/10.1016/j.asoc.2012.09.007
11. Marín, N., Sánchez, D.: On generating Linguistic descriptions of time series. Fuzzy Sets Syst. **285**, 6–30 (2016). https://doi.org/10.1016/j.fss.2015.04.014

12. Hudec, M., Bednárová, E., Holzinger, A.: Augmenting statistical data dissemination by short quantified sentences of natural language. J. Official Stat. **34**(4), 981–1010 (2018). https://doi.org/10.2478/jos-2018-0048
13. Ramos-Soto, A., Martin-Rodillab, P.: Enriching Linguistic descriptions of data: a framework for composite protoforms. Fuzzy Sets Syst. (2019). https://doi.org/10.1016/j.fss.2019.11.013
14. Hernández-Sampieri, R., Torres, C.P.M.: Metodología de la investigación. Vol. 4, McGraw-Hill Interamerican Mexico (2018)
15. Littell, J.H., Corcoran, J., Pillai, V.: Systematic reviews and meta-analysis. Oxford University Press, ISBN: 978-0-19-532654-3 (2008)
16. Donis-Díaz, Bello, R., Kacprzyk, J.: Using ant colony optimization and genetic algorithms for the Linguistic summarization of creep data. Intelligent Systems'2014, Cham, pp. 81–92 (2015). https://doi.org/10.1007/978-3-319-11313-5_8
17. Kacprzyk, J., Zadrożny, S.: Fuzzy logic-based Linguistic summaries of time series: a powerful tool for discovering knowledge on time varying processes and systems under imprecision. Wiley Interdiscip. Rev. Data Min. Knowl. Discov. **6**(1), 37–46 (2016). https://doi.org/10.1002/widm.1175
18. Wilbik, A., Kaymak, U., Dijkman, R.M.: A method for improving the generation of Linguistic summaries. IEEE International Conference on Fuzzy Systems (FUZZ-IEEE), pp. 1–6 (2017). https://doi.org/10.1109/FUZZ-IEEE.2017.8015752
19. Zadeh, L.A.: A prototype-centered approach to adding deduction capability to search engines–the concept of protoform. Intelligent Systems, Proceedings, First International IEEE Symposium, vol. 1, pp. 2–3 (2002).https://doi.org/10.1109/IS.2002.1044219
20. Kacprzyk, J., Zadrożny, S.: Linguistic database summaries and their protoforms: towards natural language-based knowledge discovery tools. Inf. Sci. **173**(4), 281–304 (2005). https://doi.org/10.1016/j.ins.2005.03.002
21. Kacprzyk, J., Zadrożny, S.: Linguistic database summaries using fuzzy logic, towards a human-consistent data mining tool. No. 20, p. 10 (2009)
22. Wilbik, A., Vanderfeesten, I., Bergmans, D., Heines, S. Mook, W.: Linguistic summaries for compliance analysis of a glucose management clinical protocol. IEEE International Conference on Fuzzy Systems (FUZZ-IEEE), pp. 1–7 (2018). https://doi.org/10.1109/FUZZ-IEEE.2018.8491449
23. Kacprzyk, J., Zadrożny, S.: Linguistic summarization of the contents of Web server logs via the ordered weighted averaging (OWA) operators. Fuzzy Sets Syst. **285**, 182–198 (2016). https://doi.org/10.1016/j.fss.2015.07.020
24. Kuhn, T.: A survey and classification of controlled natural languajes. Comput. Lingusitic **40**(1), 121–170 (2014). https://doi.org/10.1162/COLI_a_00168
25. Khedidja, B., Allel, H., Mohand, L.: Data summarization for sensor data management: towards computational-intelligence-based approaches. Int. J. Comput. Digital Systems. **9**(5), 825–833 (2020). https://doi.org/10.12785/ijcds/090505
26. Kacprzyk, J.: An interactive fuzzy logic approach to Linguistic data summaries. In: 18th International Conference of the North American Fuzzy Information Processing Society-NAFIPS (Cat. No. 99TH8397), pp. 595–599 (1999). https://doi.org/10.1109/NAFIPS.1999.781763
27. Kacprzyk, J., Zadrożny, S.: Linguistic summarization of data sets using association rules. Fuzzy Systems, FUZZ'03. In: The 12th IEEE International Conference, vol. 1, pp. 702–707 (2003)
28. Dubois, D., Prade, H.: Gradual inference rules in approximate reasoning. Inf. Sci. **61**(1), 103–122 (1992). https://doi.org/10.1016/0020-0255(92)90035-7
29. George, R., Srikant, R.: Data summarization using genetic algorithms and fuzzy logic. Genetic Algorithms and Soft Computing, pp. 599–611 (1996)
30. Pérez, I., Piñero, P.Y., García-Vacacela, R., Bello, R., Acuña, L.A.: Discovering fails in software projects planning based on Linguistic summaries. In: International Joint Conference on Rough Sets. Lecture Notes in Computer Science ISSN 0302-9743. Springer, 12179 LNAI, ISBN 978-3-030-52704-4, pp. 365–375 (2020). https://doi.org/10.1007/978-3-030-52705-1_27
31. Pérez, I., Piñero, P.Y., Bello, R., Acuña, L.A., García-Vacacela, R.: Linguistic summaries generation with hybridization method based on rough and fuzzy sets. In: International Joint

Conference on Rough Sets, pp. 385–397 (2020). https://doi.org/10.1007/978-3-030-52705-1_29
32. Kaczmarek-Majer, K., Hryniewicz, O., Dominiak, M., Święcicki, Ł.: Personalized Linguistic summaries in smartphone-based monitoring of bipolar disorder patients. (2019).https://doi.org/10.2991/eusflat-19.2019.56
33. Kacprzyk, J., Zadrożny, S., Dziedzic, M.: A novel view of bipolarity in Linguistic data summaries. In: Issues and Challenges of Intelligent Systems and Computational Intelligence. Springer, pp. 215–229 (2014). https://doi.org/10.1007/978-3-319-03206-1_16
34. Kacprzyk, J., Zadrożny, S.: Fquery for Access: Fuzzy Querying for a Windows-Based DBMS. Fuzziness in database management systems. Springer, pp. 415–433 (1995). [Online] Available: https://doi.org/10.1007/978-3-7908-1897-0_18. [Accessed: 4 May 2021]
35. Rasmussen, D., Yager, R.R.: Finding fuzzy and gradual functional dependencies with SummarySQL. Fuzzy Sets Syst. **106**(2), 131–142 (1999). https://doi.org/10.1016/S0165-0114(97)00268-6
36. Wu, D., Mendel, J.M., Joo, J.: Linguistic summarization using if-then rules. Fuzzy systems (FUZZ), IEEE Int. Conf. 1–8 (2010)
37. Wu, D., Mendel, J.M.: Linguistic summarization using IF–THEN rules and interval type-2 fuzzy sets. IEEE Trans. Fuzzy Syst. **19**(1), 136–151 (2011)
38. Pérez, I., Santos, O., García, R., Piñero, P.Y., Ramírez, E.C.: Descubrimiento de resúmenes lingüísticos para ayuda a la toma decisiones en gestión de proyecto. Revista Cubana de Ciencias Informáticas **12**, 163–175 (2018)
39. Smits, G., Nerzic, P., Pivert, O., Lesot, M.J.: Efficient generation of reliable estimated Linguistic summaries. In: IEEE International Conference on Fuzzy Systems (FUZZ-IEEE), pp. 1–8 (2018). https://doi.org/10.1109/FUZZ-IEEE.2018.8491604
40. Kacprzyk, J., Strykowski, P.: Linguistic summaries of sales data at a computer retailer via fuzzy logic and a genetic algorithm. In: Proceedings of the 1999 Congress on Evolutionary Computation-CEC99 (Cat. No. 99TH8406), vol. 2, pp. 937–943 (1999). https://doi.org/10.1109/CEC.1999.782523
41. Donis-Díaz, Muro, A.G., Bello, R., Morales, E.V.: A hybrid model of genetic algorithm with local search to discover Linguistic data summaries from creep data. Expert Syst. Appl. **41**(4), 2035–2042 (2014). https://doi.org/10.1016/j.eswa.2013.09.002
42. Dijkman, R., Wilbik, A.: Linguistic summarization of event logs—a practical approach. Inf. Syst. **67**, 114–125 (2017). https://doi.org/10.1016/j.is.2017.03.009
43. Duraj, A., Szczepaniak, P.S., Chomatek, L.: Intelligent detection of information outliers using Linguistic summaries with non-monotonic quantifiers. In: International Conference on Information Processing and Management of Uncertainty in Knowledge-Based Systems, pp. 787–799 (2020). https://doi.org/10.1007/978-3-030-50153-2_58
44. Duraj, A., Szczepaniak, P.S., Ochelska-Mierzejewska, J.: Detection of outlier information using Linguistic summarization. Flexible Query Answering Systems 2015, Cham, pp, 101–113 (2016). https://doi.org/10.1007/978-3-319-26154-6_8
45. Aguilar, C., et al.: Aplicación de la minería de datos anómalos en organizaciones orientadas a proyectos. Revista Cubana de Ciencias Informáticas **10**, 195–209 (2016)
46. Castro, G.F., et al.: Platform for project evaluation based on soft-computing techniques. pp. 226–240 (2016). [Online] Available: https://doi.org/10.1007/978-3-319-48024-4_18. [Accessed: 12 April 2021]
47. Peláez-Aguilera, M.D., Espinilla, M., Fernández, M.R., Medina, J.: Fuzzy Linguistic protoforms to summarize heart rate streams of patients with ischemic heart disease, p. 11 (2019). https://doi.org/10.1155/2019/2694126
48. Amghar, D., Amine, M.: Extracting a Linguistic summary from a medical database. Int. J. Intell. Syst. Appl. **10**(12), 16–26 (2018). https://doi.org/10.5815/ijisa.2018.12.02
49. Dua, D., Graff, C.: UCI Machine learning repository. University of California, Irvine, School of Information and Computer Sciences (2017). [Online] Available: http://archive.ics.uci.edu/ml [Accessed: 4 April 2021]

50. Rojas, Á.R.: Resúmenes lingüísticos para riego de cultivos. Tesis, Universidad Técnica Federico Santa María, Departamento de Informática, Santiago, Chile (2018). [Online] Available: https://repositorio.usm.cl. [Accessed: 5 May 2021]
51. Piñero, P., Pérez, I., García, R., Toscanini, P.: Caracterización de los estándares de gestión de proyectos y su impacto en la gestión económico financiera de las organizaciones orientadas a proyectos. Guayaquil, Ecuador: Universidad Católica de Santiago de Guayaquil, ISBN: 978-9942-82-514-8 (2020)
52. Wilbik, A., Gilsing, R., Turetken, O., Ozkan, B., Grefen, P.: Intentional Linguistic summaries for collaborative business model radars. IEEE International Conference on Fuzzy Systems (FUZZ-IEEE), pp. 1–7 (2020). https://doi.org/10.1109/FUZZ48607.2020.9177587
53. Jain, A., Popescu, M., Keller, J., Rantz, M., Markway, B.: Linguistic summarization of in-home sensor data. J. Biomed. Inform. (2019). https://doi.org/10.1016/j.jbi.2019.103240
54. Jain, A., Keller, J.M., Bezdek, J.C.: Quantitative and qualitative comparison of periodic sensor data. In: IEEE-embs International Conference on Biomedical and Health Informatics (bhi), pp. 37–40 (2016). https://doi.org/10.1109/BHI.2016.7455829
55. Díaz-Hermida, F., Vidal, J.C.: Fuzzy quantification for Linguistic data analysis and data mining. (2018). [Online] Available: https://arxiv.org/abs/1807.07389v1 [Accessed: 4 April 2021]
56. Genç, S., Akay, D., Boran, F.E., Yager, R.R.: Linguistic summarization of fuzzy social and economic networks: an application on the international trade network. Soft. Comput. **24**(2), 1511–1527 (2020). https://doi.org/10.1007/s00500-019-03982-9
57. Sanchez-Valdes, D., Alvarez-Alvarez, A., Trivino, G.: Dynamic Linguistic descriptions of time series applied to self-track the physical activity. Fuzzy Sets Syst. **285**, 162–181 (2016). https://doi.org/10.1016/j.fss.2015.06.018
58. Heble-Lahera, C., Cascallar-Fuentes, A., Ramos-Soto, A., Diz, A.B.: Empirical study of fuzzy quantification models for Linguistic descriptions of meteorological data. In: IEEE International Conference on Fuzzy Systems, pp. 1–7 (2020). https://doi.org/10.1109/FUZZ48607.2020.9177716
59. Degtiarev, K.Y., Remnev, N.V.: Linguistic resumes in software engineering: the case of trend summarization in mobile crash reporting systems. Procedia Comput. Sci. **102**, 121–128 (2016). 210.1016/j.procs.2016.09.378
60. Gilsing, R., Wilbik, A., Grefen, P., Turetken, O., Ozkan, B.: a formal basis for business model evaluation with Linguistic summaries. Enterprise, Business-Process and Information Systems Modeling. Springer, pp. 428–442 (2020), https://doi.org/10.1007/978-3-030-49418-6_29
61. Rangel, J. R., Pérez, I., García, R.C., Piñero, P.: Strategic factors in the context of project management: Manage. Perspect. Espacios, ISSN 0798-1015, vol. 39, p. 10 (2018)
62. Pérez, I., García, R., Piñero, P., Mahdi, G.S., Peña, M.: Experiencias en el uso de técnicas de softcomputing en la evaluación de proyectos de software. Revista Investigación Operacional **41**(1), 106–117 (2020)
63. Igde, E., Aydoğan, S., Boran, Akay, D.: Linguistic summarization of structured patent data (2017)
64. Kacprzyk, J., Zadrożny, S.: Reaching consensus in a group of agents: supporting a moderator run process via Linguistic summaries. Soft Computing Applications for Group Decision-making and Consensus Modeling. Springer, pp. 465–485 (2018)
65. Kacprzyk, J., Zadrożny, S.: On a fairness type approach to consensus reaching support under fuzziness via linguistic summaries. In: IEEE International Conference on Fuzzy Systems (FUZZ-IEEE) (2016). https://doi.org/10.1109/FUZZ-IEEE.2016.7737937

Chapter 2
New Linguistic Data Summarization Approach for Prediction Problems in Project Management Applications

Iliana Pérez Pupo, Pedro Y. Piñero Pérez, Salah Hassan Al-subhi, Roberto García Vacacela, Hugo A. Martínez Noriega, and Nicolas Villavicencio Bermúdez

Abstract In the proposal, multilayer networks are built where the layers represent the different elements that make up the linguistic summaries and systems with single-value triangular neutrosophic numbers are combined for the computation of the inference. In the validation, the new inference model was applied in the scenario of making-decision in project management and the "Arithmetic mean of the error in the prediction" was applied as a metric. Comparisons are made by evaluating the inference model with different summary parameters. In addition, the linguistic summaries are combined with the fuzzy cognitive maps and the extension NCM_LDS is generated. The new proposed extension is compared with other map extensions, it is shown that the proposed inference mode-lo reports better results than the rest of the maps in the diagnostic stage, but that it does not have significantly better results than the m-FCM in the decision and prognosis stages. The NCM_Indeterminacy map is shown to be the one with the worst results.

Keywords Linguistic summaries · Prediction problems · Project management

I. Pérez Pupo (✉) · P. Y. Piñero Pérez · H. A. Martínez Noriega
Grupo de Investigaciones en Gestión de Proyectos, Universidad de las Ciencias Informáticas, Carretera a San Antonio km 2 ½, La Habana, 54830 Havana, CP, Cuba
e-mail: iperez@uci.cu

P. Y. Piñero Pérez
e-mail: ppp@uci.cu

S. H. Al-subhi
Solarplex Electyrical and Electronics Contracting Co. Combined Security Group Co, Alfahahel, Kuwait

R. García Vacacela · N. Villavicencio Bermúdez
Facultad de Especialidades Empresariales, Universidad Católica Santiago de Guayaquil, Guayaquil, Ecuador
e-mail: roberto.garcia@cu.ucsg.edu.ec

N. Villavicencio Bermúdez
e-mail: nicolas.villavicencio@cu.ucsg.edu.ec

P. Y. Piñero Pérez et al. (eds.), *Artificial Intelligence in Project Management and Making Decisions*, Studies in Computational Intelligence 1035,
https://doi.org/10.1007/978-3-030-97269-1_2

1 Introduction

The discovery of behavior patterns and their interpretability is one of the latent problems in decision-making processes. This problem increases with the rapid growth of data generated by information systems and sensor networks. In general, these data are not readable or understandable to the naked eye by decision makers, an element that, together with the volume of data that is generated, hinders decision-making processes. Thus arises the called "rich data, poor information" dilemma present in decision-making [1]. In this sense, the development of new techniques and tools that allow the discovery of hidden non-trivial dependencies in the data is an object of interest for decision makers.

In this context, the linguistic data summarization (LDS) arises as one of the descriptive knowledge discovery techniques with an interesting and promising approach to produce summaries from a database using natural language [2]. Authors such as Yager and Zadeh were pioneers in the development of linguistic data summarization. His publications [3, 4] set guidelines in the conceptualization and subsequent development of the linguistic summarization of data as one of the techniques of soft computing. Different authors have characterized the techniques of linguistic data summarization as shown below:

- Kacprzyk [5] describes it as one of the paradigms of word computing that can capture the essence of data.
- Boran et al. [6] propose that it is a descriptive knowledge discovery technique capable of extracting potential, useful and abstract knowledge from both numerical and categorical data. This technique allows you to discover the patterns that summarize the relationships between attributes in a large database with more abstract conceptual representations.
- More recently, Pérez et al. [7] establish that it is an interdisciplinary area of knowledge, which is part of soft computing and involves a set of knowledge areas among which fuzzy logic, machine learning, statistics, databases and others stand out. Its objective is to help decision-making from the automatic construction of summaries in natural language that reflect the implicit knowledge and the intrinsic relationships that exist between the variables in heterogeneous databases.

As part of this research, different tertiary reviews [6, 7, 8] and works published in books, journals, theses and event reports [9, 10, 11] were reviewed. As can be seen in Fig. 1, linguistic summaries have been applied mostly to descriptive problems of phenomena, events, happenings and so on. To a lesser extent they have been used in classification or prediction problems [12].

Regarding the discovery of linguistic summaries, various techniques are identified. For example, descriptive statistics [6] generally used for the construction of summaries based on type 1 and type 2 protoforms. These techniques are characterized by generating quantitative summaries and with a limited capacity to discover new knowledge [13]. Another method is the one based on queries [14, 15], which exploits the facilities of database query languages. There are also the rule-based

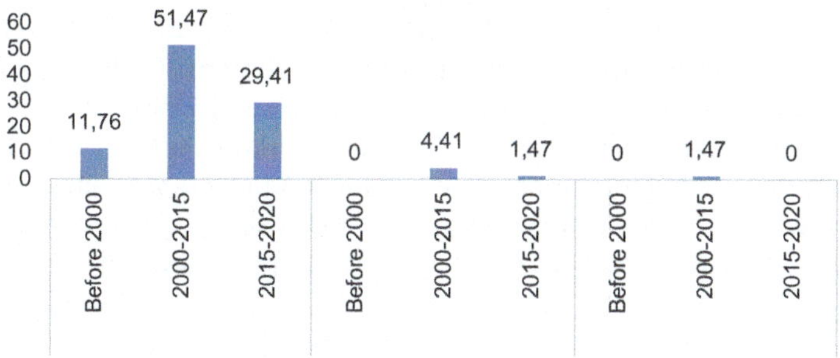

Fig. 1 Linguistic summaries applications

ones [16, 17] that are often based on the construction of a candidate rule base that are then transformed into linguistic summaries, generating many summaries that are eliminated at a later stage; this approach generally does not sufficiently exploit the information associated with the correlation of the data in the problem in question. Meta-heuristics have also been used to generate linguistic summaries [18, 19], where genetic algorithms have been the most widely used technique, but they involve a high computational cost. More recently, other authors have exploited the hybridization of different soft computing and machine learning techniques, such as rough sets [20], probabilistic models [21, 22], rule learning [23, 24] and anomalous data mining [25]. These approaches incorporate greater robustness into the algorithms and a better exploitation of information associated with the data, such as the correlation between the variables or specific aspects of the application domains.

From this review it is identified that there are many opportunities associated with the application of linguistic summarization techniques and the learning of linguistic summaries in solving decision-support problems, as shown below:

- Linguistic summaries are identified as having a group of similarities with fuzzy inference rules and therefore, new inference methods can be built from linguistic summaries inspired by fuzzy inference systems.
- Linguistic summaries of data have generally been used descriptively, but similarities are identified with fuzzy inference rules and therefore, new inference methods could be built from linguistic summaries inspired by fuzzy inference systems.
- Different algorithms have been developed for the generation of linguistic summaries, but it is identified that there are still lines open for research based on the possibility of hybridization of different techniques and a better exploitation of information associated with the data.
- In the search carried out, no applications of the linguistic data summarization techniques were found in project-oriented organizations

Given this situation, the development of an algorithm for the use of linguistic summaries in an inference system that can be used in solving prediction problems is proposed in this research.

The work is structured in the following sections. The second section presents a brief analysis associated with the structure and algorithms for learning linguistic summaries. In addition, the elements that constitute similarities between linguistic summaries and fuzzy rules are analyzed. Then, in the third section, an inference model based on linguistic summaries is presented. In the fourth section an application of the proposed inference model is presented in decision-making in project management. Finally, the conclusions and lines of future work are presented.

2 Structure of Linguistic Summaries and Contact Points with Fuzzy Inference Systems

The summaries are made up of a set of elements that express the relationships between concepts, such as the quantifiers (*Q*), the filters (*R*), the summarizers (*S*) and the quality indicators of the linguistic summaries (*T*). In the construction of the summaries, these concepts are grouped according to the structures "*Qy's are S*" and "*QRy's are S*" [3] as shown below:

- *Qy's are S*: for linguistic summaries without filter. Example "Most of the workers are punctual". In this case, the quantifier *Q* represents the linguistic term "Most", while the expression is "punctual" represents the summarizer.
- *QRy's are S*: for linguistic summaries with a filter, where *R* can be made up of one or more attributes qualifying the object. For example, "Most young workers are late" is a summary where the expression "young" constitutes a filter and "they are late" represents the summarizer.

Zadeh [4] and other authors [26] have broken down these linguistic summaries structures into stereotyped forms called protoforms, which have been extended according to the characteristics of the different application scenarios [7]. However, the most widespread are those proposed by Zadeh [27], in these hierarchies are established where one passes through the levels of abstraction as shown below:

- The type 0 protoform is the one with the lowest level of abstraction, since it is assumed that all the elements that make up the summary are known, we want to know is their degree of truth *T* [4, 26]. While in type 1 and 2 protoforms we want to discover the quantifier (*Q*).
- In the case of protoforms type 3 and 4, the quantifier (*Q*) and the structure of the summary are known and we want to discover the summarizer *S* and the filter *R* (for the case of protoform 4).
- The type 5 protoform is the one with the highest level of abstraction, since no element is known and, therefore, it is a matter of discovering everything.

In order to evaluate summaries, authors develop different *T* measures [4, 26], proposed:

- Degree of truth (T_1): evaluates the truth of a summary based on the object's membership to the summary and to the summary's quantifier.
- Degree of imprecision (T_2): calculates the summary vagueness degree and summary imprecision's degree by considering the alternative values for each summarizer.
- Degree of coverage (T_3): calculates the objects relative frequency that belongs to summarizer's fuzzy set and to the filter's fuzzy sets.
- Degree of appropriateness (T_4): measures the usefulness of the summary, combining the coincidence relative frequency between the objects and the summary with the degree of coverage. This measure reports low values with high values of coverage degrees.
- Length of summary (T_5): measure to get the summary length based on variables number implicated on it. Very large summaries with a high length are usually incompressible.

A certain analogy can be identified between the linguistic summaries with the form "*QRy's are S*" with the fuzzy rules with the structure "x_1 is A and x_2 is $B \rightarrow y$ is C". In the previous example, the fuzzy rule is formed by the antecedents "x_1 is A and x_2 is B" and the consequent "y is C" such that A, B, C are fuzzy sets. In this sense, the summarizers S may have a certain similarity with the consequent, while we can establish a certain analogy between the filters R with the antecedents "x_1 is A and x_2 is B".

On the other hand, there is a sustained development of fuzzy inference systems and their applications in many sectors of society [28]. These systems have the basis of their operation in the knowledge expressed in the fuzzy rules, there are three fundamental types of systems: Mamdani, Sugeno and Tsukamoto model [29, 30]. These systems differ fundamentally in the form of the consequent of their fuzzy rules [31] and the inference mechanisms that they establish. An important element that differentiates SIBs from classical Rule-Based Systems is that the result of the inference is obtained by applying numerous rules and reconciling their partial inferences from an aggregation procedure.

3 A New Approach for Inference Based on Linguistic Summaries

Starting from an understanding of fuzzy inference systems and the similarities between linguistic summaries and fuzzy rules, a new approach to Inference and prediction based on linguistic summaries is proposed in this section. This approach combines fuzzy theory, neutrosophic theory and linguistic data summarization theory in the inference process and is characterized by the following elements:

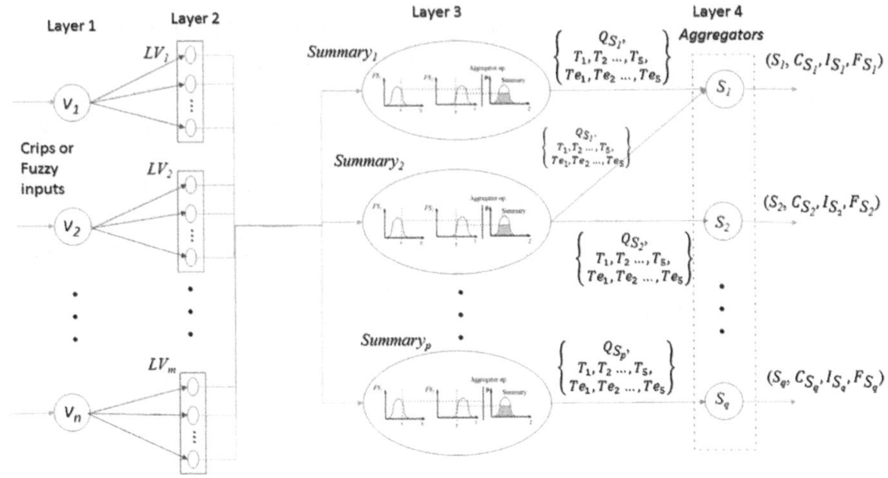

Fig. 2 Inference in LDS

- It requires as input a set of linguistic summaries previously learned with any of the algorithms described in the consulted bibliography, such as those proposed by Iliana Pérez [20, 21, 23, 25].
- The elements that make up the linguistic summaries, quantifiers (Q), filters (R) and summarizers (S) must be represented using neutrosophic fuzzy sets.
- The use of neutrosophic fuzzy sets is suggested [32] and a set of mathematical operators are established that facilitate the processes of adding neutrosophic numbers.
- A topology oriented to multilayer nodes (four layers) is established that supports the inference process, view Fig. 2.

The first layer represents the system inputs, each node corresponds to a crisp or fuzzy variable. This layer is strong connected with the second layer, each variable in the first is connected with its associated fuzzy variable. That's means, there is an edge from each node in the first layer to a set of fuzzy sets in the second one.

The second layer contains a set of linguistic variables that represents both filters A and summaries B, with $A \cap B \neq \emptyset$. But each linguistic variable consist of neutrosophic fuzzy sets (SVNS) [32, 33] view Definition 1.

Definition 1 [34]: Let X be a universe of discourse with a generic element in X denoted as x. A single-valued neutrosophic set (SVNS) A, see Fig. 3 in X is: $A = \{x, T_A(x), I_A(x), F_A(x) \mid x \in X\}$, where T_A is the certainty function, I_A is the indeterminacy function and F_A is the falsehood function. For each element x in X, $T_A(x)$, $I_A(x)$, $F_A(x)$ $\in [0, 1]$ y $0 \leq T_A(x) + I_A(x) + F_A(x) \leq 3$. For convenience the triple ($T_A(x)$, $I_A(x)$, $F_A(x)$) is called a single value neutrosophic number SVNN, see Eqs. (1), (2) and (3).

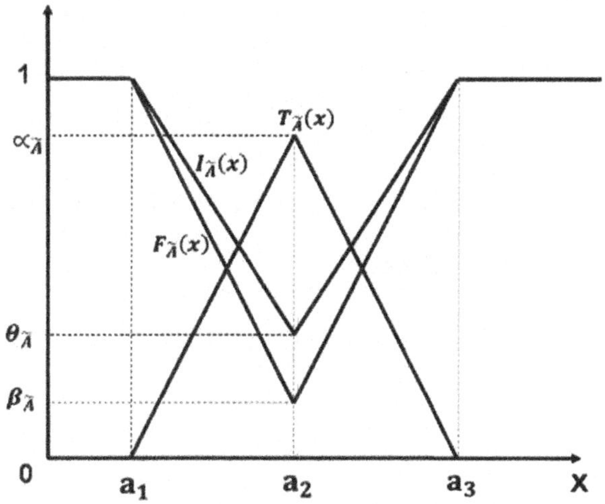

Fig. 3 Example of single-valued neutrosophic set (SVNS)

$$T_{\tilde{A}}(x) = \begin{cases} \propto_{\tilde{A}} \left(\frac{x-a_1}{a_2-a_1}\right) & (a_1 \le x < a_2) \\ \propto_{\tilde{A}} & (x = a_2) \\ \propto_{\tilde{A}} \left(\frac{a_3-x}{a_3-a_2}\right) & (a_2 < x \le a_3) \\ 0 & \text{in other case} \end{cases} \tag{1}$$

$$I_{\tilde{A}}(x) = \begin{cases} \frac{(a_2-x+\theta_{\tilde{A}}(x-a_1))}{(a_2-a_1)} & (a_1 \le x < a_2) \\ \theta_{\tilde{A}} & (x = a_2) \\ \frac{(x-a_2+\theta_{\tilde{A}}(a_3-x))}{(a_3-a_2)} & (a_2 < x \le a_3) \\ 1 & \text{in other case} \end{cases} \tag{2}$$

$$F_{\tilde{A}}(x) = \begin{cases} \frac{(a_2-x+\beta_{\tilde{A}}(x-a_1))}{(a_2-a_1)} & (a_1 \le x < a_2) \\ \beta_{\tilde{A}} & (x = a_2) \\ \frac{(x-a_2+\beta_{\tilde{A}}(a_3-x))}{(a_3-a_2)} & (a_2 < x \le a_3) \\ 1 & \text{in other case} \end{cases} \tag{3}$$

In the third layer each node represents a linguistic summary, the filters on each summary are associated with the correspondent nodes on the second layer. In order to evaluate each summary, this layer applies t-norms [31] or compensatory operators [35], to aggregate the evaluation summaries´ filters. In this layer the aggregation of filters represents the activation level of linguistic summarizers, all summarizers in the same linguistic summary receive the same activation value.

The operations between single-valued neutrosophic numbers are defined as shown below [36]. Let $a = (T_1, I_1, F_1)$ and $b = (T_2, I_2, F_2)$ be two SVNNs.

$$\mathrm{a} \oplus \mathrm{b} = (\mathrm{T}_1 + \mathrm{T}_2 - \mathrm{T}_1\mathrm{T}_2,\ \mathrm{I}_1 \cdot \mathrm{I}_2, \mathrm{F}_1 \cdot \mathrm{F}_2) \tag{4}$$

$$a \otimes b = (T_1 \cdot T_2,\ I_1 + I_2 - I_1 \cdot I_2, F_1 + F_2 - F_1 \cdot F_2) \tag{5}$$

$$k \cdot a = \left(1 - (1 - T_1)^k,\ (I_1)^k,\ (F_1)^k\right) \text{ such that } k > 0,\ k \in IR \tag{6}$$

$$a^k = \left((T_1)^k, 1 - (1 - I_1)^k, 1 - (1 - F_1)^k\right), \text{ such that } k > 0,\ k \in IR \tag{7}$$

It is also true that,

$$a \oplus b = b \oplus a \tag{8}$$

$$a \otimes b = b \otimes a \tag{9}$$

$$n_1 \cdot a \oplus n_1 \cdot b = n_1 \cdot (a \oplus b) \tag{10}$$

$$n_1 \cdot a \oplus n_2 \cdot a = (n_1 + n_2) \cdot a, \text{ such that } n_1, n_2 \in IR \tag{11}$$

$$a_1^n \otimes a_2^n = a^{(n_1+n_2)}, \text{ such that } n_{1,} n_2 \in IR \tag{12}$$

The fourth layer contain as many nodes as summarizers there are. In this layer occurs an aggregation process of the activation's levels associated to each summarizer. This layer uses the t-values of each summary as weight of each summary activation. The following algorithm resume the inference process, view Algorithm 1.

Algorithm 1 *Inference process (summaries, input_values)*

```
    Inputs
list_t_values: list of quality ítems to calculate the summary weight.
    hsummarizers = hash.new()
    for each summary in summaries do
      for each filter en summary.filters
        eval_filter = evaluate_filter(filter, input_values)
        fill_values << eval_filter.evaluation
        fill_certainty_values << eval_filter.svnn //SVNN for each activation filter
      end
     antecedent_result = aggregator.evaluate_t_norm(fill_values)
     fill_svnn_values = do_aggregate_certainty(fill_certainty_values)
     t_weight = calculate_summary_weigth(summary.t_values, list_t_values)
     activation = antecedent_result ⊗ t_weight
     for each summarizer en summary.summarizers do
       hsummarizers.add(summarizer, [activation, fill_svnn_values] )
      end
    end // cycle line 2
    Outputs = []
    for each summarizer in hsummrizers do
      value = hsummarizers [summarizer].aggregate_activation_levels
      Outputs.add(summarizer , value)
    end
    return Outputs
```

4 Application in Decision-Making in Project Management

In the validation, the database "Decision making in project management" (200821_gp_toma_decisiones) of the Project Management Research Repository [37] is used, made up of 1 011 records of project cuts. From this database a meta-algorithm is built that combines different algorithms for learning linguistic summaries proposed by Iliana Pérez and collaborators [21, 24, 25, 38, 39, 40]. From the learning process, 490 linguistic summaries were discovered.

The metric "Arithmetic mean of the error in the prediction" (ME), whose unit of measurement is numerical between 0 and 1 and is calculated using Eq. (13). The calculation of the error E_i includes the errors in the first three prediction proposals provided by each combination of the inference method used. This indicator, the smaller it is, the better the results.

$$ME = \frac{1}{N}\sum_{i=1}^{N} E_i \tag{13}$$

The following validation tests are applied based on the proposed indicators:

- Test 1: Evaluate the impact of the use of different combinations of T indicators in the inference process from the Arithmetic Mean Error metric with different groups of linguistic summaries.
- Test 2: Comparison of the proposal with other inference methods using the Arithmetic Mean Error metric.

4.1 Results of Test 1 Impact of the Use of Different Combinations of T Indicators in the Inference Process

For this test, the 490 linguistic summaries learned were grouped according to the attributes they use in different groups. Considering that A_f the set of attributes used in the filters and B_S the set of attributes used in the summarizers, then the summary groups are divided into:

1. The first two groups formed by 164 summaries associated with the diagnostic of projects and the identification of problems in projects. The first group that we will call *Diagnostic_Low* is made up of summaries where there is a clear separation of the set of attributes used in the A_f filters and attributes used in the B_S summarizers, that is, $A_f \cap B_S = \emptyset$. While the second group we call *Diagnostic_High*, it is true that $A_f \cap B_S \neq \emptyset$.
2. Two other groups made up of 225 decision-oriented linguistic summaries that are made during the decision-making stage during the cut of project. These groups are called: *Decision_Low* which satisfies that the attributes used as filters and summarizers are clearly separated from $A_f \cap B_S = \emptyset$ and *Decision_High* which fulfills that $A_f \cap B_S \neq \emptyset$.

In addition, two other groups are formed, consisting of 101 linguistic summaries that are associated with forecasts that are predicted from the decisions made. In such a way that we call *Pronostic_Low* the one that fulfills that the attributes used as filters and summarizers are clearly separated from $A_f \cap B_S = \emptyset$ and *Pronostic_High* that fulfills that $A_f \cap B_S \neq \emptyset$.

Some examples of the summaries you get are presented below:

- Summary used in the diagnostic stage (*Diagnostic_Low*): Some registries indicate that 100.0% of the time the projects that "have high data quality" are satisfied that "they have defects in the control of tasks".
- Summary used in the decision stage (*Decision_Low*): Few records indicate that in 100.0% of the time the projects that "have poor performance of human resources" and "defects in the control of tasks" are fulfilled that have "defected in scope".

- Summary used in the decision stage (*Decision_High*): Few records indicate that 100.0% of the time the projects that "have defects in the effectiveness of human resources" and "have the decision to increase quality control" it is fulfilled that "they have the decision to renegotiate the scope".

Different results of the proposed inference model are compared considering variations of the group of indicators T in the six groups of linguistic summaries constructed and 12 combinations of parameters, see Table 1.

The Table 1 identifies that the best results are obtained by the combinations that use the sets of indicators $\{T_1, T_2, T_3, T_4\}$ and $\{Te_{1a}, Te_3, Te_4\}$ with mean values of the metric "Arithmetic mean of the error in the prediction" (*ME*), of 0.06 and 0.066 respectively. Between these two combinations of T values, the set $\{Te_{1a}, Te_3, Te_4\}$ obtained the lowest standard deviation, that is, it has a more stable behavior.

4.2 Results of Test 2 Comparison of the Proposal with Other Inference Methods

For this test, the 490 linguistic summaries learned are combined with a Fuzzy Inference maps that we call *NCM_LDS*. This map is characterized by using linguistic summaries of data in the connections between concepts with an architecture similar to rule-based maps [41]. The results of the *NCM_LDS* map are compared with the *m-FCM* [42, 43] and *NCM_Indeterminacy* [44, 45] map extensions reported in the bibliography.

The comparison is made on the same database used in the previous test that represents a multi-stage sequential decision-making problem described by M. J. Rezaee, S. Yousefi and collaborators in [46]. In the database the stages of diagnostic, decision and prognosis are clearly identified. The different methods are compared from the metric "Arithmetic mean of the error in the prediction" and the results are independently evaluated in the three stages, view Table 2 and Fig. 4.

Figure 4 shows, in a network diagram, the integrated results of the comparison in the three stages, from considering the "Arithmetic mean of the error in the prediction", where it is shown that the best results are obtained by the algorithm m or *NCM_LDS*, followed by *m-FCM*, with *NCM_Indeterminacy* being the worst performer. The *NCM_LDS* map outperforms the *m-FCM* map regarding prediction in diagnostic, but does not have significant differences regarding the decision and prognosis stages.

5 Conclusions

This article proposes a new inference approach based on linguistic data summarization to solve complex predictions problems. The proposal made is inspired by the

Table 1 "Arithmetic mean of the error in the prediction" with different parameters

	Diagnostic_Low	Diagnostic _High	Decision_Low	Decision_High	Pronostic_low	Pronostic _High	*Promedio de ME*	StdDev ME
Without T	0.116	0.124	0.037	0.063	0.008	0.099	0.075	0.046
T_1, T_4	0.116	0.162	0.108	0.117	0.038	0.009	0.092	0.057
T_1	0.079	0.079	0.079	0.208	0.001	0.15	0.099	0.071
T_4	0.116	0.161	0.088	0.069	0.038	0.01	0.08	0.054
T_3	0.064	0.078	0.067	0.161	0.009	0.099	0.08	0.05
T_1, T_3, T_4	0.116	0.139	0.08	0.068	0.009	0.033	0.074	0.049
T_1, T_2, T_3, T_4	0.064	0.134	0.011	0.134	0.009	0.008	0.06	0.061
Te_{1a}	0.116	0.14	0.05	0.072	0.009	0.099	0.081	0.047
Te_3	0.116	0.077	0.108	0.072	0.038	0.038	0.075	0.033
Te_4	0.092	0.091	0.036	0.035	0.099	0.099	0.075	0.031
Te_{1a}, Te_4	0.116	0.14	0.036	0.072	0.009	0.12	0.082	0.052
Te_{1a}, Te_3, Te_4	0.042	0.14	0.036	0.072	0.009	0.099	0.066	0.048

Table 2 Descriptive analysis of the variable "Arithmetic mean of the error in prediction" in the stages of diagnostic, decision and prognosis

		NCM_Indeterminacy	*m-FCM*	*NCM_LDS*
Diagnostic	Media	0.6339	0.2687	0.0493
	StdDev	0.10377	0.29901	0.14229
Decision	Media	0.5555	0.1038	0.1039
	StdDev	0.10848	0.22754	0.22766
Prognostic	Media	0.4698	0.1095	0.1095
	StdDev	0.06311	0.22033	0.22041

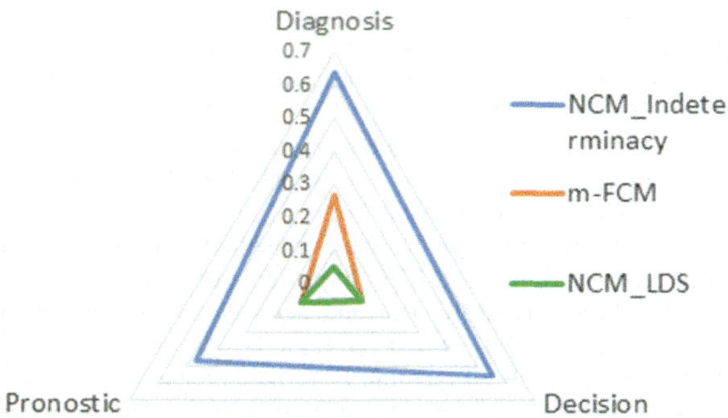

Fig. 4 Behavior of the algorithms before "Arithmetic mean of the error in the prediction"

similarities that exist between fuzzy rules and linguistic data summaries, as well as the operation of fuzzy inference systems.

From linguistic summaries, a multilayer network is built where the layers represent the different elements that make up the linguistic summaries. The use of neutrosophic fuzzy sets is established for the representation of linguistic terms. Aggregation processes take place from operations with single-value triangular neutrosophic numbers.

The validation of the new inference model is carried out by applying the same in the decision-making scenario in project management and the metric "Arithmetic mean of the error in the prediction". Comparisons of the inference model are made using different variants of quality indicators as weights that affect the inference process. It is shown that the best results are achieved using the indicator sets $\{T_1, T_2, T_3, T_4\}$ and $\{Te_{1a}, Te_3, Te_4\}$.

In addition, linguistic summaries are combined with fuzzy cognitive maps for the resolution of a multi-stage sequential decision-making problem that occurs during project evaluation. We have called this extension NCM_LDS and the inference result of this extension is compared with implementations of m-FCM and NCM_Indeterminacy maps. It is shown that the NCM_LDS reports better results than the rest of the maps in the diagnostic stage, but that it does not have significantly better results than the m-FCM in the decision and prognosis stages. The NCM_Indeterminacy map is shown to be the one with the worst results.

References

1. Wu, D., Mendel, J.M.: Linguistic summarization using IF–THEN rules and interval type-2 fuzzy sets. IEEE Trans. Fuzzy Syst. **19**(1), 136–151 (2010). https://doi.org/10.1109/TFUZZ.2010.2088128
2. Yager, R.R.: On linguistic summaries of data. Knowl. Discov. Databases, 378–389 (1991)
3. Yager, R.R.: A new approach to the summarization of data. Inf. Sci. **28**(1), 69–86 (1982). https://doi.org/10.1016/0020-0255(82)90033-0
4. Zadeh, L.A.: A computational approach to fuzzy quantifiers in natural languages. Comput. Math. Appl. **9**(1), 149–184 (1983)
5. Kacprzyk, J.: Fuzzy logic for linguistic summarization of databases. In: IEEE International Fuzzy Systems. Conference Proceedings (Cat. No.99CH36315), vol. 2, pp. 813–818 (1999). https://doi.org/10.1109/FUZZY.1999.793053
6. Boran, F.E., Akay, D., Yager, R.R.: An overview of methods for linguistic summarization with fuzzy sets. Expert Syst. **61**, 356–377 (2016). https://doi.org/10.1016/j.eswa.2016.05.044
7. Pérez, I., Piñero, P.Y., Martín, N., Bello, R.: Tendencias en la sumarización lingüística de datos. Rev. Cuba. Transform. Digit. **2**(1), 79–101 (2021)
8. Piñero, P.Y., Pérez, I., García, R., Toscanini, P.: Caracterización de los estándares de gestión de proyectos y su impacto en la gestión económico financiera de las organizaciones orientadas a proyectos. Guayaquil, Ecuador: Universidad Católica de Santiago de Guayaquil, ISBN: 978-9942-82-514-8 (2020).
9. Wilbik, A., Dijkman, R.M.: On the generation of useful linguistic summaries of sequences. IEEE International Conference on Fuzzy Systems, pp. 555–562 (2016). https://doi.org/10.1109/FUZZ-IEEE.2016.7737736
10. Kacprzyk, J.: Intelligent data analysis via linguistic data summaries: a fuzzy logic approach. Classification and Information Processing at the Turn of the Millennium. Springer, pp. 153–161 (2000). https://doi.org/10.1007/978-3-642-57280-7_17
11. Kaczmarek-Majer, K., Hryniewicz, O., Dominiak, M., Święcicki, Ł.: Personalized linguistic summaries in smartphone-based monitoring of bipolar disorder patientshttps://doi.org/10.2991/eusflat-19.2019.56
12. Chiang, D.A., Chow, L.R., Wang, Y.F.: Mining time series data by a fuzzy linguistic summary system. Fuzzy Sets Syst. **112**(3), 419–432 (2000). https://doi.org/10.1016/S0165-0114(98)00003-7
13. Khedidja, B., Allel, H., Mohand, L.: Data summarization for sensor data management: towards computational-intelligence-based approaches. Int. J. Comput. Digit. Syst. **9**(5), 825–833 (2020). https://doi.org/10.12785/ijcds/090505
14. Sojka, P., Hudec, M., Švaňa, M.: Linguistic summaries in evaluating elementary conditions, summarizing data and managing nested queries. Informatica, 1–16 (2020). https://doi.org/10.15388/20-INFOR428
15. Zadrożny, S., Kacprzyk, J.: Fuzzy Analytical Queries: A New Approach to Flexible Fuzzy Queries. IEEE International Conference on Fuzzy Systems, pp. 1–8 (2020). https://doi.org/10.1109/FUZZ48607.2020.9177556
16. Wilbik, A., Kaymak, U., Dijkman, R.M.: A method for improving the generation of linguistic summaries. IEEE International Conference on Fuzzy Systems, pp. 1–6 (2017). https://doi.org/10.1109/FUZZ-IEEE.2017.8015752

17. Smits, G., Nerzic, P., Pivert, O., Lesot, M.J.: Efficient generation of reliable estimated linguistic summaries. IEEE International Conference on Fuzzy Systems, pp. 1–8 (2018). https://doi.org/10.1109/FUZZ-IEEE.2018.8491604
18. Donis-Díaz, Muro, A.G., Bello, R., Morales, E.V.: A hybrid model of genetic algorithm with local search to discover linguistic data summaries from creep data. Expert Syst. Appl., **41**(4), 2035–2042 (2014), Part 2. https://doi.org/10.1016/j.eswa.2013.09.002
19. Donis-Díaz, Bello, R., Kacprzyk, J.: Using ant colony optimization and genetic algorithms for the linguistic summarization of creep data. Intell. Syst. 81–92 (2015). https://doi.org/10.1007/978-3-319-11313-5_8
20. Pérez, I., Piñero, P.Y., Bello, R., Acuña, L.A., García-Vacacela, R.: Linguistic summaries generation with hybridization method based on rough and fuzzy sets. International Joint Conference on Rough Sets, Lecture Notes in Computer Science ISSN 0302-9743. Springer, 12179 LNAI, ISBN 978-3-030-52704-4, pp. 385–397 (2020). https://doi.org/10.1007/978-3-030-52705-1_29
21. Pérez, I., Santos, O., Bello, R., Piñero, P.Y.: Algorithms for linguistic data summarization, help in decision-making in project-oriented organizations. XXII Ibero-American Conference on Software Engineering, CIbSE, pp. 633–640 (2019)
22. Pérez, I., Piñero, P.Y., Bello, R., García, R., Piñero, P.E., Piñero, C.M.: Aplicaciones de la sumarización lingüística de datos en la toma de decisiones en gestión de proyectos. V Conferencia Internacional en Ciencias Computacionales e Informáticas (CICCI) (2020)
23. Pérez, I., Santos, O., García, R., Piñero, P.Y., Ramírez, E.C.: Descubrimiento de resúmenes lingüísticos para ayuda a la toma decisiones en gestión de proyecto. Rev. Cuba. Cienc. Informáticas **12**, 163–175 (2018)
24. Pérez, I., López, P., Varona, E., Piñero, P.Y., García, R.: Construcción de resúmenes lingüísticos a partir rasgos de la personalidad y el desempeño en el desarrollo de software. Rev. Cuba. Cienc. Informáticas **12**, 135–150 (2018)
25. Pérez, I., Piñero, P.Y., García-Vacacela, R., Bello, R., Acuña, L.A.: Discovering fails in software projects planning based on linguistic summaries. International Joint Conference on Rough Sets. Lecture Notes in Computer Science ISSN 0302-9743, Springer, 12179 LNAI, ISBN 978-3-030-52704-4, pp. 365–375 (2020). https://doi.org/10.1007/978-3-030-52705-1_27
26. Kacprzyk, J., Zadrożny, S.: Linguistic database summaries and their protoforms: towards natural language based knowledge discovery tools. Inf. Sci. **173**(4), 281–304 (2005). https://doi.org/10.1016/j.ins.2005.03.002
27. Zadeh, L.A.: A prototype-centered approach to adding deduction capability to search engines–the concept of protoform. In: Intelligent Systems. Proceedings, First International IEEE Symposium, vol. 1, pp. 2–3 (2020). https://doi.org/10.1109/IS.2002.1044219
28. Sremac, S., Kazimieras, E., Matić, B., Kopić, M., Stević, Ž.: Neuro-fuzzy inference systems approach to decision support system for economic order quantity. Econ. Res.-Ekon. Istraživanja, **32**(1), 1114–1137 (2019). https://doi.org/10.1080/1331677X.2019.1613249
29. Blej, M., Azizi, M.: Comparison of Mamdani-type and Sugeno-type fuzzy inference systems for fuzzy real time scheduling. Int. J. Appl. Eng. Res. **11**(22), 11071–11075 (2016)
30. Sari, W.E., Wahyunggoro, O., Fauziati, S.: A comparative study on fuzzy Mamdani-Sugeno-Tsukamoto for the childhood tuberculosis diagnosis. In: AIP Conference Proceedings, vol. 1755 (2016)
31. Cox, E., O'Hagan, M.: The fuzzy systems handbook, second edition: a practitioner's guide to building, using and maintaining fuzzy systems, 2nd edn. Morgan Kaufmann, San Diego (1998)
32. Pérez, I., Piñero, P.Y., García-Vacacela, R., Bello, R., Santos, O., Vázquez, M.: Extensions to Linguistic summaries indicators based on Neutrosophic Theory, applications in Project Management Decisions. Neutrosophic Sets Syst. vol. 22 (2018)
33. Smarandache, F., Pramanik, S.: New trends in neutrosophic theory and applications. Brussels, Belgium: Pons Editions, ISBN: 978-1-59973-498-9 (2016). [Online] Available: http://arxiv.org/abs/1611.08555. [Accessed: 2 May 2021]
34. Wang, H., Smarandache, F., Zhang, Y., Sunderraman, R.: Single valued neutrosophic sets. Multispace multistructure, vol. 4, pp. 410–413 (2010)

35. Yager, R.R., Kacprzyk, J.: The ordered weighted averaging operators: theory and applications (1997). Springer. https://doi.org/10.1007/978-1-4615-6123-1
36. Liu, P., Wang, Y.: Multiple attribute decision-making method based on single-valued neutrosophic normalized weighted Bonferroni mean. Neural Comput. Appl. **25**(7), 2001–2010 (2014). https://doi.org/10.1007/s00521-014-1688-8
37. Piñero, P.Y., Pérez, I., Rivero, C.C., Rojas, C., Sosa, R., Torres, S.: Repositorio de datos para investigaciones en gestión de proyectos. Rev. Cuba. Cienc. Informáticas **13**(1), 176–191 (2019)
38. Pérez, I., Piñero, P.Y., Bello, R., Acuña, L.A., García, R.: Linguistic summaries generation with hybridization method based on rough and fuzzy sets. Rough Sets, Havana, Cuba, pp. 385–397 (2020), https://doi.org/10.1007/978-3-030-52705-1_29
39. Pérez, I., García, R., Piñero, P.Y., Mahdi, G.S., Peña, M.: Experiencias en el uso de técnicas de softcomputing en la evaluación de proyectos de software. Rev. Investig. Oper. **41**(1), 106–117 (2020)
40. Melendez, J.R., Gracia, G.E.: Corporate social responsibility: Decisions and strategic actions in the industrial scenario of oil companies. Espacios, **40**(11) (2019). [Online] Available: http://www.revistaespacios.com/a19v40n11/19401123.html. [Accessed: 14 April 2021]
41. Barriba, I., Rodríguez-Díaz, A., Castro, J.R., Sanchez, M.A.: An approach to fuzzy inference system based fuzzy cognitive maps. Computer Science and Engineering—Theory and Applications. Springer International Publishing, pp. 151–166 (2018). https://doi.org/10.1007/978-3-319-74060-7_8
42. D'Onofrio, S., Papageorgiou, E.: Using fuzzy cognitive maps to arouse learning processes in cities. Designing Cognitive Cities, Springer International Publishing, pp. 107–130 (2019). https://doi.org/10.1007/978-3-030-00317-3_5
43. Harmati, I.Á., Kóczy, L.T.: Notes on the rescaled algorithm for fuzzy cognitive maps. Computational Intelligence and Mathematics for Tackling Complex Problems, Springer International Publishing, pp. 43–49 (2020). https://doi.org/10.1007/978-3-030-16024-1_6
44. Al-Subhi, S.H., Papageorgiou, E.I., Pérez, I., Mahdi, G.S., Acuña, L.A.: Triangular neutrosophic cognitive map for multistage sequential decision-making problems. Int. J. Fuzzy Syst. (2021). https://doi.org/10.1007/s40815-020-01014-5
45. Villamar, C.M., Suarez, J., Coloma, L.D., Vera, C., Leyva, M.: Analysis of technological innovation contribution to gross domestic product based on neutrosophic cognitive maps and neutrosophic numbers. Neutrosophic Sets Syst. (2019). https://doi.org/10.5281/zenodo.3569637
46. Rezaee, M.J., Yousefi, S., Babaei, M.: Multi-stage cognitive map for failures assessment of production processes: an extension in structure and algorithm. Neurocomputing **232**, 69–82 (2017). https://doi.org/10.1016/j.neucom.2016.10.069

Chapter 3
Linguistic Data Summarization with Multilingual Approach

Iliana Pérez Pupo, Pedro Y. Piñero Pérez, Salah Hassan Al-subhi, Gaafar Sadeq Saeed Mahdi, and Rafael E. Bello Pérez

Abstract The development of information systems increases the volume of data and the need for its processing for decision-making. Algorithms are required that allow the discovery of behavior patterns and their interpretability. In this context, the linguistic summarization of data arises as one of the descriptive knowledge discovery techniques with a promising approach to produce summaries from a database using natural language, where authors such as Yager and Zadeh were pioneers and set guidelines in the development of these techniques. In this work, new algorithms are proposed for the generation of linguistic summaries from data that combine different soft computing techniques such as: rough sets, the learning of probabilistic graphs with controlled natural languages for the generation of linguistic summaries with an approach multilingual. In particular, controlled natural languages are proposed for the Spanish, English, Japanese and Arabic languages. The proposed algorithms are compared with other techniques reported in the bibliography and are subject to expert evaluation.

Keywords Artificial intelligent · Controller natural language · Linguistic data summarization · Project management

I. Pérez Pupo (✉) · P. Y. Piñero Pérez
Grupo de Investigaciones en Gestión de Proyectos, Universidad de las Ciencias Informáticas, Carretera a San Antonio km 2 ½, La Habana, 54830 Havana, CP, Cuba
e-mail: iperez@uci.cu

S. H. Al-subhi
Solarplex Electyrical and Electronics Contracting Co. Combined Security Group Co, Alfahahel, Kuwait

G. S. S. Mahdi
Business Intelligence and Consulting Group, BusinessRedmine SA, Yemen Office, Yemen

R. E. Bello Pérez
Centro de Investigación en Informática, Universidad Central Marta Abreu de las Villas, Santa Clara, Cuba
e-mail: rbellop@uclv.edu.cu

P. Y. Piñero Pérez et al. (eds.), *Artificial Intelligence in Project Management and Making Decisions*, Studies in Computational Intelligence 1035,
https://doi.org/10.1007/978-3-030-97269-1_3

1 Introduction

Several authors [1–4] assert that one of the latent problems in unstructured decision-making processes is associated with the discovery of patterns in data from information systems. In this sense, the development of techniques and tools that allow the discovery of hidden non-trivial dependencies in the data under an easily interpretable approach by decision-makers from different areas or domains constitutes an object of interest [2, 5–7].

In this context, techniques such as Linguistic Data Summarization (LDS) have been created. These techniques allow the generation of linguistic summaries from structured databases, using sentences in natural language [8].

In the consulted bibliography, several works of systematic reviews associated to LDS [9–12] stand out. These works characterize the different LDS techniques, showing that the following difficulties exist:

- The decision-making environments where these techniques have been applied to have uncertainty with phenomena such as: ambiguity, indeterminacy or imprecision [13]. However, in the techniques reported in the bibliography, it is appreciated that they only analyze the degree of certainty in the information, avoiding the treatment of indeterminacy or falsehood.
- Most of the algorithms reported in the bibliography generate summaries that reflect frequent behaviors of the data, but not those of low frequency. This affects its applicability in environments where anomalous data needs to be detected, such as fraud detection in revenue assurance, error detection in project planning [14], among others.
- Despite the high applicability of project management in dissimilar sectors of society, there is little LDS researches in this context; which is one of the motivations for this research, considering that there is a repository of databases [15] that facilitate the validation of the results.
- The algorithms reported in the bibliography, such as those based on rules [5] and metaheuristics [16], do not exploit the statistical information associated with the correlation between variables during the search process. This affects the efficiency of the algorithms because they can generate candidate summaries that do not have an actual representation in the data. These jobs generally include intermediate stages of cleaning candidate summaries.
- In the proposals made by [17] the generation of summaries with "Many" and "Most" quantifiers is favored, limiting the generation of linguistic summaries with other quantifiers. This approach does not facilitate the identification of situations with little occurrence in the database, affecting the efficiency of the algorithms in solving anomalous data detection problems, such as the detection of epidemics, failures, income assurance, among others.
- The metaheuristics reported in the bibliography [17] have high computational cost, an element that negatively influences the applicability of these algorithms in real decision-making environments.

- As a result of this analysis, it was identified that most of the works are based on the simple treatment of strings in the English language [1, 11, 12]. This situation affects the efficiency of the algorithms for the generation of the summaries regarding their interpretability and presentation in multiple languages.

The objective of this research is to improve the efficiency of the algorithms to generate summaries with quality, interpretable and that facilitate decision-making under a multilingual approach.

The work is structured in the following sections. The second section presents algorithms for the generation of linguistic summaries under a multilingual approach are presented. In the third section, the results are analyzed and the proposed algorithms are compared with others reported in the bibliography. Finally, the conclusions and future work are presented.

2 New Approach for Linguistic Summaries Generation by Using Controller Natural Language

The linguistic summaries are made up of elements set that express the relationships between concepts, such as quantifiers (*Q*), filters (*R*), summarizers (*S*) and quality indicators of linguistic summaries (*T*). In the construction of summaries, these concepts are grouped according to the structures "*Qy's are S*" and "*QRy's are S*" [18] as shown below:

- *Qy's are S*: linguistic summaries without filter, like "Most of the workers are punctual". In this summary, *Q* represents the linguistic term "Most", while the expression "punctual" represents the summarizer.
- *QRy's are S*: linguistic summaries with filter, where *R* can be made up of one or more attributes qualifying the object, like "Most young workers are late", where "young" constitutes the filter and "they are late" represents the summarizer.

Zadeh [19] and other authors [20] have formalized these linguistic summaries structures into stereotyped forms called protoforms, which have been extended according to the characteristics of the different application scenarios [9]. However, the most widespread are those proposed by Zadeh [21].

The possible values of *S* and *R*, represented by linguistic variables, are represented depending on the problem in question and constitute inputs to the algorithms.

The algorithms for generating linguistic summaries that are proposed in this section have, as an essential element, that they generate a set of candidate summaries. Each candidate summary constitutes an object *P(Q, R, S, T)* formed by the quantifier *Q*, the filters *R*, the summarizers *S* and a set of quality indicators *T*. The formation of the linguistic summaries with this data structure, facilitates its representation under a multilingual approach, using different controlled natural languages.

2.1 Definition of Controlled Natural Languages for the Construction of Multilingual Linguistic Summaries

A novel element that is proposed in this research is the construction of linguistic summaries under a multilingual approach, using Controlled Natural Languages (CNL), see Definition 1. The objective of using controlled natural languages is to achieve greater expressiveness in the linguistic summaries generated.

Definition 1 [22, 23]: A controlled natural language (CNL) is defined as a constructed language, based on a certain natural language, more restrictive in terms of lexicon, syntax and/or semantics while preserving the majority of their natural properties, so that base language speakers can intuitively and correctly understand texts in controlled natural language, at least to a substantial degree. Seen in a practical way, a CNL is defined by a grammar that establishes the syntax and a dictionary that describes the lexicon and semantics.

To characterize languages, in general we can use the PENS classification scheme [22], which is based on the following dimensions:

- Precision, which measures the degree to which the meaning of a text can be extracted from it, such that each language can be classified as: imprecise (P1), less imprecise (P2), reliable in interpretation (P3), interpreted in a way deterministic (P4) and with a certain semantics (P5).
- Expressiveness, which describes the range of propositions that a language is capable of expressing. Considering this criterion, languages can be classified as: inexpressive (E1), low expressiveness (E2), medium expressiveness (E3), with high expressiveness (E4) and with maximum expressiveness (E5).
- Naturalness, which measures how close the language is to the natural language, in terms of readability and understandability of the base natural language speakers. Under this criterion, languages are classified as: non-natural (N1), with a predominance of non-natural elements (N2), with a predominance of natural elements (N3), with natural sentences (N4) and with natural texts (N5).
- Simplicity, related to the effort required to fully implement the syntax and semantics of the language, through a formal model. Considering this criterion, languages can be classified as: very complex (S1), without exhaustive descriptions (S2), with large descriptions (S3), with short descriptions (S4) and with very short descriptions (S5).

For each dimension, the conceptual extremes of the attributes represent natural languages. For example, the English language is classified as $P_1E_5N_5S_1$, while the language of propositional logic is classified as $P_5E_1N_1S_5$. The linguistic summaries proposed in this research are generated from CNL that are not located in either of the two extremes, but in intermediate characterizations of each dimension; being the classification $P_3E_3N_4S_4$ the most appropriate to characterize them.

In particular, in this research controlled natural languages are constructed for the languages: Spanish, English, Japanese and Arabic. For each CNL authors define a grammar and a dictionary associated to domain. CNL for summaries in English: Grammar 1 constitutes an example built for English language.

CNL for summaries in English:

Grammar 1 defined for the controlled natural language proposed for the English language is presented below.

<*Linguistic summary*>::= <D> <data descriptor connector> <Q> <quantifier connector> <y’> <filters connector> <R> <summarizer connector> <S> |
<Q> <y’> <filters connector> <R> <summarizer connector> <S> |
<Q> <y’> <summarizer connector> <S>

<D>::= “almost all” | “most” | “many” | “around a half of’ | “some” | “few” | “very few”

<data descriptor connector>::= “records show, that”

<quantifier connector>::= “of times,”

<filters connector>::= “with”

<summarizer connector>::= “have”

<y’>::= <subject>

<subject>::= <simple phrase>

<Q>::= <linguistic quantifier> | <numeric quantifier>| <mixed quantifier>| <percent quantifier>

<linguistic quantifier>::= “almost all” | “most” | “many”| “around a half of’ | “some” | “few” | “very few”

<numeric quantifier>::= “more than 95% of’ | “around of 85% of’ | “around of 75% of’ | “approximately el 50%” | “close to 33%” | “less than 17%” | “less than 5%”

<mixed quantifier>::= “very few (less than 5%)” | “few (around of 15%)” | “some of (close to 33%)” | “around a half of’ | “many (close to 65%)” | “most of (around of 83%)” | “almost all of’

<percent quantifier>::= <percent connector > <numeric value in percent>

<percent connector>::= “in”

<S>::= <phrase>

<R>::= < phrase>

< phrase>::= < phrase> <logical operator> <simple phrase > | <simple phrase>

<logical operator>::= <conjunction> | <disjunction>

Example of English's summary: "Very few records show, that 94.0% of times projects with regular data quality, have bad productivity."

Where:

<D>:	Very few	<data descriptor connector>:	records show, that
<Q>:	94.0%	<quantifier connector>:	of times,
<y'>:	Projects	<filters connector>:	with
<R>:	regular data quality	<summarizer connector>:	have
<S>:	bad productivity		

CNL for summaries in Spanish:

Grammar 2 defined for the controlled natural language proposed for the Spanish language is presented below.

<Linguistic summary>::= <D> <data descriptor connector> <Q> <quantifier connector> <y'> <filters connector> <R> <summarizer connector> <S> |
<Q> <genre connector> <y'> <filters connector> <R> <summarizer connector> <S> |
<Q> <genre connector> <y'> <summarizer connector> <S>
<D>::= "casi la totalidad" | "la mayoría" | "muchos"| "aproximadamente la mitad" | "algunos" | "pocos" | "muy pocos"
<gender connector>::= "de los" | "de las"
<data descriptor connector>::= "registros en la base de datos, reportan que"
<quantifier connector>::= "de las veces,"
<filters connector>::= "que cumplen" | "que tienen" | "que son"
<summarizer connector>::= "tienen" | "se cumple que"
<y'>::= <subject>
<subject>::= <simple phrase>
<Q>::= <linguistic quantifier> | <numeric quantifier>| <mixed quantifier>| <percent quantifier>
<linguistic quantifier>::= "casi la totalidad" | "la mayoría" | "muchos" | "aproximadamente la mitad" | "algunos" | "pocos" | "muy pocos"
<numerical quantifier>::= "más del 95%" | "alrededor del 85%" | "alrededor del 75%" | "aproximadamente el 50%" | "cercano al 33%" | "menos del 17%" | "menos del 5%"
<mixed quantifier>::= "muy pocos (menos del 5%)" | "pocos (alrededor del 15%)" | "algunos (alrededor del 33%)" | "aproximadamente la mitad" | "muchos (cerca del 65%)" | "la mayoría (alrededor del 83%)" | "casi la totalidad"

Table 1 Example of application of controlled language in linguistic summaries

Parts of CNL	Summary 1: about pregnant cardiopaths	Resumen 2: about projects evaluation
<*D*>	muchos	la mayoría
<gender connector>	de los	de los
<data descriptor connector>	registros en la base de datos, reportan que en el	
<*Q*>	100.0%	90.404%
<quantifier connector>	de las veces	de las veces,
<*y'*>	las embarazadas cardiópatas	los proyectos
<filters connector>	que tienen	que tienen
<*R*>	ausencia de soplo holosistólico	muy alta calidad del dato
<summarizer connector>	tienen	tienen
<*S*>	ausencia de cianosis	muy baja cantidad de competencias con baja evaluación

<percent quantifier>::= <percent connector > <numeric value in percent>
<percent connector>::= "en el"
<S>::= <phrase>
<R>::= < phrase>
< phrase>::= < phrase> <logical operator> <simple phrase > | <simple phrase>
<logical operator>::= <conjunction> | <disjunction>

Examples of summaries obtained by using Grammar 2 (Table 1):

(1) "Muchos registros en la base de datos, reportan que en el 100.0% de las veces, las embarazadas cardiópatas que tienen ausencia de soplo holosistólico tienen ausencia de cianosis."
(2) "La mayoría de los registros en la base de datos, reportan que en el 90.404% de las veces, los proyectos que tienen muy alta calidad del dato tienen muy baja cantidad de competencias con baja evaluación."

CNL for summaries in Japanese

Grammar 3 defined for the CNL proposed for the Japanese language:

<Linguistic summary>::= <data descriptor connector> <D> <quantifier connector D> <Q> <quantifier connector Q> <R> の <y'> は <S> と報告しています。|
<S> の <y'> の <Q> は、<R>です。|
<Q> の <y'> は <S> です。
<D>::= "ほとんど" | "たくさん" | "約半分" | "いくつか" | "少数" | "非常に少数"
<Q>::= <linguistic quantifier> | <percent quantifier>
<linguistic quantifier>::= "ほとんど" | "たくさん" | "約半分" | "いくつか" | "少数" | "非常に少数"
<percent quantifier>::= <Numeric value in percent>
<data descriptor connector>::= "データベース内の"
<quantifier connector D>::= "記録が、"
<quantifier connector Q>::= " % の確率で、"
<y'>::= <subject>
<subject>::= <simple phrase>
<S>::= <phrase>
<R>::= <phrase>
<phrase>::= <phrase> | <particle> | <simple phrase>
<particle>::= "は" | "が" | "と" | "の"

Examples of Japanes summary:
Example 1:
データベース内の非常に数少ない記録が、94.0%の確率で、データ品質の悪いのプロジェクトは悪い評価と報告しています。
CNL: <data descriptor connector> <D> <quantifier connector D> <Q> <quantifier connector Q><R> の<y'> は <S> と報告しています。
Traduction: "*Very few records in the database report a 94.0% probability that projects with bad data quality have bad evaluation.*"

Where:

<R>: デ ー タ 品質の 悪 い — bad data quality

<y'>: プ ロ ジ ェ ク ト — projects

<S>: 悪い 評価 — bad evaluation

<D>: 非常に 数少な い — Very few

<Q>: 94.0%

Example 2: [illegible]

CNL : <S> [illegible] <y'> [illegible] <Q> [illegible] , <R> [illegible] . (Structure: *QRy's are S*)

Traduction: "*Most projects with bad data quality have bad evaluation*".

Where:

<*R*>*:* [illegible] bad data quality

<*y'*>*:* [illegible] projects

<*S*>*:* [illegible] bad evaluation

<*Q*>*:* [illegible] most

Example 3: [illegible]

CNL: <Q> [illegible] <y'> [illegible] <S> [illegible] . (Structure: Qy's are S)

Traduction: "*Most projects have bad evaluation.*"

Where:

<Q>: [illegible]

<y'>: [illegible]

<S>: [illegible]

CNL for summaries in Arabic

Grammar 4 defined for the controlled natural language proposed for the Arabic language is presented below.

<Linguistic summary>::= <S> <summarizer connector> <R> <filters connector> <y'> <quantifier connector> <Q> <data descriptor connector> <D> |
<S> <summarizer connector> <R> <filters connector> <y'> <Q> |
<S> <summarizer connector> <y'> <Q>

<D>::= "جدا قليل" | "القليل" | "بعض" | "ما يقارب النصف من" | "الكثير" | "أغلبية"

<data descriptor connector>::= "السجلات تظهر أن"

<quantifier connector>::= "من المرات،"

<filters connector>::= "التي لديها"

<summarizer connector>::= "لديه"

<y'>::= <subject>

<subject>::= <simple phrase>

<Q>::= <linguistic quantifier> | <numeric quantifier>| <mixed quantifier>| <percent quantifier>

<linguistic quantifier>::= "القليل" | "بعض" | "تقريباً نصف" |"الكثير من" | "أغلبية" | "تقريباً كل" | "قليل جداً من" | "من"

<numerical quantifier>::= "" | "حوالي 50%" | "حوالي 75%" | "حوالي 85%" | "أكثر من 95%" | "أقل من 5%" | "أقل من 17%" | "ما يقارب 33%"

<mixed quantifier>::= "حوالي) بعض" | "(حوالي 15%) القليل" | "(أقل من 5%) قليل جداً من" | "أغلبية" | "(ما يقارب 65%) الكثير" | "تقريباً النصف" | "(33%" | "تقريباً كل" | "(حوالي 83%)"

<percent quantifier>::= <percent connector > <numeric value in percent><percent

<percent connector>::= "في"

<S>::= <phrase>

<R>::= < phrase>

< phrase>::= < phrase> <logical operator> <simple phrase > | <simple phrase>

<logical operator>::= <conjunction> | <disjunction>

Example: قليل جداً من السجلات في قاعدة البيانات تشير إلى أن هناك نسبة 94.0% أن المشاريع التي لديها جودة بيانات سيئة لديها تقييم سيء

Traduction: "*Very few records in the database report a 94.0% probability that projects with bad data quality have bad evaluation.*"

Where:

<R>: سيئة بيانات جودة — bad data quality

<y'>: مشاريع — projects

<S>: سيء تقييم — bad evaluation

<D>: جدا قليل — Very few

<Q>: 94.0%

3 New Algorithms for Generation of Multilingual Linguistic Summaries

This section discusses different approaches that combine linguistic data summarization with controlled natural languages to generate summaries that can be expressed in multiple languages. Algorithms are proposed following two fundamental approaches:

- The generation of linguistic summaries combining probabilistic graphs.
- The generation of summaries applying the theory of rough sets.

In the analysis of the algorithms, the variables are used: *n*—size of the database, *p*—number of variables. *K*—mean size of the summaries, *g*—generations, *d*—number of individuals and *z*—number of summaries.

In the definition of the parameters for the experimental runs, it is established as a principle that the parameters that are common to several algorithms are set with a value to mitigate bias in the experimentation. The parameters are used:

- A_{LV} a set of linguistic variables is constructed for each database that is used by all the algorithms.
- α the value 0.33 is set and is applied equally by all algorithms.
- $\alpha_{kI} = 0.9$ and $\alpha_{k2} = 0.8$, they are used by the *RST_LDS* algorithm and are considered moderately high values, so as not to restrict too much the extent of the search process.
- *U:* dataset
- *T:* list of *T* quality values used to sort summaries
- *CNL:* natural language controlled defined
- LV_Q: linguistic fuzzy variable established for quantifiers Fig. 1, Eq. (1)

$$\mu_Q(x) = \begin{cases} 1 & if\ x \le b\ and\ Q\ left\ function \\ a \cdot e^{-\frac{(x-b)^2}{2c^2}} & other\ functions \\ 1 & if\ x > b\ and\ Q\ right\ function \end{cases} \tag{1}$$

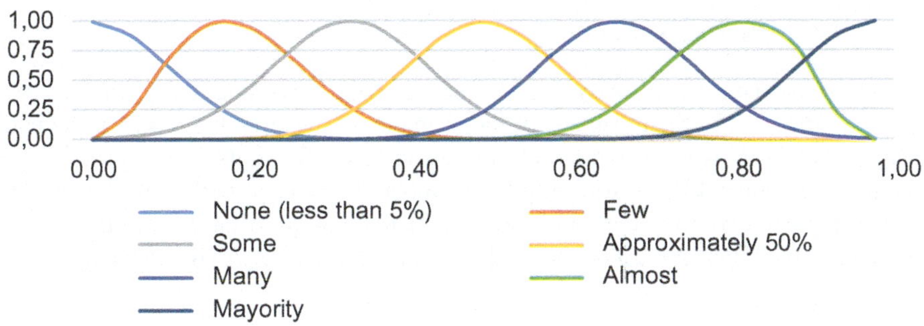

Fig. 1 Linguistic variable LV_Q

3.1 LPA_LDS Algorithm Based on Probabilistic Graphs

In this algorithm the input data set is transformed into boring values applying the principle of maximum membership. Then probabilistic graph learning techniques [24, 25] are applied that allow the discovery of relationships between variables. From the learning of the graphs, a set of trees is generated that are used for the construction of candidate summaries. Then, the candidate linguistic summaries are humanized allowing the generation of linguistic summaries with a multilingual approach, see Algorithm 1.

Algorithm 1 LPA_LDS.

Inputs

$U, A_{LV}, LV_Q, T, CNL, \alpha$

1. *Candidates* = $\varnothing$
2. *UB = transform_fuzzy(U,* A_{LV}*)*
3. *prob_graph = do_probabilistic_graph(*U_B*)*
4. *for each prob_graph*$_i$ *tree in prob_graph do*
5. *if prob_graph*$_i$ *has more than one vertex*
6. *candidate_summaries*$_i$ *= do_candidate_from_branchs (prob_graph*$_i$*)*
7. *Candidates* $\leftarrow$ *candidate_summaries*$_i$

   End of step 5

   End of step 4

8. *CandidateSummaries = generate_summaries(U, Candidates,* A_{LV}*,* LV_Q*, T)*
9. *calculate_quantifiers(CandidateSummaries,* LV_Q*)*
10. *calculate_T(CandidateSummaries,* α*)*
11. *Summaries = humanize_summaries (CandidateSummaries,* A_{LV}*, LVQ, CNL)*
12. *ranking_summaries = sort_summaries(Summaries)*
13. *Retorn ranking_summaries*

In step 2, a data set (U) is received as input and the set of linguistic variables A_{LV} that describe each of the attributes of the problem in question. Each one of the data of the input set U is transformed into linguistic labels by applying the principle of maximum membership, considering the membership of each data to the fuzzy sets of the linguistic variable that represents the attribute in question. From this process the transformed data set U_B is obtained. This step has a complexity $O(p{\cdot}n{\cdot}m)$.

Then in step 3, an algorithm for learning the probabilistic model [26] that best approximates the behavior of the data is applied to the transformed data set. Examples of algorithms that can be used in this step are listed below:

- *Chow Liu*: algorithm developed in 1968, initially proposed for the construction of trees from data [25].
- *Rebane-Pearl Algorithm* [27] extends the Chow Liu algorithm and enables the learning of poly-trees, allowing the description of higher-order interactions [25].
- *Polytree Aproximation Algorithm* (PA) 1998 that bases learning on the calculation of mutual marginal information and mutual conditional information.
- *Learning Polytree Algorrithm (*LPA): this algorithm is the one used in experimentation and the one recommended in this research, it was proposed in [28] and its main contribution is the improvement of efficiency with respect to the algorithms that precede it [26]. This algorithm calculates the marginal and conditional dependencies using calculation methods based on the concept of entropy. The complexity of this algorithm is $O(n^3)$ [26].

The algorithms in step 3 that learn the probabilistic models generate a poly-tree, where each tree can generate a candidate summary. Then in step 4, a candidate object is generated for each of the trees, in such a way that the node at the root of the tree is identified as the summarizer attribute and the nodes in the branches are identified as filters. The number of trees that can be generated is bounded at the top by the total number of attributes p and the number of elements in filters and maximum summarizers is $p-1$. Therefore, the complexity of this step is bounded by $p(p-1)$, where $O(p^2)$ regarding the order of relationships [26].

In steps 5 through 8, candidate summaries are generated from the list of *Candidates* by transforming each object into a summary object whose attributes are filters and summarizers.

In step 9, for each of the candidate summaries, we proceed to determine the quantifier that best adapts to the summary with an algorithm that has a complexity $O(n^2 \cdot p)$.

In step 10, the T indicators are calculated to evaluate the quality of the linguistic summaries obtained. The computational complexity of this step is bounded superiorly by $O(n \cdot p)$ and is given by the complexity of the methods for calculating the indicators. At the end of this step, you have a set of candidate summaries (*Candidates*), each candidate with the structure $P(Q, R, S, T)$.

Then in step 11, the final summaries (*Summaries*) are generated from humanizing the candidate summaries *humanize_summaries(Candidates,* A_{LV}, LV_Q, *LNC)* using Algorithm 3. with a complexity $O(n \cdot m \cdot p^2)$.

In step 12 the summaries are ordered according to T values. The computational complexity of this step is bounded superiorly by $O(n \log n)$.

3.2 Algorithm for the Generation of Linguistic Summaries Based on Rough Sets (RST_LDS)

The theory of rough sets is also presented as a valid alternative for the generation of linguistic summaries. This theory was proposed in 1986 by Z. Pawlak for application in data inconsistencies [29–31].

Let an information system be $IS = (U, A \cup D)$ where U is a set of data described by the sets of attributes A and D, usually referred to in rough set theory as predictive attributes and decision attributes respectively, H is called the total set of attributes of the problem in question, such that $H = A \cup D$, $X \subseteq U$ a set of objects and $B \subseteq A$ a selected set of attributes. From the information contained in B, the variable X can be approximated as follows:

The lowest approximation of X with respect to B is:

$$B * (X) = \{x \in U : B(x) \subseteq X\} \tag{2}$$

The top approximation of X with respect to B is:

$$B^*(X) = \{x \in U : B(x) \cap X \neq \Phi\} \tag{3}$$

The border region can be defined as:

$$BN_B(X) = B * (X) - B * (X) \tag{4}$$

The negative region of decision d with respect to B is:

$$NEG_B(X) = U - B * (X) \tag{5}$$

Indiscernibility relation: defines the equivalence relation IND_B [32, 33] and this relation is denoted by:

$$IND_B(B) = \{(x, y) \in U \times U : a(x) = a(y) \text{for each } a \in B\} \tag{6}$$

The positive region of decision d with respect to B is:

$$POS_B(d) = \grave{E}\{B_*(X) : X \in U/IND(d), d \in D\} \tag{7}$$

Another useful concept is the dependence in k degree, which is explained in the next paragraph Definition 2.

Definition 2 Intuitively, a set of decision attributes D depends entirely on a set of attributes B, denoted by $B \Rightarrow D$, if all the values of the attributes D are uniquely determined by the values of the attribute in B. In other words, D is totally dependent on B, if there is a functional dependence between the values of D and B [32]. D depends on B to k degree, where $k \in [0, 1]$ and is denoted by $B \Rightarrow_k D$, see Eq. (8). If $k = 1$ then D depends totally on B, whereas if $k < 1$ then D depends on partially by B.

$$\boldsymbol{k} = \frac{|\boldsymbol{POS_B(D)}|}{|\boldsymbol{U}|} \tag{8}$$

$$POS_B(D) = \bigcup_{X \in U/D} B_*(X) \tag{9}$$

Algorithm 2 is based on the theory of rough sets for the generation of linguistic summaries. This algorithm extends the work proposed in [34] with the multilingual approach.

In the first step of this algorithm, the variables of the input data set are transformed into linguistic variables, applying the principle of maximum membership. Two sets of attribute groups are constituted that represent the filters and summarizers of the linguistic summaries to be constructed. These sets are initialized with all the variables of the data set as independent subsets, during the search process they are transformed. Subsequently, the dependence in k degree is calculated between groups of attributes that represent filters and summarizers; those that exceed the established threshold will be candidate summaries. Then, the candidate linguistic summaries are filtered and humanized using a controlled natural language, allowing the generation of the final linguistic summaries.

Algorithm 2 *RST_LDS*

Inputs

U, A_{LV}, LV_Q, T, CNL, α

Information system $IS = (U, A \cup D)$

A, D: information system attributes.

α_{k1}: alpha cut off k_1 establishes the upper limit of the dependence in k degree in the generation of summaries.

α_{k2}: alpha cut off k_2 establishes the lower limit of the dependence in k degree in the generation of summaries

1. *Transform $SI = (U, A \cup D)$ in linguistic dataset considering fuzzy sets and maximum membership principles.*
2. *A_ItemSet = attributes_to_sets($A \cup D$)*
3. *D_ItemSet = A_ItemSet*
4. *Stackset.push(A_ItemSet)*
5. *CS = {}*
6. *While Stackset is not empty*

6.1. *A_ItemSet = Stackset.pop*

6.2. *for each $B \subseteq$ A_ItemSet*

6.3. *for each $X \subseteq$ D_ItemSet: $B \cap X = \varnothing$*

6.4. $POS_B(D) = \bigcup_{X \in U/D} B_*(X)$, $k(X,B) = \frac{|POS_B(D)|}{|U|}$, $B \Rightarrow_k D$, $k(B, X) \geq \alpha_{k2}$

6.5. *for each $O_t(B,X) \in POS_B(D)$*

6.6. *if ($k(B, X) \geq \alpha_{k1}$) then*

6.7. *$CS = CS \cup O_t(B,X)$*

6.8. *else*

6.9. *if ($k(B, X) \geq \alpha_{k2}$) then*

6.10. *Stackset.push({$B \cup X$})*

End of step 6.9

End of step 6.6

End of step 6.5

End of step 6.3

End of step 6.2

End of step 6

7. *CandidateSummaries = summaries_from_candidates(U, CS, A_{LV}, LV_Q, T)*
8. *calculate_quantifiers(CandidateSummaries, LV_Q)*
9. *calculate_T(CandidateSummaries, α)*
10. *Summaries = humanize_summaries(CandidateSummaries, A_{LV}, LV_Q , CNL)*
11. *ranking_summaries = sort_summaries(Summaries)*
12. *Retornar ranking_summaries*

In step 1, the information system (IS) is transformed into a set of linguistic data considering the fuzzy sets and the principles of maximum membership, it has complexity $O(p{\cdot}n{\cdot}q)$.

In step 2, each element $a \in A \cup D$ is transformed into a set $\{a\}$ and inserted into a set *A_ItemSet*, Example: $\{a_1, a_2, a_3\} \geq \{\{a_1\}, \{a_2\}, \{a_3\}\}$. Then in step 3 and 4 the *D_ItemSet* variables and the *Stackset* stack are initialized. The complexity of these steps is $O(p)$.

Step 6 is the heart of this algorithm. In steps 6.2 and 6.3 we create pairs of sets B and X that represent possible filters and summarizers respectively. Then, in step 6.4, the positive region $POS_B(D)$ is calculated using Definition 2 and Eqs. (1.21), (1.22). The condition $k(B, X) \geq \alpha_{k2}$, helps to find summaries between attributes of partial dependency, but considering the sets of attributes that have a minimum level of relationship. Then, in step 6.6, $O_t \in U$, $O_t(B, X)$ represents the values of the attributes $a \in B \cup X$ in the O_t object; if there is a dependency greater than the threshold α_{k1} in the O_t context, then the values of the $O_t(B, X)$ will be used as candidate summaries. If the dependency found $\alpha_{k2} < \mathrm{k}(B, X) < \alpha_{k1}$, the set $\mathrm{Z} = B \cup X$ is generated and this new set is added to the stack to continue the search process where this new set will become part of filters in new summaries. The complexity of the cycle started in step 6 is bounded superiorly by $O(2^p)$.

In step 7 the pairs of sets that are in *CS* are traversed; each set represents a possible linguistic summary. Therefore, the complexity of this step is bounded by $O(n)$. Then steps 8 to 12 are applied exactly the same as steps 8–12, proposed in Algorithm 1.

At the end of step 9, the algorithm generate a set of candidate summaries (*CandidateSummaries*); then in step 10 of Algorithm 3 is applied to generate the linguistic summaries with a computational complexity $O(n{\cdot}m{\cdot}p^2)$.

Finally, the complexity of the *RST_LDS* algorithm is $O(max(n^2{\cdot}p,\ n{\cdot}m{\cdot}p^2))$. This algorithm, by its nature, allows the discovery of linguistic summaries that represent situations with a very low frequency of appearance in the database and with a high dependency between attributes. These summaries allow the identification of anomalous data, which broadens the applicability of this algorithm in different contexts, for example, in scenarios associated with income assurance [14] or computer security.

3.3 Algorithm for the Humanization of Linguistic Summaries Using Controlled Natural Languages.

Algorithm 3 describes how these CNL are used to construct linguistic summaries from candidate summaries.

This algorithm traverses the set of candidate summaries obtained by the algorithms proposed in the previous subsections, where each summary has the structure $P(Q, R, S, T)$. Between steps 4 and 6, the different elements that make up the summary are translated using the $CNL_{Dicctionary}$ dictionary defined for the CNL in the specific domain of application.

Once all the elements that make up the summary have been translated into the specified CNL, in step 7 we proceed to join or concatenate the parts of the summary, so that the sentence is finally constructed according to the grammar specified in $CNL_{Grammar}$.

Algorithm 3 *humanize_summaries (CandidateSummaries,* A_{LV}, LV_Q, *CNL)*

Inputs
CandidateSummaries: list of candidate summaries
A_{LV}, LV_Q, *T, CNL*
1. *summaries = []*
2. *For each P(Q, R, S, T) in CandidateSummaries*
3. *translations = []*
4. *For each element of P(Q, R, S, T)*
5. *element_translation=search_dictionary(element,* A_{LV}, LV_Q, $CNL_{Dicctionary}$*)*
6. *translations ← element_translation*
   End of step 4
7. *sentence = generate_sentence(translations,* $CNL_{Grammar}$*)*
8. *summaries ← sentence*
   End of step 2
9. *Retorn summaries*

The complexity of this algorithm is $C(n, p, m) = n\cdot(p + 1)\cdot m\cdot(p + 1) = n\cdot m\cdot(p^2 + 2p + 1)$, given because the cycle in step 2 is executed at most n times, the cycle in step 4 is executed at maximum $p + 1$ times and the complexity of the dictionary search in step 5 is bounded by $O(m\cdot(p + 1))$. Therefore, Algorithm 3 has a complexity of $O(n\cdot m\cdot p^2)$.

4 Analysis of Results and Validation of the Proposed Algorithms

In this section, the proposed algorithms are validated by comparing them with the *AG_LDS* and *ACO_LDS* algorithms based on genetic algorithms and ant colony respectively [17]. The algorithms are compared with 10 databases and with respect to the following indicators:

- Group of indicators for the evaluation of the summaries described in [35] that we will call traditional indicators T_1, T_2, T_3, T_4 and T_5.
- Extended indicators Te_{1a}, Te_{1b}, Te_3, Te_4 and Te_5 proposed in [36].
- The indicator "Strength of discovered dependencies" proposed in [17].
- The indicator "Coverage of different situations in database" that measures the capacity of the algorithms to generate linguistic summaries that cover different

phenomena in the data. This indicator constitutes a measure of the versatility of the algorithms for their application in different contexts.

To calculate the indicator "Coverage of different situations in database", the relative frequency of the appearance of the following situations is analyzed:

- Very few records in database
- Few records in database
- Some records in database
- About half of the records in database
- Many of the records in database
- Most of the records in database

Then, for each database, the dispersion of the relative frequencies is calculated using the standard deviation. What is desired in this case is that there be a balance in amount of summaries generated, covering each of the different situations described.

In algorithm comparison tests, parametric and non-parametric statistical tests are used as appropriate, using the *IBM SPSS Statistics* version 25 tool [37]. It works with a significance of 0.05 and confidence intervals of 99%. The following 10 databases are used in the algorithm comparison.

Databases associated with the evaluation and decision-making in project management, taken from the repository published in [15, 38]:

1. BD_gp_eval_proy_hard: database with 229 records that represent real cases of project evaluation based on 8 indicators in different cuts.
2. BD_gp_eval_proy_integral: database with 8430 records that represent the comprehensive evaluation of projects based on 19 attributes.
3. BD_gp_indicadores: database made up of 174 records associated with the evaluation of project programs based on 14 attributes.
4. BD_gp_competencias. database made up of 20,574 instances and 12 attributes that represent the evaluation of the competences of the participants in projects.
5. BD_gp_rrhh: database of 3 931 instances and 9 attributes associated with the evaluation of human resources in project outages.
6. BD_gp_requerimientos: database associated with the management of project requirements, made up of 1 310 instances and 7 attributes.
7. BD_gp_toma_decisiones: database associated with decision-making in project management and made up of 1 011 instances and 17 attributes.

Database associated with medical decision-making. The first two associated with clinical records of pregnant women with heart disease, offered by medical experts from the "Ramón González Coro" Gyneco-obstetric Teaching Hospital and the Institute of Cardiology and Cardiovascular Surgery in Havana. The third database is on covid cases in Cuba, published by the Ministries of Public Health.

- *BD_med_congenito:* database created from clinical records on symptoms, diagnoses, treatments and prognoses of pregnant women with heart disease with congenital diseases. It is made up of 464 instances and 20 attributes.

- *BD_med_valvular*: database created from clinical records on symptoms, diagnoses, treatments and prognoses of pregnant women with heart disease with valvular diseases. Made up of 304 instances and 17 attributes.
- *BD_med_covidCuba*: database published by the Cuban Ministry of Public Health associated with covid cases in Cuba that records the sex, age, country of origin of the cases, municipality, province and type of contagion of some patients by covid-19. It includes 18 151 instances and 6 attributes.

4.1 Comparison of the Proposed Algorithms with Others Reported in the Bibliography

In this section, all algorithms are analyzed based on their global behavior, considering the indicators explained above. As part of the analysis of the overall performance of the algorithms, the *Page's L Test* method [39], was used, which shows, in various ways, the position or ranking of the algorithms with respect to each database and the proposed indicators.

4.1.1 Procedure for the Application of the Page's L Test

1. Selection of the criteria for establishing the ranking, in our case they are the indicators T_1, T_2, T_3, T_4, T_5, Te_{1a}, Te_{1b}, Te_3, Te_4, Te_5, "Coverage of different situations in database", "Strength of discovered dependencies" and runtime.
2. Assignment of the ranking values, assigning a position to each algorithm with respect to its result in the comparison tests carried out previously, considering the following elements:
    a. For each indicator, each algorithm is assigned a ranking value that corresponds to the position where it was located. The best algorithm is ranked first and then the rest are ranked based on the results.
    b. If *n* algorithms, with respect to the same indicator, were located in the same place, they are assigned the same value corresponding to the midpoint of the n positions. For example, if two algorithms are in the first place with respect to a certain indicator, each one is assigned the value 1.5.
3. Ordering of the samples from the sums of the rankings.
4. Application of the non-parametric *Page's L Test*. In all cases, the level of statistical significance is 0.05, which means that the null hypothesis will be rejected as long as the result is ≤ 0.05 the Statex tool [40] is used in this step.

4.1.2 Application of the Page's L Test

For the application of *Page's L Test*, the global summary of the position of each algorithm with respect to each indicator analyzed in the comparisons was used as input data.

$H_0 : m_1 = m_2 = ...$ Null hypothesis

$H_1 : m_1 < m_2 < ....$ Alternative hypothesis.

The results showed that there are significant differences: $L = 1058, Z = 3.632446$, $P(> Z) = 0.00014$, therefore, the null hypothesis is rejected and it is shown that the data comply with the order $m_1 < m_2 < ....$

As part of the confirmation of these results, the Statex computer tool itself applies Chi-Square techniques and Monte Carlo simulation, generating the following results.

Chi-Square Results:

$X^2 = 13.194662$, $\frac{P(>X)}{2} = 0.00014$. The null hypothesis is rejected and it is shown that the data comply with the order ($m1 < m2 < ...$).

It is shown that, regarding the analyzed indicators, the algorithms are ordered in the following ranking: *RST_LDS, GA_LDS, LPA_LDS* and *ACO_LDS.*

In the global analysis, it is concluded that the proposed *RST_LDS* algorithm obtains the best results while the *ACO_LDS* algorithm is the one that obtains the lowest evaluation.

In the experimentation runs, all the algorithms were executed on the same computer, with the following characteristics: I5, 4 cores and 8 GB RAM; in this way possible biases in the measurement due to the use of different technologies are avoided.

Regarding the execution time indicator, the algorithms are located in the following order: *RST_LDS, LPA_LDS, GA_LDS* and *ACO_LDS* see Table 2.

Table 2 shows the computational complexity of the algorithms used in the validation: those proposed in the research and those reported in the bibliography. In it, it is observed that the proposed algorithms improve the computational complexity of those reported in the bibliography. There is a correspondence between the computational complexity of the algorithms and the execution time.

With the previous analysis, it is shown that the proposed algorithms improve the efficiency with respect to the algorithms based on metaheuristics used in the experimentation.

Table 2 Complexity of algorithms

Algorithms	Complexity	Mean time (seconds)
RST_LDS	$O(n^2 \cdot p)$	2.11
LPA_LDS	$O(n^3)$	52.76
GA_LDS	$O(kgnd)$	1212.62
ACO_LDS	$O(zgnd^2)$	19,692.03

4.2 Validation of the Algorithms in Their Ability to Generate Summaries Under a Multilingual Approach

For the validation of the algorithms under a multilingual approach, algorithms are applied to the case study on Project Evaluation at the Center for Consulting and Development of Business Architectures of the University of Computer Science. This case study is chosen because it has enough data collected in the information system for project management BusinessRedmine [41, 42] between the years 2010–2020. In this scenario, the algorithms were applied to analyze:

- The relationship that exists between the personality traits of human resources and their performance in project results.
- Decision making during project outages.
- The detection of errors during the planning processes.

Five experts participated in the evaluation of the linguistic summaries in Spanish and their applicability in the use case, 4 of them PhDs and a master's degree in project management. The average of 29.6 years of experience being 13 the minimum value of years of experience of the experts. The experts evaluated the summaries obtained with respect to the following criteria: interpretability of summaries and ease of decision-making. In general, the following results are obtained:

- Regarding the "interpretability of summaries", a high evaluation was obtained with a low dispersion of opinions.
- Regarding the "ease of decision-making", most of the criteria were evaluated with high values while presenting low dispersion in opinions.

The R_{WG} correlation coefficient (Benavente Reche 2009) was also applied for the analysis of the agreement between the experts, whose value was $R_{WG} = 0.95$, showing that there is agreement between them.

To evaluate the multilingual approach, summaries were generated in Spanish, English, Japanese and Arabic. The summaries generated were evaluated by another set of experts in the specified languages (see Table 3).

The linguistic summaries were evaluated by the experts regarding the following criteria: grammar, simplicity, readability and spelling. For the aggregation of the experts' preferences, the computation technique with 2-tuples words [43, 44] is applied with the following set of linguistic terms LBTL = {Very low, Low, Medium, High, Very high}. In Fig. 2 the aggregation of the evaluations assigned by the experts in these languages to the linguistic summaries generated in the study case is represented, oscillating between "high" and "very high".

Figure 2 shows that the criteria best evaluated globally were simplicity and spelling. Arabic and Japanese were rated better than English, but in all three languages, most of the criteria scored high or very high. In general, the experts consulted were satisfied with the results, but it is identified as an open line of research, to continue working on readability.

Table 3 Experts that reviewed the summaries in different languages

Experts that reviewed the summaries in Arabic			
	1st language	2nd language	Time
Expert 1	Arabic	Spanish	18 years of Spanish proficiency
Expert 2	Arabic	Spanish	20 years of mastery of Spanish
Experts that reviewed the summaries in Japanese			
	1st language	2nd language	Time
Expert 3	Japanese	Spanish	More than 15 years of command of Spanish
Experto 4	Japanese	Spanish	13 years of Spanish proficiency
Experts that reviewed the summaries in English			
	1st language	2nd language	Time
Expert 5	English	Spanish	Know both languages since childhood
Expert 6	Spanish	English	More than 30 years of English proficiency

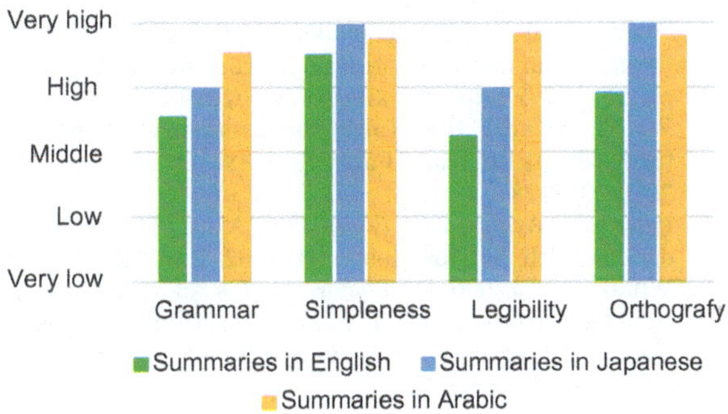

Fig. 2 Result of the evaluation assigned by the experts, to the linguistic summaries in the English, Japanese and Arabic languages

5 Conclusions

In the bibliography analyzed, it is identified that there is consensus among the different researchers regarding the elements that make up the linguistic summaries. Furthermore, it was identified that most of the protoforms and algorithms reported in the bibliography are designed for the English language, affecting their ability to represent summaries in multiple languages.

In the experiments carried out, it is shown that the *RST_LDS* algorithm surpasses the rest of the algorithms with which it is compared, reported in the bibliography. The applicability of the algorithms proposed for decision-making is demonstrated from a project management case study.

The interpretability of the summaries is validated from a multilingual approach and the results obtained are submitted to the analysis of experts in the Spanish, English, Japanese and Arabic languages; considering the grammar, simplicity, legibility and spelling criteria. In most of the criteria, the evaluation was between high and very high, with spelling the criterion for the best evaluation and the readability of the summaries in English, the criterion that obtained the lowest evaluation.

As future work, new forms of construction of linguistic summaries and methods for their visualization will be developed, covering a greater number of languages. In addition, it is intended to generate new algorithms based on protoforms that enhance the integration of multiple summaries in the form of long texts that support decision-making.

References

1. Dijkman, R., Wilbik, A.: Linguistic summarization of event logs—a practical approach. Inf. Syst. **67**, 114–125 (2017). https://doi.org/10.1016/j.is.2017.03.009
2. Amghar, D., Amine, Chikh, M.: Extracting a Linguistic summary from a medical database. Int. J. Intell. Syst. Appl., **10**(12), 16–26 (2018). https://doi.org/10.5815/ijisa.2018.12.02
3. Kaczmarek-Majer, K., Hryniewicz, O., Dominiak, M., Święcicki, Ł.: Personalized linguistic summaries in smartphone-based monitoring of bipolar disorder patients.https://doi.org/10.2991/eusflat-19.2019.56
4. Peláez-Aguilera, M.D., Espinilla, M., Fernández, M.R., Medina, J.: Fuzzy Linguistic Protoforms to Summarize Heart Rate Streams of Patients with Ischemic Heart Disease. Hindawi, pp. 11 (2019). https://doi.org/10.1155/2019/2694126
5. Wu, D., Mendel, J.M.: Linguistic summarization using IF–THEN rules and interval type-2 fuzzy sets. IEEE Trans. Fuzzy Syst. **19**(1), 136–151 (2010). https://doi.org/10.1109/TFUZZ.2010.2088128
6. Igde, E.Y., Aydoğan, S., Boran, F.E., Akay, D.: Linguistic summarization of structured patent data. Int. Sch. Sci. Res. Innov. **11**, 1062–1065 (2017)
7. Wilbik, A., Barreto, D., Backus, G.: On relevance of Linguistic SUMMARIES–a case study from the agro-food domain. International Conference on Information Processing and Management of Uncertainty in Knowledge-Based Systems, pp. 289–300 (2020). https://doi.org/10.1007/978-3-030-50146-4_22
8. Yager, R.R.: On Linguistic Summaries of Data. Knowl. Discov. Databases, pp. 378–389 (1991)
9. Pérez, I., Piñero, P.Y., Martín, N., Bello, R.: Tendencias en la sumarización lingüística de datos. Rev. Cuba. Transform. Digit. **2**(1), 79–101 (2021)
10. Boran, F.E., Akay, D., Yager, R.R.: An overview of methods for Linguistic summarization with fuzzy sets. Expert Syst. Appl. **61**, 356–377 (2016). https://doi.org/10.1016/j.eswa.2016.05.044
11. Marín, N., Sánchez, D.: On generating Linguistic descriptions of time series. Fuzzy Sets Syst. **285**, 6–30 (2016). https://doi.org/10.1016/j.fss.2015.04.014
12. Ramos-Soto, A., Martin-Rodillab, P.: Enriching linguistic descriptions of data: A framework for composite protoforms. Fuzzy Sets Syst., pp. 26 (2019). https://doi.org/10.1016/j.fss.2019.11.013
13. García, J.A., Peña, A., Piñero, P.Y., Pérez, R.B.: Project control and computational intelligence: trends and challenges. Int. J. Comput. Intell. Syst. **10**(1), 320–335 (2017). https://doi.org/10.2991/ijcis.2017.10.1.22
14. Castro, G.F.: Modelo para el aseguramiento de ingresos en organizaciones orientadas a proyectos basado en minería de datos anómalos. Doctoral, Universidad de las Ciencias Informáticas, La Habana, Cuba (2017)
15. Piñero, P.Y., Pérez, I., Hechavarría, C., Rojas, C., Sosa, R., Torres, S.: Repositorio de datos para investigaciones en gestión de proyectos. Rev. Cuba. Cienc. Informáticas **13**(1), 176–191 (2019)
16. Donis-Díaz, Muro, A.G., Bello, R., Morales, E.V.: A hybrid model of genetic algorithm with local search to discover linguistic data summaries from creep data. Expert Syst. Appl., **41**(4), part 2, 2035–2042 (2014). https://doi.org/10.1016/j.eswa.2013.09.002

17. Donis-Díaz, Bello, R., Kacprzyk, J.: Using ant colony optimization and genetic algorithms for the linguistic summarization of creep data. Intelligent Systems'2014, pp. 81–92 (2015). https://doi.org/10.1007/978-3-319-11313-5_8
18. Yager, R.R.: A new approach to the summarization of data. Inf. Sci. **28**(1), 69–86 (1982). https://doi.org/10.1016/0020-0255(82)90033-0
19. Zadeh, L.A.: A computational approach to fuzzy quantifiers in natural languages. Comput. Math. Appl. **9**(1), 149–184 (1983)
20. Kacprzyk, J., Zadrożny, S.: Linguistic database summaries and their protoforms: towards natural language based knowledge discovery tools. Inf. Sci. **173**(4), 281–304 (2005). https://doi.org/10.1016/j.ins.2005.03.002
21. Zadeh, L.A.: A prototype-centered approach to adding deduction capability to search engines–the concept of protoform. In: Intelligent Systems. Proceedings, First International IEEE Symposium, vol. 1, pp. 2–3 (2002). https://doi.org/10.1109/IS.2002.1044219
22. Kuhn, T.: A Survey and Classification of Controlled Natural Languages. Comput. Linguist. ISSN 0891–2017, **40**(1), 121–170, doi: https://doi.org/10.1162/COLI_a_00168 (2014).
23. Kittredge, R.I.: Sublanguages and Controlled Languages. The Oxford Handbook of Computational Linguistics, 2nd edn, ISBN 9780199573691, pp. 430–447 (2003). https://doi.org/10.1093/oxfordhb/9780199573691.013.015
24. Pérez, I., Santos, O., Bello, R., Piñero, P.Y.: Algorithms for linguistic data summarization, help in decision-making in project-oriented organizations. XXII Ibero-American Conference on Software Engineering, CIbSE, pp. 633–640 (2019)
25. Chow, C., Liu, C.: Approximating discrete probability distributions with dependence trees. IEEE Trans. Inf. Theory **14**(3), 462–467 (1968). https://doi.org/10.1109/TIT.1968.1054142
26. Soto, M.: A single connected factorized distribution algorithm and its cost of evaluation. Doctorado, Instituto de Cibernética, Matemática y Física. Departamento Matemática Interdisciplinaria, Universidad de La Habana, Cuba (2003)
27. Rebane, G., Pearl, J.: The Recovery of Causal Poly-Trees from Statistical Data. Third Conference on Uncertainty in Artificial Intelligence (2013)
28. Soto, M., Ochoa, A.: A factorized distribution algorithm based on polytrees. Proceedings of Congress on Evolutionary Computation, vol. 1, pp. 232–237.https://doi.org/10.1109/CEC.2000.870300
29. Pawlak, Z., Skowron, A.: Rough sets: some extensions. Inf. Sci. **177**(1), 28–40 (2007). https://doi.org/10.1016/j.ins.2006.06.006
30. Wang, G.: Extension of rough set under incomplete information systems. IEEE World Congress on Computational Intelligence, Conference on Fuzzy Systems. FUZZ-IEEE'02, vol. 2, pp. 1098–1103 (2002). https://doi.org/10.1109/FUZZ.2002.1006657
31. Zhai, J., Zhang, S., Zhang, Y.: An extension of rough fuzzy set. J. Intell. Fuzzy Syst. **30**(6), 3311–3320 (2016). https://doi.org/10.3233/IFS-152079
32. Pawlak, Z.: Rough sets. Int. J. Comput. Inf. Sci. **11**(5), 341–356 (1982). https://doi.org/10.1007/BF01001956
33. Bello, R., Verdegay, J.L.: Rough sets in the soft computing environment. Inf. Sci. ISSN 0020-0255, **212**, 1–14 (2012). https://doi.org/10.1016/j.ins.2012.04.041
34. Pérez, I., Piñero, P.Y., Bello, R., Acuña, L.A., García-Vacacela, R.: Linguistic summaries generation with hybridization method based on rough and fuzzy sets. International Joint Conference on Rough Sets. Lecture Notes in Computer Science ISSN 0302-9743, Springer, ISBN 978-3-030-52704-4, pp. 385–397 (2020). https://doi.org/10.1007/978-3-030-52705-1_29
35. Kacprzyk, J., Zadrożny, S.: Linguistic data summarization: a high scalability through the use of natural language? Scalable Fuzzy Algorithms for Data Management and Analysis: Methods and Design. IGI Global, pp. 214–237 (2010)
36. Pérez, I., Piñero, P.Y., García, R., Bello, R., Santos, O., Leyva, M.Y.: Extensions to Linguistic summaries indicators based on Neutrosophic theory: applications in project management decisions. Neutrosophic Sets Syst. Univ. N. M. **22**, 87–100 (2018)
37. McCormick, K., Salcedo, J.: SPSS Statistics for Data Analysis and Visualization. John Wiley & Sons, ISBN: 978-1-119-00355-7 (2017)

38. Pérez, I.: Repositorio de investigaciones en gestión de proyectos (2021). [Online] Available: https://gespro.uci.cu/projects/repositorio-de-investigaciones-en-gestion-de-proyectos/dmsf. [Accessed: 2 March 2021]
39. Bartz-Beielstein, T., et al.: Benchmarking in optimization: Best practice and open issues. ArXiv200703488 Cs Math Stat (2020)
40. Mandel, I.: Expert system for statistics STATEX—30 years after. Social Science Research Network, Rochester, NY, SSRN Scholarly Paper ID 3664990 (2020)
41. López, P.: Procedimiento para la aplicación de test de personalidad como apoyo a la gestión de recursos humanos en proyectos informáticos. Máster, Universidad de las Ciencias Informáticas, La Habana, Cuba (2017)
42. Pérez, I., García, R., Piñero, P.Y., Mahdi, G.S., Peña, M.: Experiencias en el uso de técnicas de softcomputing en la evaluación de proyectos de software. Rev. Investig. Oper. **41**(1), 106–117 (2020)
43. Herrera, F., Martínez, L.: A 2-tuple fuzzy linguistic representation model for computing with words. IEEE Trans. Fuzzy Syst. **8**(6), 746–752 (2000). https://doi.org/10.1109/91.890332
44. Zhang, S., Wei, G., Alsaadi, F.E., Hayat, T., Wei, C., Zhang, Z.: MABAC method for multiple attribute group decision making under picture 2-tuple linguistic environment. Soft Comput. **24**(8), 5819–5829 (2020). https://doi.org/10.1007/s00500-019-04364-x

Chapter 4
Project to Improve Offensive Phase Finalization of Futsal Teams by Using Linguistic Data Summarization Techniques

Gregorio Morales González, Beatriz Sánchez Córdova, Magda Mesa Anoceto, Pedro Y. Piñero Pérez, and Iliana Pérez Pupo

Abstract In this paper authors collect and analyze data from 2018 futsal season of the University of Informatics Sciences team. In order to obtain linguistic summaries, authors apply linguistic data summarization techniques. The linguistic summaries are storage in database and later use for coaches for decision-making on 2019 season. To validate the research, authors analyze the variables involved in the offensive phase finalization in the "*Universidad de las Ciencias Informáticas (UCI)*" futsal team with respect to the 2018/2019 seasons, from observation and SPSS statistical analysis, significant differences are evident in the variable goal and positive passes, managing to score 33 goals in 2019 and 13 goals in 2018; for this, 85 positive shots were produced averaging 2.57 shots to reach goal. In the plays of static positional strategies there are significant differences in the variables goal and positive shots. Ten more goals are scored in 2019 as every 3.69 positive shot opportunities the team scores a goal. While in 2018 three goals are scored and an average of 7.66 positive shots are used to score a goal. Positional transitions on the move improve in 2019 where eight goals are scored and on average 2.12 shots are produced to score a goal. Counterattacking transitions present a stable behavior; this variable depends on the way of defending to close and counterattack. In free-kicks two goals are scored in

G. Morales González
Departamento de Deportes, Grupo de Investigaciones en Gestión de Proyectos, Universidad de las Ciencias Informáticas, Habana, Cuba
e-mail: grego@uci.cu

B. Sánchez Córdova · M. Mesa Anoceto
Universidad del Deporte, Habana, Cuba

P. Y. Piñero Pérez · I. Pérez Pupo (✉)
Grupo de Investigaciones en Gestión de Proyectos, Universidad de las Ciencias Informáticas, Habana, Cuba
e-mail: iperez@uci.cu

P. Y. Piñero Pérez
e-mail: ppp@uci.cu

P. Y. Piñero Pérez et al. (eds.), *Artificial Intelligence in Project Management and Making Decisions*, Studies in Computational Intelligence 1035,
https://doi.org/10.1007/978-3-030-97269-1_4

2019, improving in this regard. The free-kick is improved in 2019 since no goals were scored in 2018 and two goals were scored this year.

Keywords Artificial intelligence · Futsal finalization · Futsal offensive · Linguistic data summarization techniques

1 Introduction

The offensive finalization phase is the key to success in the game of futsal. One of the problems to be solved in Cuban futsal is the definition in the resolution of the plays. Therefore, the behavior of these situations in the game must be studied to find technical tactical solutions in their execution. For De Bortoli et al. [1], "the achievement index is the best predictor of performance in futsal"; that is, the successful completion of offensive actions marks the end result in this sport. Therefore, the constant study of the variables of shooting on goal and the efficiency translated into a goal continues to be a problem of permanent study in futsal.

Tenroller [2] states that the shot "is the action of hitting the ball while standing or moving, deflecting it or giving it a trajectory. It is a technical foundation that precedes the goal" (p. 67).

For Tenroller [2]; Lozano [3] Lima [4]; Bolsonaro [5]; López [6] and Lago [7] the shot or shot on goal is the culmination of the offensive game and is considered the most decisive technical action when establishing a positive result in the match. Any type of hitting the ball is valid if this is a goal. The shot is usually fired quickly and explosively.

The finalization is one of the offensive phases given by the individual technical action preceded by a series of situations that can be individual, group and collective that make all the elaboration carried out culminate in a goal, that is, the finalization is the final result of the tactical decisions of the whole team or a part of it.

The offensive phase finalization is the essence of the game itself; it is the one that culminates the actions to attack from the openings in their different manifestations. All the attacks, be these fast-positional attacks and the transitions with their variants are directed to look for the goal. Finalization is the end product of the game that is accomplished with goals and causes teams to win competitive events.

For the authors Morales et al. [8] the study of the theoretical-methodological foundations for the improvement of offensive actions in the finalization phase in the futsal game, indicates that the structure of its internal logic from the technical and tactical components with the correct psychological preparation are the basis of counterattacks, quick attacks and positional attacks that improve the finalization phase.

Morales et al. [9] state that the study and planning of games in reduced spaces, becomes a methodological resource that causes the tactical improvement of the support and breaking clearances and in turn, raises the conditional and coordinate motor skills making the player be exposed to a wide range of cognitive learning strategies where they can choose those that fit their style of play.

It is important to take into account the study of the overall behavior of the team in competition. This can help restructure future planning and be able to provide answers in future competitions. Seeing the shortcomings in 2018 in the elaboration of offensive plays and the lack of completion of these, the training sessions are restructured based fundamentally on the game as a fundamental method of training. For this purpose, these are broken down into tactical blocks based on adapted games, conditioned games and modeled games. These include important variables for their improvement in the offensive phase finalization.

The objective of this article is to analyze the information from futsal team in 2018 season by using linguistic data summarization techniques. With the linguistic summaries obtained, the coaches improve his game strategies for 2019 season.

The paper consists of three sections. First section is the introduction to the futsal training problem. In second section authors presents the methodology applied in the research. Third section present the comparison results between 2018 and 2019 seasons. The SPSS program is used in the search for significant differences between the 2018 and 2019 seasons of the university futsal team in the first category provincial and university championships in the Cuban capital. Finally, authors present the paper conclusions.

2 Discovering Linguistic Summaries Deal with Futsal Team Weaknesses

The methodology for the observational categorical selection according to the research criteria, part of studies validated by Anguera and Hernández-Mendo [10] who support it from a nomothetic follow-up (the number of times a behavior is repeated is studied) and multidimensional (both proxemic and gestural behaviors are explained).

The instruments used to collect the data are based on the use of video cameras and note sheets registered on the technological support of the data. Ten observers are used that make up the observation protocol for data collection. TP, goals, static strategy plays, positional transitions in movements and transitions of counterattacks and their series in time are analyzed. From the collection of these in annotation sheets and from the observation of videos of the observed variables, these results are contrasted to have a better assessment of what was verified in both seasons.

The information about the 2018 season including both official games and training games, was collected by considering the following rules:

- Collect information about the "goals", "fails shots" and "positive shots" associate to each player.
- Positional transitions in movements for each player.
- Transitions of counterattacks and their series in time are analyzed.
- Information about static strategy plays.

- Name of player.

We collected the following data of 2018 season:

- Information from five game sections by 42 weeks. These game sections consist of both official games and training sections.
- In each game sections coaches get around 24 measures deal with:
  - Positional transitions in movements for each player.
  - Transitions of counterattacks and their series in time are analyzed.
  - Information about static strategy plays.

With collected information we create a database with 5040 rows.

In order to analyze the collected information, authors applied artificial intelligence techniques. In this case authors applied linguistic data summarization techniques [11] and they obtain linguistic summaries that represent the player's performance and the main mistakes in each game situation. Authors apply the following steps:

1. To apply algorithms, the discover linguistic summaries from data that cover all protoforms describes by Zadeh [12] and Kacprzyk [13]
2. Storage of the summaries in a database.
3. Recover linguistic summaries from database by coaches.
4. Decision-making by consider linguistic summaries' information.

In the first step, authors apply the algorithm *RST_LDS* [14] to discover linguistic summaries from data. This algorithm permits to discover summaries with the structures: "Q*y*'s are *S*" and "Q*Ry*'s are *S*".

Q represents the quantifiers ("very few", "few", "approximately half of situations", "much", "very much").

R and *S* are summary filters and summarizers respectively:

- Static positional strategy plays: static positional strategy plays on offense are those game situations that are started from set pieces to start or restart the game. They are broken down into: kickoff, restart kickoff in the event of a goal, goal kickoffs, corner kicks and kickoffs (for the investigation they are divided into kickoff in the defensive zone and kickoff in the offensive zone)
- Positional transitions in motion: the positional transitions in movement are assumed in the investigation as those recoveries of the ball in the action of the game that start fundamentally from the interceptions and the anticipations that are generated in the defense-attack transition, becoming the manager of the offensive elaboration when the surprise in the counterattack is lost. Its essential characteristic is the art of keeping the ball from short possessions of up to five passes, averages of between five and seven passes and longer possessions of seven onwards in the quest to end game situations in shot to goal.
- Counter-attack transitions: It is the transition that is made as soon as the ball is recovered and to attack with numerical superiority to carry out the finalization. Its essence is to surprise the opponent by taking advantage by having one or more players in attack than the opponent's defenders.

- Goals scored (GOAL) and their series in time: the goal is established as regulated in this type of game by the International Football Federation for futsal sport (FIFA-FS). The series in time indicates what happened during the eleven games by seasons.
- Positive shots (PS): a positive shot is one that is made in the direction of the goal and that its trajectory is located within the interior range of the goalposts. Any type of shot that does not meet this parameter is not considered in the analysis.
- Failure shots (NS): a negative shot.
- Name of players

Example of different summaries obtained are:

- *Much* of players from "*Positional transitions in motion*" get "Failure shots".
- *Approximately half of situations* players from "Static positional strategy plays" get "Failure shots".

In the second step, all linguistic generated summaries are storage in database. All summaries in database could be recovered from queries that involve filters or summarizers.

Later in third step the coaches can recover linguistic summaries by present different queries to summaries database to with the structure:

- "*Q*" players' get "*S*": in this case different *S* results was given like parameters and the algorithms sought quantifier *Q*
- "*Q*" of players' from position "*R*" get "*S*": in this case "*S*" and "*R*" situations were given and the algorithms sought quantifier "*Q*" too.
- "*Q*" of players' from position "*R*" get "*S*"*:* in this summary were given to the algorithm "*Q*" and "*R*" information and the retrieve algorithms return the values in "*S*".
- "*Q*" of players' from position "*R*" get "*S*"*:* but in this case the algorithm does not receive any parameter and every element were discovered.

The coaches analyzed the information provided by the linguistic summaries and based on them, took measures to improve the results, which are shown in next section.

3 Results and Discussion

The methodology is shown to a group of experts in soccer, futsal and other sports of collective ball games, who evaluate it considering the Viability dimension of the methodology with respect to the indicators "formal quality" and "social unit of the methodology":

1. A survey is applied that collects the agreement of the experts regarding 13 criteria aimed at evaluating training methods in Cuba by applying the multi-criteria analysis technique.

2. For the aggregation of the experts' preferences, the computation technique with 2-tuples words is applied [15, 16] with the following set of linguistic terms *LBTL = {Not adequate, Little adequate, suitable, fairly suitable, very suitable}*. It is decided to apply this technique because it has no loss of information, unlike other computing techniques with words such as ordinal scale and the extension principle.
3. The results are analyzed considering the aggregate preferences of the experts and their agreement in the evaluation.
4. From the analysis that was initially performed, the LDS algorithm was applied.
5. According to the methodology, a series of games is applied as a training method in confined spaces to perfect the offensive tactical phase of completion.

During validation, 31 experts in soccer, futsal and other sports of collective ball games participated in the application of the multi-criteria analysis method. These experts completed an evaluation form where the methodology is evaluated against 13 criteria. The criteria collected in questions 1, 2 and 3 are aimed at measuring the "Formal quality" indicator. The rest of the questions (from 4 to 13) evaluate the "social unit of the methodology".

From the analysis of the experts' criteria, the results of Fig. 1 are obtained. This figure shows the aggregated preferences of the experts in the stages before and after the methodology has been applied. In it, it is evident that, in all the questions, both those related to "Formal quality" and those related to "social unit of the methodology", there is a high evaluation with a low dispersion in opinions; both before and after the methodology has been applied. But a better result is obtained from the evaluation of the criteria, after the application of the methodology.

The R_{WG} correlation coefficient [17] was also applied for the analysis of the agreement between the experts in the initial stage, before applying the methodology, whose value was $R_{WG} = 0.96$ and in the final stage, after applying the methodology,

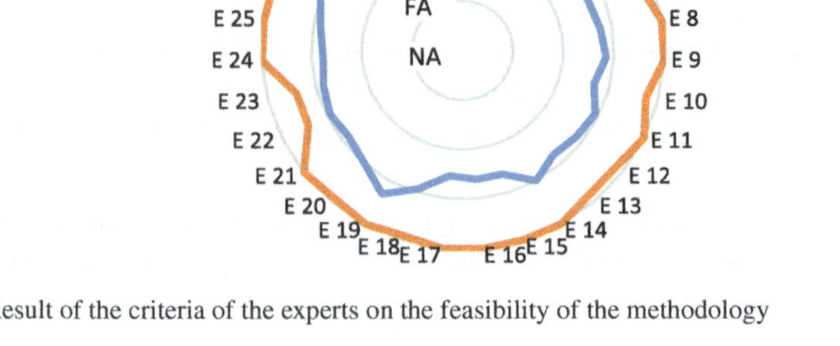

Fig. 1 Result of the criteria of the experts on the feasibility of the methodology

whose value was $R_{WG} = 0.99$. In both cases a high agreement is identified in the opinion of the experts, but in the final stage there is greater agreement.

For the evaluation of the blocks of specific aspects in the games in confined spaces with completion, both in the initial stage (preparatory stage) and in the final stage (competitive stage), tests were carried out to compare the results in both stages. For this, well-founded statistical tests were carried out and the IBM SPSS Statistics version 25 tool [18] was used. The following steps were followed with a significance of 0.05 and confidence intervals of 99%:

Step 1. A normality test is applied (using the Shapiro–Wilk and Kolmogorov–Smirnov tests) and descriptive analysis.
Step 2. In case the samples comply with the normal distribution, parametric tests are applied using Student's *t*-test test for two related samples.
Step 3. In case the samples do not comply with the normal distribution, non-parametric tests are applied using Friedman for n related samples.
Step 4. If Friedman indicates that there are no significant differences in the n related samples, then it is concluded that there are no significant differences.
Step 5. If Friedman indicates that there are significant differences in the *n* related samples, then a post-hoc Wilcoxon Test is applied for two related samples.

In this analysis it is identified that all the variables have a normal distribution (see Table 1), since all the variables have a significance greater than 0.05. That is why the parametric *t*-student test of two-to-two comparison is applied, see Table 2.

When applying the parametric Student's *t*-test, it is evident that all the variables report a higher mean in the competitive stage and that there are significant differences in the pairs except in variable 6, associated with giving away the coverage of support and control in movement and completion.

Table 3 is shown below, which expresses the behavior of the variables goal and positive shots made by the team under study in the provincial first division and university championships in 2018 and 2019. It summarizes all heading levels.

Table 1 Normality tests Kolmogorov-Smirnov

	Statistical	gl	Sig		Statistical	gl	Sig
V1	0.175	13	0.200*	After V1	0.143	13	0.200*
V2	0.125	13	0.200*	After V2	0.152	13	0.200*
V3	0.221	13	0.083	After V3	0.089	13	0.200*
V4	0.121	13	0.200*	After V4	0.209	13	0.123
V5	0.184	13	0.200*	After V5	0.193	13	0.200*
V6	0.161	13	0.200*	After V6	0.184	13	0.200*
V7	0.169	13	0.200*	After V7	0.185	13	0.200*
V8	0.152	13	0.200*	After V8	0.159	13	0.200*
V9	0.175	13	0.200*	After V9	0.198	13	0.174

*Lower bound of the true significance, according to the SPSS statistic tool

Table 2 Statistical tests of paired samples

		Mean	N	Desv. Deviation	Desv. Mean error	Sig. (bilateral)
Pair 1	V1	0.5154	13	0.11616	0.03222	0.002
	After V1	0.6731	13	0.16281	0.04515	
Pair 2	V2	0.4454	13	0.12928	0.03585	0.008
	After V2	0.5907	13	0.10240	0.02840	
Pair 3	V3	0.5346	13	0.12647	0.03508	0.001
	After V3	0.6863	13	0.12716	0.03527	
Pair 4	V4	0.4333	13	0.15943	0.04422	0.007
	After V4	0.5721	13	0.14679	0.04071	
Pair 5	V5	0.3946	13	0.13265	0.03679	0.003
	After V5	0.5691	13	0.15818	0.04387	
Pair 6	V6	0.4954	13	0.19959	0.05536	0.070
	After V6	0.5801	13	0.20275	0.05623	
Pair 7	V7	0.4423	13	0.12722	0.03529	0.000
	After	0.7021	13	0.12698	0.03522	
Pair 8	V8	0.4557	13	0.19258	0.05341	0.002
	After V8	0.5834	13	0.19547	0.05421	
Pair 9	V9	0.3172	13	0.14698	0.04077	0.003
	After V9	0.4769	13	0.12848	0.03563	

Table 3 General total of goals with respect to positive shots in the 2018 and 2019 seasons

11 played games	Goal	Positive	Media	StdDev
2018	14	62	1.27	0.78
2019	33	85	3	1.41

In the eleven games played in 2018, 14 goals were scored from 62 positive shots on goal. The average in terms of goals and positive shots per game behaved for 1.27% and 5.45%, respectively. The standard deviation in the goal variable is 0.78 and in positive shots it is 1.29.

In eleven games played in 2019, 33 goals were scored from 85 positive shots, for an average of 3 goals where the standard deviation in relation to the goal represents 1.41, averaging eight shots per game.

Table 4 shows the behavior of the four offensive tactical variables involved in goal scoring. Their detailed study can be the basis of the elaboration of decisive plays in the attacks in the game of futsal.

Finally, Table 5 describes the results of the SPSS software application, which is applied to the variables defined in the observation and statistical description from the paired samples in the 2018 and 2019 championships, respectively.

Table 4 Total goals with respect to positive shots, by offensive tactical variables in the finalization phase in the 2018 and 2019 seasons

Static positional strategy plays	POSIT	GOAL	**Counter-attack transitions**	POSIT	GOAL
Est2018	23	3	Con2018	12	9
Est2019	48	13	Con2019	17	10
Positional transitions in motion			**Free kicks**		
TranP2018	21	2	Free2018	1	0
TranP2019	20	8	Free2019	6	2

Table 5 Statistical result of the variables observed in the offensive phase finalization in the 2018 and 2019 championships

Paired samples test					
		Paired differences	t	Df	Sig. (2-tailed)
Pair 1	Free2018_Gol—Free2019_Gol	−0.24131	−3.684	10	0.004
Pair 2	Free2018_Pos—Free2019_Pos	−0.84833	−5.333	10	0.000
Pair 3	Est2018_Gol—Est2019_Gol	−0.14531	−3.708	10	0.004
Pair 4	Est2018_Pos—Est2019_Pos	−0.45919	−4.114	10	0.002
Pair 5	TranP2018_Gol—TranP2019_Gol	0.53516	−1.456	10	0.176
Pair 6	TranP2018_Pos—TranP2019_Pos	1.22376	−0.410	10	0.690
Pair 7	Con2018_Gol—Con2019_Gol	0.53558	−0.803	10	0.441
Pair 8	Con2018_Pos—Con2019_Pos	0.95557	0.000	10	1.00

3.1 *Variable Goal in the 2018/2019 Seasons*

Regarding the goal variable, Fig. 2 shows the behavior over time for eleven games for each season. In 2018, 14 goals were scored and in 2019, 33 goals were scored.

Fig. 2 Behavior over time of the goals of the UCI futsal team in the 2018 and 2019 seasons

The final average of goals per game in 2018 is 1.27. The goals are distributed as follows: three goals are scored in a match, two are scored in two matches, while in seven matches a single goal is scored in each of them, failing to score goals in one match. This means that the number of goals does not behave in the same way in the eleven games as shown by the standard deviation.

In 2019 the team's scoring capacity is increased compared to the previous season. The average for matches is three goals. The 33 scored goals being distributed as follows: five goals are scored in one match, while four goals are scored in each of four matches; in three games two goals are scored in each one, in two games three goals are scored adding six goals and in one game one goal is scored; there is a 0 goal game. The standard deviation is also scattered. This responds to the changing characteristics and demands of the game of futsal considering players and opponents.

There are significant differences in the goal variable from one season to another as described in Table 5, scoring 19 more goals than in the 2018 season.

3.2 Variable Positive Shots in 2018/2019 Seasons

Figure 3 shows the behavior over time of the positive shots made in both championships. When comparing the positive shots in the provincial championships of 2018 and 2019, the following evaluations of aforementioned behavior are reached.

In the 2018 season, by matches, six positive shots were fired on average, while in 2019, nine shots were taken on average in the eleven matches made. From the third game to the fifth game it decreases in both years, in this aspect when analyzing together with the technical direction and the players it is agreed that it is the stage of closing studies of the first semester at the university and the sporting results tend to decrease when not have the whole team in those games.

There is an inseparable relationship between goal and positive shot. This is corroborated by the significant differences between the positive shots from one season to the next, as shown in Table 5. In 2018, 62 positive shots are needed for every 4.42 chances to score a goal; while in 2019 of 85 positive shots made, 2.57 chances are needed to score a goal.

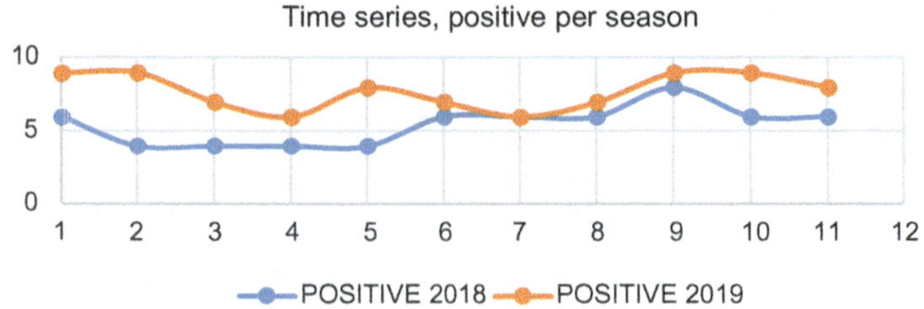

Fig. 3 Behavior over time, positive shots of the UCI futsal team in the 2018 and 2019 seasons

3.3 Static Positional Strategy Plays in 2018/2019 Seasons with Respect to Goal and Positive Shots

In 2018, only three goals were scored from positional static plays (one from a corner kick and two sidekicks, one in the defensive zone and the other in the offensive zone). 23 shots were a shoot resulting in goal.

In 2019, 13 goals were scored distributed in the following indicators. Out of ten positive shots from goal kick, three goals are achieved; out of ten positive corner kicks, three goals are scored; out of nine positive offensive kick-offs three goals are scored; out of seven positive defensive lateral kicks, two goals are scored and out of six positive direct free kicks two goals are scored.

Table 5 corroborates the significant differences in the variables goal and positive shots from one season to another. Ten more goals were scored in 2019 as every 3.69 positive shooting opportunity, the team scores a goal. While in 2018 7.66 positive shots are used on average to score a goal.

3.4 Positional Transitions in Motion in 2018/2019 Seasons Regarding Goal and Positive Shots

Regarding positional transitions in motion, two goals were scored in 2018, one of them by interception and the other from an anticipation.

In 2019 season eight goals are scored, this makes the team performance better. The distribution of the goals is as follows: in interceptions three goals are scored using five positive shots, from anticipations three goals are achieved from six positive shots and in rapid attack two goals from six positive shots.

Although in 2018 and 2019 seasons the positive shots are quantitatively similar (20 and 21 respectively), improvements are achieved in relation to positive shots. In 2018 the team makes 10.5 positive shots to score a goal. While in 2019 he executes an average of 2.12 positive shots to score a goal, the behavior of the team being more effective.

This behavior can be given by the improvement in the conservation of the ball from directed, conditioned games and games modeled for the pressure output that were trained with a view to the 2019 season. Conservation movements, support unmarking and rupture that allow to be more forceful.

The Counter-attack transitions behave in 2018 as the variable in that season with the highest goal production. Nine goals from 17 positive shots were scored. They are scored in counterattacks of two against one, four goals using eight positive shots. In three-on-one situations, three goals are scored from six positive shots. In three-on-two situations, a goal is scored from two positive shots and in a four-on-two situation, the only opportunity presented in the competition is seized.

The behavior in 2019 was better with the minimum difference that one more goal is scored. Ten goals are scored in 16 positive shots, although this variable improves, its behavior remains stable. They are scored in counterattacks of two against four goals using seven positive shots. In three-on-one situations, four goals are scored from six positive shoots. In three-on-two situations, two goals are scored from three positive shots and in the four-on-two situation no goals are scored.

4 Conclusions

The generation of linguistic summaries form 2018 season data, permits to coaches improve the training plans. The futsal team get 2019 results better than 2018 season.

There are significant differences in the goal variable from one season to another, where the team shows better effectiveness in 2019 scoring 33 goals, achieving a difference of 19 goals compared to 2018 where 13 goals are scored.

In terms of positive shots there are significant differences. In 2018, 62 positive shots are taken needing 4.42 shots to score a goal; while in 2019 out of 85 positive shots, it takes 2.57 chances to score a goal.

In the static positional strategy plays, significant differences are corroborated in the variables goal and positive shots from one season to another. Ten more goals are scored in 2019. For every 3.69 positive shot opportunities the team scores a goal. Whereas in 2018, three goals are scored and an average of 7.66 positive shots are used to score a goal.

Positional transitions on the move improve in 2019 over 2018 in terms of scoring and positive shots, where eight goals are scored and an average of 2.12 shots are taken to score a goal.

The counterattacking transitions improve from one year to another, although the differences are not significant, improvement is achieved; this variable depends on the way the team defends and counterattacks.

The free-kick is improved in 2019 since no goals were scored in 2018 and two goals were scored this year.

References

1. De Bortoli, R., et al.: Comparación entre posición declarada y eficacia subjetiva en fútbol sala. Red: Revista de Entrenamiento Deportivo (España) **16**(1), 23–26 (2002)
2. Tenroller, C.: Futsal: Ensino e Práctica, 1 ed. 67, p. 115. ULBRA, Canoas (2004)
3. Lozano, J.: Curso internacional FIFA. Entrenadores de máximo nivel. Cuba, La Habana (2009)
4. Lima, M.: Perfil das finalizacoes no Futsal: un estudo do XXII Jogos da Juventude do Parana. Universidade Federal de M inas Gerais, Monografia (2010)
5. Bolsonaro, J.R.: Analise das finalizacoes na fase final da Liga Futsal 2012. Revista Brasilera de Futsal e Futebol. San Pablo 7(24), 148–152 (2015)
6. López, J.V.: Manual de la UEFA para entrenadores de futsal. Union des Associations Européennes de Football (UEFA), Suiza UEFA, Nyon (2017)

7. Lago, C.: Optimización del rendimiento físico–deportivo en fútbol sala femenino: desde la epidemiología lesional a una propuesta de intervención (tesis doctoral). Faculty of Education and Sports Sciences, University of Vigo Campus A Xunqueira, Spain (2019)
8. Morales, G., Perez, A., Paizan, O.: Reduced games for the improvement of the support and rupture demarcation in futsal. Serie Científica de la Universidad de las Ciencias Informáticas **13**(2), 55–65 (2020). ISSN: 2306-2495
9. Morales, G., Sánchez, B., Mesa, M.: Methodology for the improvement of the offensive training of finalization in university futsal. I Conference Online Sport Science Applied COCAR (2020)
10. Anguera, M.T., Hernández–Mendo, A.: La metodología observacional en el ámbito del deporte. J. Sport Sci. **9**, 135–160 (2013)
11. Pérez, I., Piñero, P.Y., et al.: Tendencias en la sumarización lingüística de datos. Rev. Cuba. Transform. Digit. **2**(1), 79–101 (2021)
12. Zadeh, L.A.: A prototype-centered approach to adding deduction capability to search engines—the concept of protoform. In: Intelligent Systems Proceedings, First International IEEE Symposium, vol. 1, pp. 2–3 (2002). http://doi.org/10.1109/IS.2002.1044219
13. Kacprzyk, J., Zadrozny, S.: Protoforms of linguistic data summaries: towards more general natural-language-based data mining tools. HIS, pp. 417–425 (2002)
14. Pérez, I., Piñero, P.Y., Bello, R., Acuña, L.A., García-Vacacela, R.: Linguistic summaries generation with hybridization method based on rough and fuzzy sets. In: International Joint Conference on Rough Sets. Lecture Notes in Computer Science, pp. 385–397. Springer. ISSN 0302-9743, 12179 LNAI, ISBN 978-3-030-52704-4. http://doi.org/10.1007/978-3-030-52705-1_29 (2020)
15. Herrera, F., MartÃnez, L.: A 2-tuple fuzzy linguistic representation model for computing with words. IEEE Trans. Fuzzy Syst. **8**(6), 746–752. http://doi.org/10.1109/91.890332 (2000)
16. Zhang, S., Wei, G., Alsaadi, F.E., Hayat, T., Wei, C., Zhang, Z.: MABAC method for multiple attribute group decision making under picture 2-tuple linguistic environment. Soft Comput. **24**(8), 5819–5829. http://doi.org/10.1007/s00500-019-04364-x (2020)
17. Asuero, A.G., Sayago, A., Gonzalez, A.G.: The correlation coefficient: an overview. Crit. Rev. Anal. Chem. **36**(1), 41–59. http://doi.org/10.1080/10408340500526766 (2006)
18. Mccormick, K., Salcedo, J.: SPSS Statistics for data analysis and visualization. Wiley, S.l. ISBN: 978-1-119-00355-7. ISBN 978-1-119-00355-7 (2017)

Chapter 5
Algorithms for Linguistic Description of Categorical Data

Carlos R. Rodríguez Rodríguez, Denis Sergeevich Zuev, and Marieta Peña Abreu

Abstract The paper proposes a method that comprises five algorithms for producing composite linguistic summaries from categorical data. The generated composite summaries reflect Evidence, Contrast, or Emphasis relations between at least two constituent summaries. The constituent summaries are instances of the LDS classical protoforms created, in this case, with frequent L1 item sets and association rules obtained from applying an association rule mining algorithm. In order to verify the feasibility of implementing the method, we performed a use case with a dataset of 2128 cases of the Economic Chamber of the Provincial People's Court of Havana. The results were consistent with expectations, obtaining 18 Evidence relations, 11 Contrast relations, and 16 Emphasis relations. Furthermore, we evaluated the interpretability of the composite summaries obtained in the use case. Specifically, we measured the accuracy of identifying the relation type implicit in the summary and their understandability. In both cases, the results were positive.

Keywords Association rules · Linguistic data summarization · Linguistic descriptions of data

1 Introduction

Linguistic data summarization (LDS) is a Data Science approach proposed by Yager [1], widely extended [2–11] and applied [12–18], which enables users to easily

C. R. Rodríguez Rodríguez (✉) · M. Peña Abreu
Vicedecanato de Investigación y Posgrado, Facultad 3, Universidad de las Ciencias Informáticas, Habana, Cuba
e-mail: crodriguezr@uci.cu

M. Peña Abreu
e-mail: mpabreu@uci.cu

C. R. Rodríguez Rodríguez · D. S. Zuev
Kazan Federal University, Kazan, Tatarstan, Russia

P. Y. Piñero Pérez et al. (eds.), *Artificial Intelligence in Project Management and Making Decisions*, Studies in Computational Intelligence 1035,
https://doi.org/10.1007/978-3-030-97269-1_5

understand data and identify its meaning and usefulness through natural language [10].

The are several types of protoforms for linguistic summaries, but the predominant ones are the so-called classical protoforms [1, 19, 20] defined as Eq. (1) or Eq. (2):

$$T(Qy' \; are/have \; S) \quad (1)$$

$$T(QFy' \; are/have \; S) \quad (2)$$

where S is a summarizer (*e.g.*, amounts less than \$ 414,300); Q is a quantity in agreement given as a fuzzy linguistic quantifier (*e.g., most*); and T is the truth degree of the summary (e.g. 0.92), which takes values in [0, 1]. In (Eq. 2), a qualifier F (*e.g., paid*) is added. F is a filter to get a specific data subset. The following is a summary like (Eq. 2): *T (Most paid lawsuits have amounts less than 414,300 pesos) = 0.92.*

The authors of [11] point out two main limitations of LDS. The first one concerns the inappropriateness of protoforms to be provided directly to users. The second one is related to the lack of expressiveness of protoforms beyond standard quantified sentences. They also emphasize that protoforms have been mainly addressed individually without considering relations among different instanced protoforms. Moreover, the mainstream of LDS consists of modeling numerical variables using fuzzy logic, constraining its application in real problems whose primary data are categorical [9].

Many approaches attempt to improve the expressiveness of summaries [3, 6, 11, 21] and facilitate working with mixed data [5, 7, 22]. However, there are still opportunities to develop proposals that aim at the same time:

- Enhance the expressiveness of summaries by exploiting the relations between them and language refinement.
- Support the addressing of problems with categorical (qualitative) data.

In order to deal with these challenges, a model for generating three types (Evidence, Contrast and Emphasis) of composite relations from categorical data is proposed in [9]. These relations are inspired by Rhetorical Structure Theory (RST) [23, 24] and consist of at least two constituent summaries created from association rules.

This paper aims to propose the algorithms that provide an implementable version of the model defined in [9]. For this purpose, Sect. 2 describes the activities that encapsulate such algorithms. Section 3 presents the design and results of a use case on economic lawsuits. Section 4 evaluates the interpretability of the composite summaries obtained in the use case. Finally, the conclusions of the research are presented.

2 Method for Generating Composite Linguistic Summaries

The method (see Fig. 1) consists of three phases: extracting association rules, building constitutive summaries and creating composite relations. The following sections

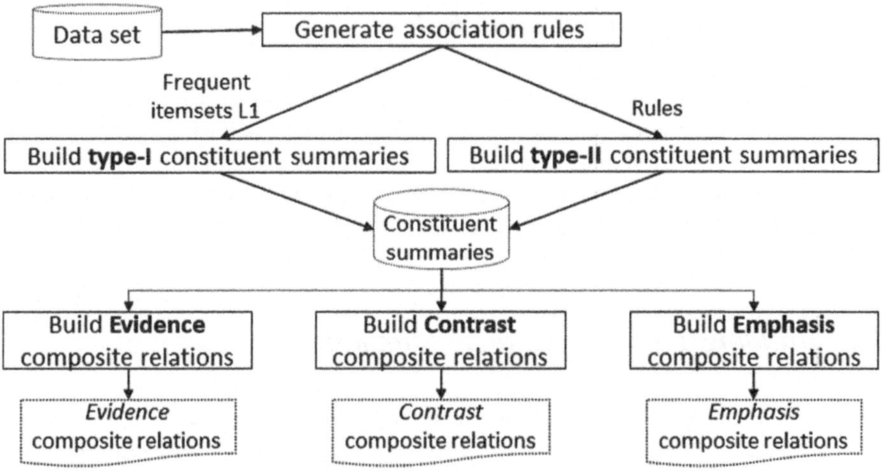

Fig. 1 Method overview diagram

describe the activities of the method and present the algorithms that enable its implementation. Table 1 list the symbols used; the antecedent includes the object and the qualifiers and the consequent comprises the summarizer.

Table 1 Main symbols used in the algorithms for the creation of composite relations

Symbol	Meaning	Example
S	Set of all constituent summaries	$S = \{s_i \mid i \in (1, 2, \ldots, k)\}$
s	Constituent summary (type-I or II)	Most paid lawsuits have amounts less than 414,300 pesos
Q	Set of linguistic quantifiers	{About half, Many, Most, Almost all}
A	Antecedent of s	paid lawsuits
B	Consequent of s	amounts less than 414,300 pesos
T	Truth degree of s	Any value in [0, 1]
C	Coverage degree of s	Any value in [0, 1]
r	Composite relation	See Examples 1–6 below
$T(r)$	Truth degree of r	Any value in [0, 1]
$S(r)$	Relation strength of r	Any value in [0, 1]
$C(r)$	Coverage degree of r	Any value in [0, 1]

2.1 Generation of Association Rules

The first activity consists of generating the association rules that will be used to create the constituent summaries. This activity provides two outputs: the frequent 1-itemsets (FI^1) and the association rules themselves. An association rule mining algorithm that uses support and confidence and directly finds frequent item sets should be applied to perform this activity, for example: *Apriori* [25], *Partition* [26], *Eclat* [27].

2.2 Building Type-I Constituent Summaries

Algorithm 1 built the type-I constituent summaries from FI^1 whose support values exceed a given threshold. It invokes algorithms *SetLingQuantifier* and *BuildSummary*.

Algorithm 1 *LDS_Type-I_CS*

Inputs:
  $FI^1 = \{[it_1^1, sup_1], \ldots, [it_n^1, sup_n]\}$: *frequent itemsets L1*
  $minConf$: *minimum accepted value of confidence*
  $minT$: *minimum accepted value of truth degree*
  Q: *set of linguistic quantifiers*
Begin
  1. A = *"objects of database"*
  2. $S = \emptyset$, *set of constituent summaries*
  3. *For each* it_i^1 *in* FI^1 *do*
  4. $sup_i = sup_i$ / *totalTransactions*
  5. *if* $sup_i \geq minConf$ *then*
  6. Q_s = *SetLigQuantifier*($sup_i, minT, Q$)
  7. $B = it_i^1$
  8. $T = \mu_{Q_s}(sup_i)$
  9. $C = sup_i$
  10. s = *BuildSummary*(Q_s, A, B, T, C)
  11. *Add s to S*
  12. *end*
  13. *End*
  14. *Return S;*
End

2.3 Building Type-II Constituent Summaries

For each rule obtained in activity 2.1, a type-II constituent summary is created using Algorithm 2. It also invokes algorithms *SetLingQuantifier* and *BuildSummary*.

Algorithm 2 *LDS_Type-II_CS*

Inputs:
Rules: set of association rules
minT: minimum accepted value of truth degree
Q: set of linguistic quantifiers
Begin
1. $S = \emptyset$, set of constituent summaries
2. For each rule in Rules do
3. Q_s = SetLigQuantifier($conf_{rule}$, minT, Q)
4. A = rule.antecedent
5. B = rule.consecuent
6. $T = \mu_{Q_s}(conf_{rule})$
7. $C = sup_{rule}/totalTransactions$
8. s = BuildSummary(Q_s, A, B, T, C)
9. Add s to S
10. End
11. Return S
End

Remark The *SetLingQuantifier* algorithm evaluates the confidence of the rule in the membership functions of each of the quantifiers Q, assigning it to the most specific one for which the summary retains a truth value higher than a given threshold ($minT$). The *BuildSummary* algorithm joins the antecedent and the consequent of the rule with the quantifier. It also assigns the values of truth (T) and coverage (C).

2.4 Building the Evidence Composite Relations

Algorithm 3 creates the Evidence composite relations [9]. It uses the results of activities 2.2 and 2.3 to generate statements consisting of one constituent summary that acts as a nucleus and one or more satellites. The satellites are the non-overlapping constituent summaries that provide the highest possible support for the nucleus statement.

Algorithm 3 *LDS_Evidence_CR*

Inputs:
$S = \{s_i | i \in (1, 2, \ldots, k)\}$, *all possible constituent summaries*
Begin
$R^e = \emptyset$, *set of Evidence relations*
For each s_i *in* S *do*
  $L_s \neq \emptyset$, *set of possible satellites*
  For each s_j *in* S, *such as* $j \neq i$ *do*
    if $\left(\left(B_{s_j} = B_{s_i}\right) \wedge \left(A_{s_j} \subset A_{s_i}\right)\right)$ *then*
      Add s_j *to* L_s
  End
  if $(L_s \neq \emptyset)$ *then*
    G = *set of subsets in* L_s *having the same attributes*
    U^* = *subset in* G *with the highest support value*
    r^e = *BuildRelation(, "since",* U^**)*
    Calculate $T(r^e)$, $S(r^e)$, *and* $C(r^e)$
    Add r^e *to* R^e
  end
End
Return R^e
End

2.5 *Building the Contrast Composite Relations*

To generate the Contrast composite relations [9], Algorithm 4 uses the outputs of activities 2.2 and 2.3 to create statements that involve two constituent summaries having the same attributes but at least for one of them, the values are different.

Algorithm 4 *LDS_Contrast_CR*

Inputs:
  $S = \{s_i | i \in (1, 2, \ldots, k)\}$, *all possible constituent summaries*
Begin
  $R^c = \emptyset$, *set of Contrast relations*
  For each s_i *in* S *do*
    For each s_j *in* S, *such as* $j \neq i$ *do*
      if

$$\left(\begin{array}{l}\left(\left(\left(A_{s_i} = A_{s_j}\right) \vee \left(A_{s_j} \subset A_{s_i}\right)\right) \wedge \left(att_{B_{s_i}} = att_{B_{s_j}}\right) \wedge PredDif\left(B_{s_i}, B_{s_j}\right)\right) \\ \vee\left(\left(\left(att_{B_{s_i}} = att_{B_{s_j}}\right) \wedge \left(att_{A_{s_i}} = att_{A_{s_j}}\right)\right) \wedge \left(PredDif\left(B_{s_i}, B_{s_j}\right) \vee PredDif\left(A_{s_i}, A_{s_j}\right)\right)\right) \\ \vee\left(\left(\left(att_{B_{s_i}} = att_{A_{s_j}}\right) \wedge \left(att_{A_{s_i}} = att_{B_{s_j}}\right)\right) \wedge \left(PredDif\left(B_{s_i}, A_{s_j}\right) \vee PredDif\left(A_{s_i}, B_{s_j}\right)\right)\right)\end{array}\right)$$

      then
        if R^c *not contain the relation* (s_j *but* s_i) *then*
          r^c = *BuildRelation*(s_i, "*but*", s_j)
          Calculate $T(r^c)$, $S(r^c)$, *and* $C(r^c)$
          Add r^c *to* R^c
        end
      end
    End
  End
  Return R^c
End

2.6 Building the Emphasis Composite Relations

Emphasis composite relations [9], are produced by Algorithm 5. It uses the outcomes of activities 2.2 and 2.3 to generate statements consisting of a constituent summary that states an assertion (the nucleus) and another (the satellite) that highlights the predominant attribute in the cases covered by the nucleus.

Algorithm 5 *LDS_Emphasis_CR*

```
Inputs:
  $S = \{s_i \mid i \in (1,2,\ldots,k)\}$, all possible constituent summaries
Begin
  $R^h = \emptyset$, set of Emphasis relations
  For each $s_i$ in $S$ do
    $L_s = \emptyset$, set of possible satellites
    For each $s_j$ in $S$, such as $j \neq i$ do
      if $\left((A_{s_i} \cup B_{s_i}) = A_{s_j}\right)$ then
        Add $s_j$ to $L_s$
    End
    if $(L_s \neq \emptyset)$ then
      $s^*$ = summary in $L_s$ with the highest support value
      $r^h$ = BuildRelation($s_i$, "especially", $s^*$)
      Calculate $T(r^h)$, $S(r^h)$, and $C(r^h)$
      Add $r^h$ to $R^h$
    end
  End
  Return $R^h$
End
```

Remark The equations to calculate *T(r)*, *S(r)*, and *C(r)* for each type of composite relation are formally defined in [9].

3 Use Case

In order to check if the model is implementable and produces results that are consistent with expectations, we conduct a use case using a dataset of real lawsuits.

3.1 Design and Implementation

The dataset used consists of 2128 cases collected from 2013 to 2015 in the Economic Chamber of Provincial People's Court of Havana. It comprises five nominal attributes without missing values:

- Plaintiff's institution: head institution of the plaintiff entity (*e.g., MICONS*)
- Defendant's institution: head institution of the defendant entity (*e.g., MITRANS*)
- Cause of lawsuit: delay, not-payment, shortage, refund, quality

- Amount of lawsuit: actual value discretized into six labels (*e.g., '(-inf-414300]'*)
- Paid: indicates if within the next six months the lawsuit was paid (Yes/No).

In order to obtain the composite summaries, we developed a Java implementation of the proposed method with the following settings:

- For generating the association rules, the Apriori algorithm (Weka version) were invoked using the following configuration: *numRules = 100, metricType = Confidence, minMetric = 0.5, delta = 0.05, minSupport = 0.1.*
- Parameters $minT$ (Algorithm 2) and $minConf$ (Algorithm 1) were set to 0.5.
- For computing $T(r)$, the t-norm minimum was used.
- The set of fuzzy linguistic quantifiers used were modeled with trapezoidal fuzzy sets as follows: *{About_half = [0.42, 0.48, 0.52, 0.58], Many = [0.52, 0.58, 1, 1], Most = [0.72, 0.78, 1, 1], Almost_all = [0.92, 0.98, 1, 1]}.*

  In order to create Contrast relations, the contrast degree between each pair of labels (l_i, l_j) for attributes *plaintiff*, *defendant*, *cause* and *paid* is $\mu_{R_{(l_i,l_j)}} = 1$. For attribute *amount*, the values of $\mu_{R_{(l_i,l_j)}}$ are balanced vectors. The following vector represents the contrast degrees of the label l_3 with respect to $l_n | n = 1, 2, \ldots, 6$: $\mu_{R_{(l_3,l_n)}} = \{x_{31} = 0.5; x_{32} = 0.25; x_{33} = 0; x_{34} = 0.25; x_{35} = 0.5; x_{36} = 1\}$.
- We discarded the relations with values of $S(r) < 0.5$.

3.2 Results and Examples

The 18 obtained Evidence relations reached for the metrics $T(r)$, $S(r)$ and $C(r)$ the values of mean and standard deviation (Mean; StdDev) expressed by the pairs (0.9050; 0.1395), (0.8433; 0.1303) and (0.3083; 0.1806), respectively. Example 1 shows a "full" evidence relation, i.e., the satellites provide evidence for all cases covered by the nucleus, so $S(r) = 1$. Example 2 shows a relation having only one satellite which satisfies 96% of the cases described by the nucleus.

Example 1 *Many* lawsuits have amounts less than $ 414,300, *since*
most paid lawsuits have amounts less than $ 414,300,
many unpaid lawsuits have amounts less than $ 414,300

$$T(r) = 1; S(r) = 1; C(r) = 0.73$$

Example 2 *Most* lawsuits against MITRANS entities are paid, *since*
most lawsuits against MITRANS entities with amounts less than $ 414,300 are paid

$$T(r) = 1; S(r) = 0.96; C(r) = 0.11$$

The use case produced 11 Contrast relations whose values of Mean and StdDev for the $T(r)$, $S(r)$ and $C(r)$ metrics were (0.8382; 0.1578), (0.5973; 0.1536) and

(0.4973; 0.1854), respectively. Examples 3 and 4 each contain two equal attributes. In Example 3, the attributes have different values in each constituent summary, so $S(r) = 1$. In Example 4, only the attribute "plaintiff's institution" has different values (MICONS/MINDUS), so $S(r) = 0.5$.

Example 3 *Many* unpaid lawsuits are for non-payment, *but*
many lawsuits for delay are paid afterwards

$$T(r) = 1; S(r) = 1; C(r) = 0.48$$

Example 4 *Almost all* lawsuits of MICONS entities have amounts less than $ 414,300, *but*
most lawsuits of MINDUS entities also have amounts less than $ 414,300

$$T(r) = 0.68; S(r) = 0.5; C(r) = 0.3$$

We also obtained 16 Emphasis relations whose values of Mean and StdDev for $T(r)$, $S(r)$ and $C(r)$ were (0.9119; 0.1298), (0.7425; 0.1434) and (0.2944; 0.1868), respectively. Examples 5 and 6 show how the nuclei provide a description and the satellites emphasize the main feature among cases covered by the nuclei. The values of $S(r)$ denote the degree of such dominance.

Example 5 *Most* lawsuits of MICONS entities are for not-payment, *especially*
almost all of they have amounts less than $ 414,300

$$T(r) = 0.88; S(r) = 0.8; C(r) = 0.27$$

Example 6 *Most* paid lawsuits have amounts less than $ 414,300, *especially*
about half of they are for not-payment

$$T(r) = 1; S(r) = 0.5; C(r) = 0.44$$

4 Evaluating the Interpretability of Relations

We also conducted an empirical evaluation to verify the interpretability of the summaries produced in the use case.

4.1 Design

The criteria defined to evaluate the interpretability of the summaries were as follow:

- C1: Identification of the relation type implicit in the summary
- C2: Understanding of the information provided in the summary.

The research questions and hypotheses that guided the study were the following:

- RQ1: Do the generated linguistic summaries describe a relationship between their constituent sentences, which can be identified by users? The null hypothesis (H_{01}) is that users can identify the relation types implicit in the generated summaries.
- RQ2: Do the generated linguistic summaries provide information that is understandable by users? The null hypothesis (H_{02}) is that users can understand the information contained in the generated summaries.

The above criteria were measured using the metrics M1 and M2, both can take values in [0, 100] and from 50% onwards, we consider it a satisfactory result.

- M1: For C1, we measured the accuracy of the subjects when they identified the relation type implicit in the summaries. For this purpose, we calculate the percentage of correct responses.
- M2: For C2, we measured the subjects' perception about the understandability of summaries using an ordered linguistic scale and calculated the percentage of positive (*"understandable", "quite", "very"*) and negative (*"not", "little"*) ratings.
- In addition, we use statistical tests with a significance threshold of 0.05 and 99% confidence intervals. For C1, we applied Cochran's Q test to the results (right/wrong) of the relation types identification. For C2, we used Friedman and Wilcoxon tests with the results (ordinal values) of the understanding assessment.

The subjects participating were intentionally selected. Initially, 27 subjects were convened and informed of the objective and characteristics of the study and expressed their willingness to participate. Finally, 23 of them completed the designed instrument, representing an 85.19% of adherence rate. The group of participating subjects had the following characteristics: (1) they are not specialists in the application area of the use case, (2) they have no previous experience working with linguistic summaries, (3) they have university degrees and (4) their ages range from 26 to 38 years old.

4.2 Instrument

The designed instrument was a survey consisting of one closed-ended and other open-ended questions. The first one comprises six sections, each containing a summary (manually verbalized), with a balance of the relations types (2 of Evidence, 2 of Contrast and 2 of Emphasis). For each summary, the subjects had to fulfill two activities:

- Activity 1: *Which of the following statements better describes this summary?* This activity is related to the evaluation of criterion C1 and subjects should select one of the following options:

- *Evidence* (an assertion is made, which is then supported by one or more additional assertions)
- *Emphasis* (an assertion is made and then an element is highlighted about it)
- *Contrast* (two opposing assertions are made)
- *Other* ________

- Activity 2: *How do you assess the understandability of the summary?* This activity is related to the evaluation of criterion C2 and subjects should select one of the following options: *"Not understandable"*, *"Little understandable"*, *"Understandable"*, *"Quite understandable"*, *"Very understandable"*.

The second question of the survey allowed subjects to formulate comments, queries, or suggestions. Besides, we requested the subjects' age and profession.

5 Results and Discussion

Subjects received the survey online without any supplementary material and submitted the responses within 48 h.

Verification of hypothesis H_{01}. It suggests that the subjects can identify the relation type implicit in each summary. For its verification, we analyze the responses to activity 1 of question 1 of the survey. Overall, the results show that the correct identification of the relation type reached 67.39%. Further analysis reveals that:

- The classifications show an unbalanced distribution among the subjects since seven of them made 46.67% of the mistakes by misidentifying three or more relations (see Fig. 2). The other errors spread among 10 subjects (one or two each). While the remaining six subjects correctly classified all relations. Given the characteristics of the subjects described above, these results seem to indicate that the accuracy in identifying the relation type depended on the previous notions that the subjects had about the terms "evidence", "contrast" and "emphasis".

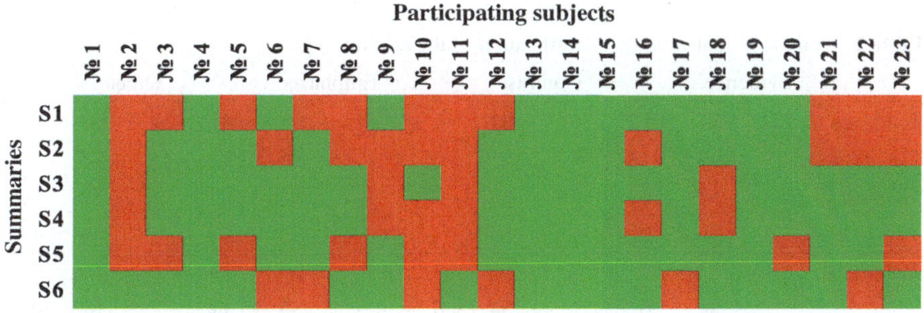

Fig. 2 Distribution of the relation type classification (summaries/users). Green indicates the right answers and red the wrong answers

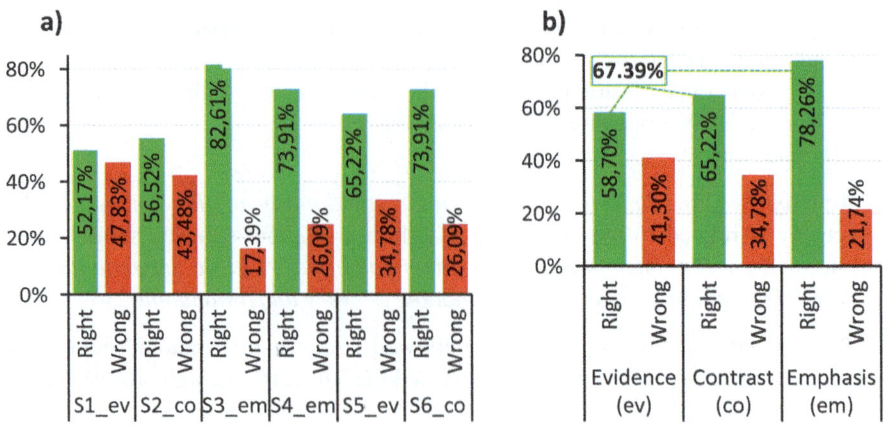

Fig. 3 Criterion: classification of relation type [individually (**a**) and by relation types (**b**)]

- Subjects most accurately identified the Emphasis relations, then the Contrast ones and were even less effective at categorizing the Evidence ones (see Fig. 3). Although Cochran's Q test found no statistically significant differences between these data (*p-value* = *0.130*), we point out some interesting behaviors. The 57.89% of the misclassified Evidence relations were mistaken for Emphasis relations, mainly in S5. At the same time, 80% of the misclassified Emphasis relations were mistaken for Evidence relations, mainly in S4 (see Table 2). Such behavior may be due to the relative similarity between the notions of "evidence" and "emphasis", which may lead to confusion.
- Many (68,75%) misclassifications of Contrast relations were categorized as Emphasis (mainly in S2), while four subjects classified S6 with the label "Other". Although the notions of "contrast" and "emphasis" are not similar, the confusion, in this case, may be linked to the content of S2 and S6. Each of them has the same consequents in their constituent summaries, which could be understood as an intention to emphasize those predicates.

Table 2 Confusion matrix in the identification of the relation type

	Evidence			Contrast			Emphasis			Other		
Evidence	27	S1	12	6	S1	6	11	S1	4	2	S1	1
		S5	15		S5	0		S5	7		S5	1
Contrast	0	S2	0	30	S2	13	11	S2	9	5	S2	1
		S6	0		S6	17		S6	2		S6	4
Emphasis	8	S3	2	2	S3	2	36	S3	19	0	S3	0
		S4	6		S4	0		S4	17		S4	0

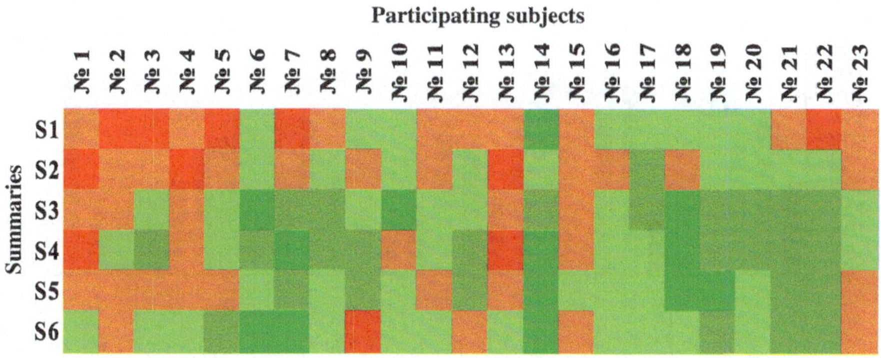

Fig. 4 Distribution of the understandability assessment (summaries/users). The colors have the same meaning as in Fig. 5.

- In all cases, the descriptions of the relation types given in the survey could have influenced the classification errors. In order to minimize the risks of confusion, they should be refined in future experiments.
- Overall, the semantics of the connectors of each relation type could also affect the classification. Therefore, it is advisable that, in real-life application environments, the users themselves define the most appropriate connectors.

Verification of hypothesis H_{02}. It states that the subjects understand the information provided in the summaries. For its confirmation, the responses to activity 2 of question 1 of the survey were analyzed. The statistical analysis shows that 63.77% of answers are positives (see Fig. 5c). Further analysis shows that:

- The evaluations show an unbalanced distribution among the subjects. Nine of them accounted for 76% of the negative ratings (*"Not understandable"* and *"Little understandable"*), assigning these labels to three summaries or more. Other nine subjects negatively evaluated one or two summaries each. The five remaining assessed all summaries positively (see Fig. 4).
- Positive evaluations have an increasing trend according to the summary number (see Fig. 5a). Thus, the order of the summaries seems to have influenced the evaluation, which could indicate that the subjects understood the summaries better as they moved through the survey. Although the differences between {S3, S4, S5, S6} are not statistically significant according to the Friedman test (*p-value* = *0.586*).
- Grouped by relation type, the Wilcoxon test found statistically significant differences (*p-value* = *0.000*) between the evaluations of Evidence relations and Emphasis relations (see Fig. 5b). This result could be related to the size and verbalization.
  - On the one hand, Emphasis relations are shorter as they contain only two constituent summaries. Moreover, after verbalization, they have a more

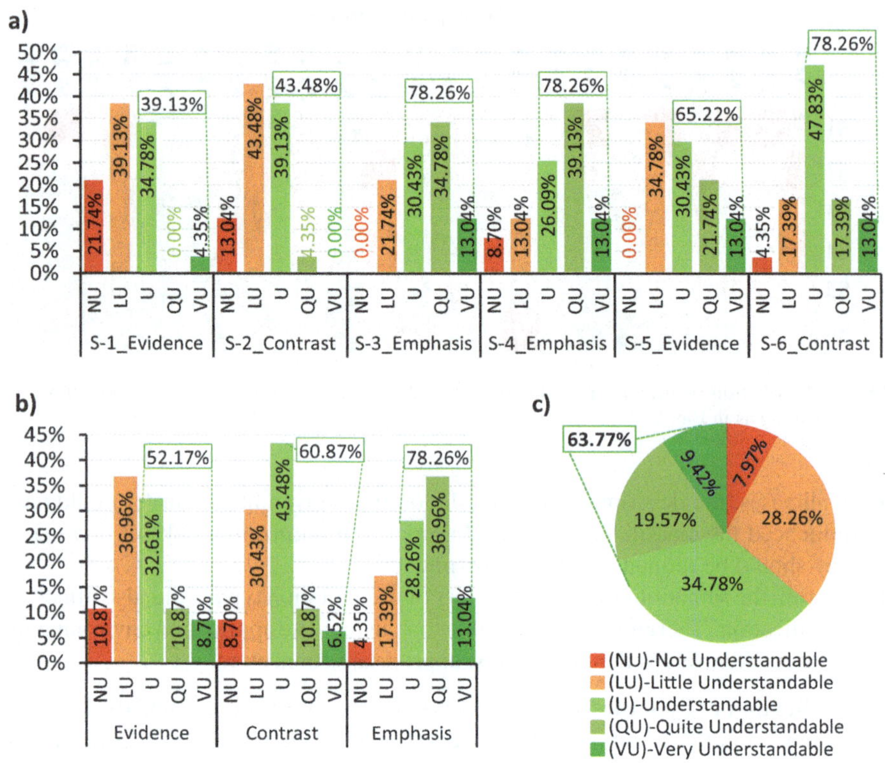

Fig. 5 Criterion: understanding [individually (a), by relations type (b), and globally (c)].

compact structure that does not repeat the nucleus information (see Examples 5 and 6).

- However, evidence relations can be consisting of two or more constituent summaries, which makes them potentially more extensive. Moreover, the verbalization repeats both the antecedent and consequent of the nucleus, although they are common for all constituent summaries (see Example 1).

- Contrast relations reached very different values among them, which are statistically significant (*p-value* = *0.003*) according to the Wilcoxon test. Both summaries (S2 and S6) contain the same number of predicates in their constituent summaries and represent the third expanded protoform of the Contrast relation [9]. Moreover, the same keywords were used in their verbalization. Therefore, none of these factors (size, structure and verbalization) seem to have influenced such a difference. However, the labels of the attributes may have affected it. These results also suggest a connection between the order of the summaries and their understanding.
- Overall, the results suggest that improved verbalization of relations could contribute to increasing their understanding.

6 Conclusions

The paper proposes an implementable version of the model presented in [9]. The proposal aims at improving the expressiveness of summaries and facilitating their generation from categorical data. The first should be the main goal of the LDS community, i.e., to provide users with more familiar and useful summaries. To this end, we chose to identify—inspired by RST—the evidence, contrast and emphasis relations underlying a set of candidate summaries. On the other hand, working with categorical data has received less attention from researchers, even though there are domains, such as the legal, where words are the primary data.

The use case showed that the proposed algorithms produce composite summaries, which agree with the protoforms defined in [9]. The obtained summaries are potentially useful for ruling new lawsuits and managing the work of the referred courtroom. At the same time, we consider that this method can be used in numerical domains by adjusting the activities related to the building of constituent summaries. For instance, the summaries produced in [18] from numerical attributes can be used as constituent summaries to create composite summaries on planning errors in project management.

The interpretability evaluation results are consistent across the two criteria analyzed. The best performing relations were the Emphasis ones, followed by the Contrast ones and finally the Evidence ones. These results seem to be related to the participating subjects' previous experience and concentration, the similarity between some concepts descriptions, the connectors used in the relations, the order of the summaries, their size and their verbalization. However, the impact of these factors on the interpretability of the summaries needs to be further investigated. Nevertheless, both criteria exceeded the defined threshold (50%) and are considered satisfactory given that the participating subjects do not belong to the application area (Law). This suggests that the generated summaries would be better interpreted by jurists specialized in commercial law. We intend to verify the latter in future research.

References

1. Yager, R.R.: A new approach to the summarization of data. Inf. Sci. **28**, 69–86 (1982)
2. Fontenla-Seco, Y., Bugarin, A., Lama, M.: Fuzzy temporal protoforms for the quantitative description of processes in natural language. In: IEEE International Conference on Fuzzy Systems, pp. 1–6 (2021). http://doi.org/10.1109/FUZZ45933.2021.9494444
3. Cornejo, M.E., Medina, J., Rubio-Manzano, C.: Linguistic descriptions of data via fuzzy formal concept analysis. In: Computational Intelligence and Mathematics for Tackling Complex Problems, pp. 119–125. Springer, Berlin (2022). http://doi.org/10.1007/978-3-030-74970-5_14
4. Duraj, A., Szczepaniak, P.S.: Linguistic summaries using interval-valued fuzzy representation of imprecise information—an innovative tool for detecting outliers. In: International Conference on Computational Science (ICCS 2021), pp. 500–513. Springer, Berlin (2021). http://doi.org/10.1007/978-3-030-77980-1_38
5. Nguyen, C.H., Pham, T.L., Nguyen, T.N., Ho, C.H., Nguyen, T.A.: The linguistic summarization and the interpretability, scalability of fuzzy representations of multilevel semantic structures of word-domains. Microprocess. Microsyst. (2021). https://doi.org/10.1016/J.MICPRO.2020.103641
6. To, N.D., Reformat, M.Z., Yager, R.R.: Question-answering system with linguistic summarization. In: IEEE International Conference on Fuzzy Systems, pp. 1–8 (2021). http://doi.org/10.1109/FUZZ45933.2021.9494389
7. Pérez, I., Piñero, P.Y., Bello, R., Acuña, L.A., García, R.: Linguistic summaries generation with hybridization method based on rough and fuzzy sets. In: International Joint Conference on Rough Sets, pp. 385–397. Springer, Berlin (2020). http://doi.org/10.1007/978-3-030-52705-1_29
8. Kacprzyk, J., Yager, R., Merigo, J.M.: Towards human-centric aggregation via ordered weighted aggregation operators and linguistic data summaries: a new perspective on Zadeh's inspirations. IEEE Comput. Intell. Mag. **14**, 16–30 (2019). https://doi.org/10.1109/MCI.2018.2881641
9. Rodríguez, C.R., Peña, M., Zuev, D.S.: Extracting composite summaries from qualitative data. In: VII International Workshop on Artificial Intelligence and Pattern Recognition. Springer, Berlin (2021). http://doi.org/10.1007/978-3-030-89691-1_26
10. Yager, R.R., Reformat, M.Z., To, N.D.: Drawing on the iPad to input fuzzy sets with an application to linguistic data science. Inf. Sci. **479**, 277–291 (2019). https://doi.org/10.1016/J.INS.2018.11.048
11. Ramos-Soto, A., Martin-Rodilla, P.: Enriching linguistic descriptions of data: a framework for composite protoforms. In: Fuzzy Sets and Systems, vol. 407, pp. 1–26 (2021). http://doi.org/10.1016/j.fss.2019.11.013
12. Heble-Lahera, C., Cascallar-Fuentes, A., Ramos-Soto, A., Diz, A.B.: Empirical study of fuzzy quantification models for linguistic descriptions of meteorological data. In: IEEE International Conference on Fuzzy Systems (2020). http://doi.org/10.1109/FUZZ48607.2020.9177716
13. Peláez-Aguilera, M.D., Espinilla, M., Fernández Olmo, M.R., Medina, J.: Fuzzy linguistic protoforms to summarize heart rate streams of patients with ischemic heart disease. Complexity (2019). http://doi.org/10.1155/2019/2694126
14. Genç, S., Akay, D., Boran, F.E., Yager, R.R.: Linguistic summarization of fuzzy social and economic networks: an application on the international trade network. Soft. Comput. **24**, 1511–1527 (2019). https://doi.org/10.1007/S00500-019-03982-9
15. Wilbik, A., Gilsing, R., Turetken, O., Ozkan, B., Grefen, P.: Intentional linguistic summaries for collaborative business model radars. In: 2020 IEEE International Conference on Fuzzy Systems, pp. 1–7 (2020). http://doi.org/10.1109/FUZZ48607.2020.9177587
16. Moreno-Garcia, J., Jimenez-Linares, L., Rodriguez-Benitez, L.: Automatic generation of linguistic descriptions of electricity consumption in the buildings of a large institution. In: Computational Intelligence and Mathematics for Tackling Complex Problems, vol. 3, pp. 101–109. Springer, Berlin (2022). http://doi.org/10.1007/978-3-030-74970-5_12

17. Moreno-Garcia, J., Jimenez-Linares, L., Liu, J., Rodriguez-Benitez, L.: Generation of linguistic descriptions for daily noise pollution in urban areas. In: IEEE International Conference on Fuzzy Systems, pp. 1–6 (2021). http://doi.org/10.1109/FUZZ45933.2021.9494388
18. Pérez, I., Piñero, P.Y., García, R., Bello, R., Acuña, L.A.: Discovering fails in software projects planning based on linguistic summaries. In: International Joint Conference on Rough Sets, pp. 365–375. Springer, Berlin (2020). http://doi.org/10.1007/978-3-030-52705-1_27
19. Zadeh, L.A.: A prototype-centered approach to adding deduction capability to search engines—the concept of protoform. In: Annual Conference of the North American Fuzzy Information Processing Society—NAFIPS (2002). http://doi.org/10.1109/NAFIPS.2002.1018115
20. Kacprzyk, J., Zadrozny, S.: Linguistic database summaries and their protoforms: towards natural language based knowledge discovery tools. In: Information Sciences, vol. 173 (2005). http://doi.org/10.1016/j.ins.2005.03.002
21. Trivino, G., Sugeno, M.: Towards linguistic descriptions of phenomena. Int. J. Approximate Reasoning **54**, 22–34 (2013). https://doi.org/10.1016/j.ijar.2012.07.004
22. Smits, G., Nerzic, P., Pivert, O., Lesot, M.J.: Efficient generation of reliable estimated linguistic summaries. In: IEEE International Conference on Fuzzy Systems (2018). http://doi.org/10.1109/FUZZ-IEEE.2018.8491604
23. Mann, W.C., Thompson, S.A.: Rhetorical structure theory: toward a functional theory of text organization. Text **8**, 243–281 (1988)
24. Hou, S., Zhang, S., Fei, C.: Rhetorical structure theory: a comprehensive review of theory, parsing methods and applications. In: Expert Systems with Applications, vol. 157 (2020). http://doi.org/10.1016/j.eswa.2020.113421
25. Agrawal, R., Srikant, R.: A fast algorithm for mining association rules. In: Proceedings of the 20th International Conference on Very Large Databases, VLDB, pp. 487–499 (1994)
26. Savasere, A., Omiecinski, E., Navathe, S.: An efficient algorithm for mining association rules in large databases. In: Proceedings of the 21st International Conference on Very Large Data Bases, pp. 432–443, Zurich (1995)
27. Zaki, M.J.: Scalable algorithms for association mining. IEEE Trans. Knowl. Data Eng. **12**, 372–390 (2000). https://doi.org/10.1109/69.846291

Chapter 6
New Indicators for the Assessment of Linguistic Summaries Considering a Rough Sets Approach

Iliana Pérez Pupo, Pedro Y. Piñero Pérez, and Rafael E. Bello Pérez

Abstract In this work, different approaches to the evaluation of linguistic summaries are analyzed. In the design of the indicators, elements of rough set theory and the neutrosophic theory are combined. Several of the proposed indicators expand the traditional indicators reported in the bibliography. For comparisons, the authors design a set of 29 test cases that represent different equivalence classes of project management decision-making scenarios. The indicators reported in the bibliography are compared with the new ones proposed in the 29 test cases. For each indicator, the authors present different graphs that allow the indicators to compare their advantages and disadvantages. Finally, the authors present the conclusion of this work and future work.

Keywords Artificial intelligent · Linguistic data summarization · Linguistic summaries assessment · Project management

1 Introduction

The remarkable development of information technologies is generating numerous data from different business management processes. Unfortunately, raw data is often not useful and does not provide enough information because it is poorly readable or understandable to human decision makers. This is one of the latent problems in

I. Pérez Pupo (✉) · P. Y. Piñero Pérez
Grupo de Investigaciones en Gestión de Proyectos, Universidad de las Ciencias Informáticas, CP 54830 La Habana, Cuba
e-mail: iperez@uci.cu

P. Y. Piñero Pérez
e-mail: ppp@uci.cu

R. E. Bello Pérez
Centro de Investigación en Informática, Universidad Central Marta Abreu de las Villas, Santa Clara, Cuba
e-mail: rbellop@uclv.edu.cu

P. Y. Piñero Pérez et al. (eds.), *Artificial Intelligence in Project Management and Making Decisions*, Studies in Computational Intelligence 1035,
https://doi.org/10.1007/978-3-030-97269-1_6

decision-making processes, closely associated with the discovery of behavior patterns and their interpretability.

A particular scenario of this situation is constituted by project-oriented organizations. In this type of organization, despite efforts to improve efficiency and effectiveness in management, many projects fail or should be renegociates. Among the fundamental causes of project failure are: poor management, inadequacies in planning, control and monitoring processes. These causes are associated with risks that can be mitigated if the information systems have techniques for the discovery of knowledge and the analysis of historical summaries of the data in linguistic form. Summaries that can be understood by decision makers and that can be used to facilitate tasks such as decision analysis, prediction or forecasting [1–4]. In this sense, knowledge discovery techniques have been classified mainly into two classes: predictive and descriptive. The goal of predictive techniques is to predict the value of a specific attribute (known as a dependent variable) based on the values of other attributes (commonly known as an explanatory or independent variable). On the other hand, the objective of descriptive techniques is to explore patterns (correlations, trends, groups, trajectories and anomalies) that summarize the underlying relationships in the data [5].

In this context, linguistic data summarization arises, as one of the descriptive knowledge discovery techniques with an interesting and promising approach to produce summaries from numerical data using natural language. With it, organizations could solve the dilemma of "rich data, poor information" present in decision-making. But, Decision-making environments in project management are characterized by being environments under uncertainty with the presence of phenomena such as: ambiguity, indeterminacy or imprecision [6]. However, the linguistic data summarization techniques reported in the bibliography show that they only analyze the degree of certainty in the information, ignoring the treatment of indeterminacy or falsehood.

In this work, different techniques for linguistic data summarization are applied for decision-making in project management. In particular, new approaches are proposed for the evaluation of linguistic summaries that combine elements of the neutrosophic theory with elements of the theory of rough sets.

This work is structured in three sections. The first section analyzes and describes a set of indicators reported in the bibliography for the evaluation of linguistic summaries. The second section suggests a set of new indicators that complement or extend the traditional ones. Then, in the third, the proposed extensions are compared with the traditional indicators, based on evaluating the behavior of both groups, in a set of test classes. Finally, the conclusions of the investigation are presented.

2 Traditional Indicators for the Evaluation of Linguistic Summaries

There are different works that address the evaluation methods of linguistic summaries. Among the first works in this sense, those published by Yager [7], Zadeh [8] and Kacprzyk and Zadrożny [5, 9] stand out. In these works, methods for calculating the truth value of a summary are proposed.

Then in other works Kacprzyk et al. [10, 11] present other indicators to evaluate the summaries such as the degree of imprecision, coverage, appropriateness and the length of a summary. But, quality indicators, for the evaluation of linguistic summaries, most used are resume by Kacprzyk in [9]:

- Degree of truth (T_1), which measures the percentage of truthfulness of a summary with respect to the data.
- Degree of imprecision (T_2), a validity criterion that complements the degree of truth, since an imprecise linguistic summary, even if it has a high degree of truth, is not very useful.
- Degree of coverage (T_3), which indicates how many objects in the database are covered or represented by the evaluated summary.
- Degree of appropriateness (T_4), which measures the relevance of a summary. That is, the ability to identify highly correlated relationships between variables, even when there is little occurrence of events that link these variables.
- Length of a summary (T_5), which is the degree of quality of the summary with respect to the number of variables it involves.

Some of these indicators are described in detail below.

2.1 *Degree of Truth*

The "degree of truth T_1" determines the validity of a summary, a criterion introduced by Yager in [7]. Equation 1 shows how to calculate this indicator for summaries without a filter, and Eq. 2 for those that do.

$$T\big(Qy's\,are\,S\big) = \mu_Q\left[\frac{1}{n}\sum_{i=1}^{n}\mu_S(y_i)\right] \tag{1}$$

$$T\big(QRy's\,are\,S\big) = \mu_Q[r] \tag{2}$$

$$r = \frac{\sum_{i=1}^{n}\mu_R(y_i)\cap\mu_S(y_i)}{\sum_{i=1}^{n}\mu_R(y_i)} \tag{3}$$

$$\mu_Q(x) = \begin{cases} 1 & for\, x \geq 0.8 \\ 2x - 0.6 & for\, 0.3 < x < 0.8 \\ 0 & for\, x \leq 0.3 \end{cases} \tag{4}$$

where

$\mu_R(y_i)$	is the degree of membership of the filter of a summary.
$\mu_S(y_i)$	is the degree of membership of the summarizer of summary.
$\cap$	represents a t-norm, which could be the "minimum" operation or a productive one.
n	is the number of objects in the database.
$\mu_Q(r)$	is the membership function represented by the linguistic quantifier Q.

The "degree of truth" measures the percentage of veracity of a summary with respect to the data; if this value were close to 1, the summary could be important for making a decision. But it could also be close to 0, as this indicator considers all cases with a degree of truth greater than zero, which is why it also values very low-certainty summaries. However, in many real scenarios where the degree of indeterminacy or falsity of the summary could also influence a decision to be made, characterized by the influence of numerous specialists, who often present neutral or opposite criteria to the variables and/or events of a scenario in question, this variant would not be viable, since it does not consider elements of indeterminacy treated in the neutrosophic theory [12].

2.2 *Degree of Imprecision* T_2

The "degree of imprecision (T_2)" is an obvious and important validity criterion [9], as a very imprecise linguistic summary with a high degree of truth is not very useful. Example, the summary "on almost every winter day, the temperature is quite cold" has a very high truth value (the T_1 would tend to 1); however, it is very imprecise, as it is not useful or does not generate valuable information. To calculate this indicator, Eqs. 5 and 6 are used. As can be seen in Eq. 6, this indicator depends on the summarizer of the summary and not on the database; that is, to calculate it, the records are not traversed, only the cardinality of each linguistic set of the summarizer is considered, as well as its domain

$$T_2 = 1 - \sqrt[m]{\prod_{j=1,\ldots,m} in(S_j)} \tag{5}$$

$$in(S_j) = \frac{card\{x \in X_j : \mu_{S_j}(x) > 0\}}{card(X_j)} \tag{6}$$

where

m is the number of fuzzy sets of the $S = \{S_1, S_2, \ldots S_j, \ldots, S_m\}$.
$card$ denotes the cardinality of the corresponding non-fuzzy set.
X_j is the domain, which represents all the finite values of the fuzzy set of the summarizer S_j.

2.3 Degree of Coverage T₃

The "degree of coverage (T_3)" indicates how many objects in the database that meet the *R* filter are covered by the summary whose summarizer is *S*. Its interpretation is simple, for example, if the value were 0.15, then it means that 15% of the objects are consistent with the summary in question. The value of this indicator clearly depends on the content of the database [9] and is calculated as shown by the Eqs. 7, 8 and 9.

$$T_3 = \frac{\sum_{i=1}^{n} t_i}{\sum_{i=1}^{n} h_i} \tag{7}$$

$$t_i = \begin{cases} 1 \; if \; \mu_S(y_i) > 0 \, and \, \mu_R(y_i) > 0 \\ 0 \qquad in \; other \; case \end{cases} \tag{8}$$

$$h_i = \begin{cases} 1 \; if \; \mu_R(y_i) > 0 \\ 0 \; in \; other \; case \end{cases} \tag{9}$$

This indicator is based on the conditional probability of the appearance of the summarizers given that the filters take place, and on the concept of support set treated in fuzzy logic. In this research, it is considered that it may be relevant to know how much a summary covers the database. In this sense, a proposal to extend this indicator is made in Sect. 3.

2.4 Degree of Appropriateness T₄

The "degree of appropriateness (T_4)" is the most relevant indicator according to Kacprzyk and Zadrożny [9], see the Eqs. 10–13. Suppose that the summary contains a description (fuzzy set) $S = (S_1, S_2, \ldots, S_m)$, which is partitioned into m summarizers composed of attributes $V_1, V_2, \ldots, V_m$ such that each summarizer corresponds to a fuzzy set, then it is denoted:

$$S_j(y_i) = \mu_{S_j}(V_j(y_i)) \tag{10}$$

$$T_4 = abs\left(\prod\nolimits_{j=1,\ldots,m} r_j - T_3\right) \tag{11}$$

$$r_j = \frac{\sum_{i=1}^{n} h_i}{n} \quad \text{for } j = 1, \ldots, n \tag{12}$$

$$h_i = \begin{cases} 1 \; if \; S_j(y_i) > 0 \\ 0 \; in \; other \; case \end{cases} \tag{13}$$

Exemplifying it with a database on workers, if 50% of them are under 25 and 50% are highly skilled, then employees who are under 25 and highly skilled can be expected to be close to 25%. However, if the degree of appropriateness is, for example, 0.39 (that is, 39% are under 25 years of age and highly qualified), then the summary found reflects a captivating relationship, not expected in our data. The T_4 is a linear combination of the T_3 and other elements.

2.5 *Length of a Summary* T5

The "length of a summary (T_5)" is an important indicator because the longer the summary, the more difficult it will be to understand, see Eq. 14.

$$T_5 = 2\left(0.5^{card(S)}\right) \tag{14}$$

To visualize how this indicator behaves, a random data sequence is constructed, which simulates the size of linguistic summaries, taking values between 1 and 8. These values are evaluated in Eq. 14. Subsequently, using the Excel tabulator, a graph is constructed of lines, where the size of the summaries is represented on the 'x' axis, and the T5 values are represented on the 'y' axis. It is obtained that this indicator has a behavior like the one shown in Fig. 1.

This is an indicator that responds to an exponential function, as represented in Fig. 1. For summaries of size 1, the T_5 reaches its maximum value, this being a not very adequate summary size, especially if it depends on taking a decision. However,

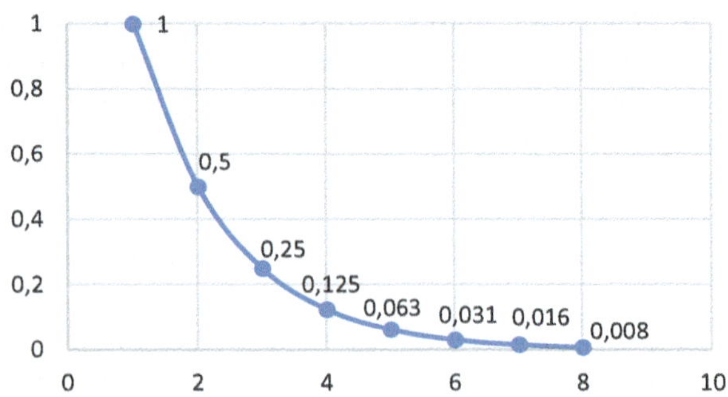

Fig. 1 T_5 graphical representation "x" means the linguistic summaries length

for summaries involving 2–4 variables, the T_5 value falls sharply, this being the most expected range of values for a summary. A summary of size 2 is good because it involves more than one variable from the database; in turn, a summary with many components is no longer legible or understandable.

These indicators, like other LDS techniques, consider the certainty of the information but do not consider the indeterminacy or falsehood in the concepts.

3 New Extensions of T Indicators to Evaluate Linguistic Summaries

In this section, a group of extensions to the traditional linguistic summaries' evaluation indicators is proposed, combining elements of the rough set theory [13, 14] and the neutrosophic theory [12].

For a better understanding of the proposals, a set of definitions and notations used to formulate the proposed extensions is introduced in next subsection. In the rest of the subsection's extensions are proposed for indicators T_1, T_2, T_3, T_4 and T_5 respectively.

3.1 Definitions and Notations Used in the Proposed Extensions

Definition 1 Neutrosophic set: let X be a universe of discourse, a neutrosophic set A over X is an object of the form: $A = \{\langle \mu_A(x), \tau_A(x), \sigma_A(x)\rangle : x \in X\}$, see Fig. 2.

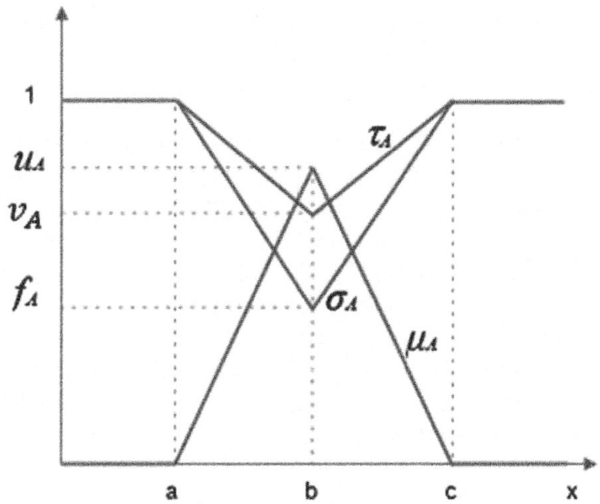

Fig. 2 Set with truth-membership (μ_A), indeterminacy-membership (τ_A) and falsity-membership (σ_A)

- $\mu_A(x) \in [0, 1]$ is a membership function that represents the degree of certainty of the membership of the value x to the set A, see Eq. 15.
- $\tau_A(x) \in [0, 1]$ membership function that represents the degree of indeterminacy of the value x to the set A, see Eq. 16.
- $\sigma_A(x) \in [0, 1]$ membership function that represents the degree of non-membership (or falsehood) of the value x to the set A, see Eq. 17.
- $0 \le \mu_A(x) + \tau_A(x) + \sigma_A(x) \le 3 \ \forall \ x \in X$

$$\mu_A(x) = \begin{cases} (x-a)u_A/(b-a) & (a \le x < b) \\ u_A & (x = b) \\ (c-x)u_A/(c-b) & (b < x \le c) \\ 0 & in\ other\ case \end{cases} \tag{15}$$

$$\tau_A(x) = \begin{cases} (b-x+v_A(x-a))/(b-a) & (a \le x < b) \\ r_A & (x = b) \\ (x-b+v_A(c-x))/(c-b) & (b < x \le c) \\ 1 & in\ other\ case \end{cases} \tag{16}$$

$$\sigma_A(x) = \begin{cases} (b-x+f_A(x-a))/(b-a) & (a \le x < b) \\ f_A & (x = b) \\ (x-b+f_A(c-x))/(c-b) & (b < x \le c) \\ 1 & in\ other\ case \end{cases} \tag{17}$$

Being $y_i \in Y$ such that, Y represents the database of n objects, $|Y| = n$ and each object y_i is made up of k attributes. For each attribute z of an object y_i there is a linguistic variable $V_z = \{A_1, A_2, \ldots, A_g\}$ formed by "g" fuzzy sets.

Definition 2 V_z being a linguistic variable associated with the z-th attribute of the objects in database Y; the membership of an object $y_i \in Y$, to the fuzzy set $A \in V_z$, is denoted $\mu_A(y_i) \in [0, 1]$ and evaluates the z-th attribute of the object y_i in the fuzzy set A.

Definition 3 The degree of belonging of the $y_i \in Y$ object to a SUMMARY linguistic summary, denoted $\mu_{SUMMARY}(y_i)$, constitutes a measure in which the object is covered or represented by a linguistic summary. SUMMARY being a linguistic summary in its generalized form "*QRy's are S*" and formed by the family of fuzzy sets of Q, $R = \{R_p\}$ and $S = \{S_1, S_2, \ldots, S_m\}$, where Q constitutes a fuzzy set of the linguistic variable of the quantifier Q, R constitutes the family of fuzzy sets that make up the filters and S the family of fuzzy sets that represent the summarizers. The degree of membership is calculated from the application of a t-norm to the set of membership values of the object's attributes and i to the family of fuzzy sets that form SUMMARY, see Eqs. 18, 19 and Definition 2.

$$\mu_{SUMMARY}(y_i) = \mathrm{TNorm}_{j \in \{1,2,\ldots,m\}}\left(\mu_{S_j}(y_i)\right) : y_i \in Y \tag{18}$$

SUMMARY has structure "*Qy's are S*", $R = \varnothing$

$$\mu_{SUMMARY}(y_i) = \text{TNorm}_{\substack{k \in \{1, 2, \ldots, \text{p}\} \\ j \in \{1, 2, \ldots, \text{m}\}}} \left(\mu_{R_k}(y_i), \mu_{S_j}(y_i)\right) : y_i \in Y \quad (19)$$

SUMMARY has structure "*QRy's are S*".

Where $\mu_{R_k}(y_i)$ represents the evaluation of the object y_i in the membership function of the fuzzy sets of R and $\mu_{S_j}(y_i)$, represents the evaluation of the object y_i in the membership function of the fuzzy sets of S_j.

Definition 4 Let SUMMARY be a linguistic summary in its generalized form "*QRy's are S*", it is defined as a lower approximation in degree α of SUMMARY and $SUMMARY_{\alpha *}(Y)$ is denoted to the set of objects $y_i \in Y$, such that Y represents the database, whose degree of membership in SUMMARY is greater than a α-cut $\alpha \in [0, 1]$, Eq. 20.

$$SUMMARY_{\alpha \rightleftarrows *}(Y) = \{y_i \in Y : \mu_{SUMMARY}(y_i) \geq \alpha\} \quad (20)$$

Definition 5 Let SUMMARY be a linguistic summary in its generalized form "*QRy's are S*", it is defined as a superior approximation of SUMMARY and $SUMMARY^*(Y)$ is denoted to the set of objects $y_i \in Y$, such that Y represents the database, whose degree of membership in SUMMARY is greater than 0. See Eq. 21.

$$SUMMARY^*(Y) = \{y_i \in Y : \mu_{SUMMARY}(y_i) \geq 0\} \quad (21)$$

From these approximations the limit region $SUMMARY_{BN}$ is defined as shown in Eq. 22.

$$SUMMARY_{BN}(Y) = \left(SUMMARY^*(Y) - SUMMARY_{\alpha *}(Y)\right) \quad (22)$$

These last three equations are inspired by the concepts of inferior approximation and superior approximation of the Rough Sets Theory [14]. The objects that belong to the lower approximation $SUMMARY_{\alpha *}(Y)$ represent the objects $y_i \in Y$, where there is no doubt that they belong to the SUMMARY summary considering the α-cut; whereas those that belong to the upper approximation $SUMMARY^*(Y)$ are the objects of $y_i \in Y$ that belong to the linguistic summary to some degree greater than 0, which includes both the objects of the lower approximation and those of the boundary region $SUMMARY_{BN}$. The objects that belong to the boundary region represent objects about which there is some doubt about the level of representativeness of a certain summary about it.

Definition 6 Let $y_i \in Y$ be an object and SUMMARY a linguistic summary, it is defined as $\tau_{SUMMARY}(\text{y}_\text{i})$ the degree of indeterminacy associated with the membership of the y_i object to the SUMMARY. It is calculated for linguistic summaries with structure "*Qy's are S*" as shown in Eq. 23, and for linguistic summaries with structure "*QRy's are S*" as shown in Eq. 24.

$$\tau_{SUMMARY}(y_i) = \text{TNorm}_{j\in\{1,2,\ldots,m\}}\left(\tau_{S_j}(y_i)\right) \tag{23}$$

$$\tau_{SUMMARY}(y_i) = \text{TNorm}_{\substack{k\in\{1,2,\ldots,p\}\\ j\in\{1,2,\ldots,m\}}}\left(\tau_{R_k}(y_i), \tau_{S_j}(y_i)\right) \tag{24}$$

Definition 7 Let $y_i \in Y$ be an object and SUMMARY a linguistic summary, it is defined as $\sigma_{SUMMARY}(y_i)$ the degree of non-membership of the object y_i to the SUMMARY. It is calculated for linguistic summaries with structure "*Qy's are S*" as shown in Eq. 25, and for linguistic summaries with structure "*QRy's are S*" as shown in Eq. 26.

$$\sigma_{SUMMARY}(y_i) = \text{TNorm}_{j\in\{1,2,\ldots,m\}}\left(\sigma_{S_j}(y_i)\right) \tag{25}$$

$$\sigma_{SUMMARY}(y_i) = \text{TNorm}_{\substack{k\in\{1,2,\ldots,p\}\\ j\in\{1,2,\ldots,m\}}}\left(\sigma_{R_k}(y_i), \sigma_{S_j}(y_i)\right) \tag{26}$$

3.2 Extensions for Calculating the Degree of Truth Te_{1a}

The Te_{1a} indicator is also a measure of the certainty of the summary, it complements the T_1 indicator, it does not replace it. Both T_1 and Te_{1a} consider the quantifier during the calculation. Equation 27 shows how the indicator Te_{1a} should be calculated.

The main difference between T_1 and Te_{1a} is that T_1 considers in the number all objects with a membership of the summary greater than zero; while Te_{1a} only considers, in the calculation of the numbering, those objects with a membership of the summary greater than the α-cut established. Furthermore, in the case of summaries with filters, T_1 places the cardinality of the database in the denominator, while Te_{1a} counts only those objects with a membership greater than zero in the denominator. In this way, the aim is to propose an indicator more focused on evaluating the accuracy of the summary, which is capable of better discriminating summaries and objects with a higher degree of certainty from those that do not have it.

$$\text{Te}_{1a} = \mu_Q\left(\frac{|SUMMARY_{\alpha*}(Y)|}{|SUMMARY^{*}(Y)|}\right), \text{ such that } \text{Te}_{1a} \in [0,1] \tag{27}$$

where: μ_Q is the fuzzy set that represents the quantifier of the summary.

On the other hand, the Te_{1b} extension is proposed by complementing the values of T_1 and Te_{1a}. It is calculated from the mean of the certainty values of objects whose membership in the summary exceeds the α-cut, see Eq. 28. This indicator has as its main difference with T_1 and Te_{1a}, which only includes the membership evaluation of filters and summarizers without considering the quantifier. In this way, it is a question of better separating, summaries with a high frequency of joint appearance of filters and summarizers, from those that represent many objects but with a low evaluation of belonging to their respective filters and summarizers.

$$\mathrm{Te}_{\mathrm{1b}} = \frac{\sum_{y_i \in SUMMARY_{\alpha*}(Y)} \mu_{SUMMARY}(y_i)}{|SUMMARY_{\alpha*}(Y)|}, \text{ such that } \mathrm{Te}_{\mathrm{1b}} \in [0,1] \tag{28}$$

3.3 Extensions to Degree of Imprecision

Te_{2a}, Te_{2b}, Te_{2c} and Te_{2d} extensions are proposed that complement T_2, not replace it. The main contribution of these indicators is the incorporation of elements of the neutrosophic theory. Unlike T_2, in the calculation of these extensions the database objects are used.

These extensions are proposed with the aim of calculating with greater precision the measures of certainty, indeterminacy and non-membership, considering both an integrated approach to these measures and their influence independently. In this way, the calculation of the quality of the summaries is facilitated, in scenarios with the presence of neutrality and indeterminacy phenomena in decision-making.

In the case of Te_{2a}, the uncertainty is calculated considering the degrees of membership, non-membership and indeterminacy of each object whose membership in the summary is greater than the α-cut established, as shown in Eq. 29.

$$\mathrm{Te}_{\mathrm{2a}} = \frac{\sum_{y_i \in SUMMARY_{\alpha*}(Y)} \left(\frac{1-\mu_{SUMMARY}(y_i)+\tau_{SUMMARY}(y_i)+\sigma_{SUMMARY}(y_i)}{3} \right)}{|SUMMARY_{\alpha*}(Y)|} \tag{29}$$

where: $\mathrm{Te}_{\mathrm{2a}} \in [0, 1]$.

The indicator Te_{2b}, calculates the uncertainty, but only considering the degrees of non-membership and indeterminacy of the objects whose membership in the summary is greater than the established α-cut, Eq. 30.

$$\mathrm{Te}_{\mathrm{2b}} = \frac{\sum_{y_i \in SUMMARY_{\alpha*}(Y)} \left(\frac{\tau_{SUMMARY}(y_i)+\sigma_{SUMMARY}(y_i)}{2} \right)}{|SUMMARY_{\alpha*}(Y)|} \tag{30}$$

where: $\mathrm{Te}_{\mathrm{2b}} \in [0,1]$.

The Te_{2c} indicator constitutes the mean of the indeterminacy of the objects whose membership in the summary is greater than the established α-cut, see Eq. 31.

$$\mathrm{Te}_{2c} = \frac{\sum_{y_i \in SUMMARY_{\alpha*}(Y)} \tau_{SUMMARY}(y_i)}{|SUMMARY_{\alpha*}(Y)|}, \text{ such that } \mathrm{Te}_{2c} \in [0, 1] \quad (31)$$

The Te_{2d} indicator constitutes the mean of the non-membership in the summary of the objects that exceed the α-cut, Eq. 32.

$$\mathrm{Te}_{2d} = \frac{\sum_{y_i \in SUMMARY_{\alpha*}(Y)} \sigma_{SUMMARY}(y_i)}{|SUMMARY_{\alpha*}(Y)|}, \text{ such that } \mathrm{Te}_{2d} \in [0, 1] \quad (32)$$

3.4 *Extension to the Calculation of the Degree of Coverage* $\mathbf{Te_3}$

The Te_3 indicator is proposed as an extension, it can replace the traditional T_3 indicator. It is built in order to determine more strictly the degree of coverage of a linguistic summary. The traditional T_3 considers all objects with membership greater than zero in the coverage calculation, while this extension gives a higher evaluation to those summaries that guarantee the coverage of a greater number of objects, but stimulating those that also achieve high membership values. Equation 33 summarizes the way to calculate this indicator.

$$\mathrm{Te}_3 = \frac{|SUMMARY_{\alpha*}(Y)|}{n}, \text{ such that } \mathrm{Te}_3 \in [0,1] \quad (33)$$

3.5 *Extension to the Calculation of the Degree of Appropriateness* $\mathbf{Te_4}$

The Te_4 extension is proposed as a complement to the traditional T_4, not as a replacement. This extension is proposed to measure the appropriateness in decision-making scenarios with the presence of phenomena such as indeterminacy or neutrality. In this sense, this indicator incorporates elements of the neutrosophic theory and calculates the appropriateness from combining the degrees of certainty, indeterminacy and non-belonging of the objects to the summary.

It is an extension that promotes the identification of peculiar relationships between variables, relationships that are not present in most of the database objects but that are strong in the objects where they appear. It establishes an inversely proportional relationship between the degree of belonging of the objects to the summary and the number of objects that belong to the superior approximation. Te_4 assigns a higher value to those summaries that represent objects with membership greater than the α-cut and not to summaries that represent many objects in the interval [0, α-cut]; that is, not to summaries with many objects in the top approximation. Another element that differentiates this extension from the traditional $T_4$is that the degree of membership also includes the degree of non-membership and indeterminacy, see Eq. 34.

$$\mathrm{Te}_4 = \frac{\sum_{y_i \in SUMMARY_{\alpha*}(Y)} \left(\frac{(1+\mu_{SUMMARY}(y_i) - \tau_{SUMMARY}(y_i) - \sigma_{SUMMARY}(y_i))}{2} \right)}{|SUMMARY^{*}(Y)|} \tag{34}$$

where: $\mathrm{Te}_4 \in [0, 1]$.

3.6 Extension to the Evaluation of the Length of Te$_5$ *Summaries*

Te_5 is proposed as an extension that replaces the traditional T_5 indicator. The main difference between these indicators is the mathematical function on which they support their calculation. This indicator is proposed with the objective of achieving a greater smoothness in the evaluation values in the face of the increase in the size of the summaries, improving the abrupt jumps that are experienced in the traditional T_5. Te_5 is based on the Gaussian curve, Eq. 35, and assigns a greater value to summaries that have lengths between two and three, therefore the parameter $b = 2.5$, while the value $k = 0.125$. These two values will be produced from trial and error until the desired smoothness is achieved.

$$Te_5 = e^{-k(\mathrm{x}-\mathrm{b})^2}, \text{ such that } Te_5 \in [0, 1], \quad k = 0.125 \text{ and } b = 2.5 \tag{35}$$

4 Comparison of Traditional and Extended Indicators

For the comparison of traditional and extended indicators, a test environment is established where 29 possible scenarios are simulated, see Table 1. Each of the scenarios is represented by a test class, where each test class represents a combination of possible filter and summary evaluation values, covering all possible scenarios in which linguistic summaries can be expressed.

The test scenarios are characterized by the following elements:

- Two prototype summaries are used. The first prototype has the structure "*Qy's R1 and R2 are S1 and S2*" formed by two filters R1 and R2, and two summarizers S1 and S2. The second prototype with the "*Qy's are* $S1$ and $S2$" structure, that is, without filters and with two summarizers S1 and S2.
- Each test class will represent a combination of possible values of quantifier-filters-summarizers when evaluating an object in a linguistic summary. To represent the

Table 1 Tests class for test scenarios

	Quantificator	Filter	Summarizer
Class ER1	Few	Low	Low
Class ER2	Few	Low	Middle
Class ER3	Few	Low	High
Class ER4	Few	Middle	Low
Class ER5	Few	Middle	Middle
Class ER6	Few	Middle	High
Class ER7	Few	High	Low
Class ER8	Few	High	Middle
Class ER9	Few	High	High
Class ER10	Around the 50%	Low	Low
Class ER11	Around the 50%	Low	Middle
Class ER12	Around the 50%	Low	High
Class ER13	Around the 50%	Middle	Low
Class ER14	Around the 50%	Middle	Middle
Class ER15	Around the 50%	Middle	High
Class ER16	Around the 50%	High	Low
Class ER17	Around the 50%	High	Middle
Class ER18	Around the 50%	High	High
Class ER19	The majority	Low	Low
Class ER20	The majority	Low	Middle
Class ER21	The majority	Low	High
Class ER22	The majority	Middle	Low
Class ER23	The majority	Middle	Middle
Class ER24	The majority	Middle	High
Class ER25	The majority	High	Low
Class ER26	The majority	High	Middle
Class ER27	The majority	High	High
Class ER28	The majority	Porous	Porous
Class ER29	The majority	Porous	Porous

ranges of values of each class of test, the terms low, medium, high are used, which represent the intervals [0, 0.33), [0.33, 0.66) and [0.66, 1] respectively.

- Each test class is made up of a set of 10 instances, such that the first 7 represent objects that correspond to the characteristics of the test class in question and the other 3 are generated with random values. An example of the objects evaluated in a summary for the test class ER1 is shown in Table 2.
- To represent the quantifier in the test scenarios, a linguistic variable is used made up of three fuzzy sets represented. Only 3 linguistic sets are defined in order not

Table 2 Object example of test class ER1

	$\mu_M(x)$ Pertenencia				$\tau_M(x)$ Indeterminación				$\sigma_M(x)$ No pertenencia			
Object 1	R1	R2	S1	S2	R1	R2	S1	S2	R1	R2	R1	S2
	0.15	0.23	0.23	0.17	0.80	0.72	0.72	0.78	0.72	0.65	0.65	0.71

to introduce unnecessary complexity in the experimentation with an increase in classes of proof.

- Test classes ER28 and ER29 represent porous classes, that is, classes whose objects, when evaluated in the summary, have many zero values in filters and summarizers.

For example, the test class ER1, see Table 1, represents the scenario of summaries that have little quantifier and the set of objects that when evaluated in the summary are evaluated as low with respect to the filters and as low with respect to the summarizers. Table 2 shows an example of an object that belongs to test class ER1. This table represents a summary made up of two filters (R1, R2), two summarizers (S1, S2) and object 1, which, when evaluated in this summary, obtains a low membership evaluation in both filters and summarizers, because the evaluation values are in the interval [0, 0.33).

4.1 *Analysis of the Behavior of the Degree of Truth Indicator and Its Extensions*

This subsection characterizes the behavior of the traditional T_1 indicators and the Te_{1a} extension. Figure 2 graphs these two indicators against summaries of the type "*QRy's are S*" while Fig. 3 shows the behavior before summaries with structure "*Qy's are S*", that is, summaries with filter and without filter respectively.

For the analysis of the behavior of the indicators, we concentrate on the logical analysis of their expected response according to their semantics and the real response to the objects of the test class.

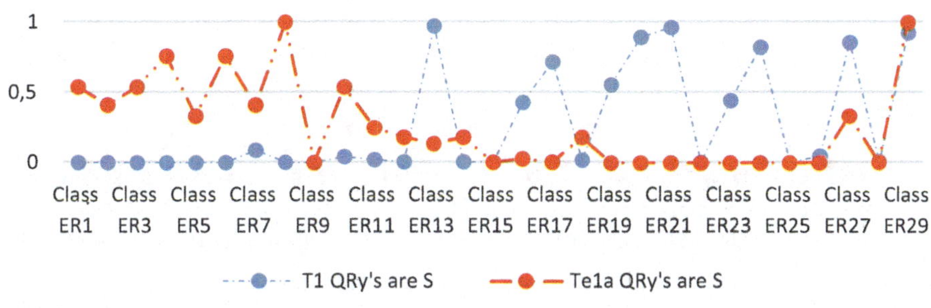

Fig. 3 Behavior of T_1 and Te_{1a} in summaries with filters

From the analysis of Fig. 2 it is identified that the linguistic summaries re-presented by the test classes from ER1 to ER8, have as semantics "few objects that comply with the filters R1 and R2 have the summarizers S1 and S2", Therefore, the T_1 and Te_{1a} indicators are expected to reach medium or high values. This is inferred because it is true that few objects meet these conditions, as indicated by the quantifier; since in these test classes, the objects take low or medium values of certainty in the fuzzy sets of the variables represented by both the filters and the summarizers. Therefore, it could safely be said that it is true that few objects meet the expression. In this context, note how Te_{1a} reports better behavior than T_1.

On the other hand, the semantics of the linguistic summaries represented by the test classes from ER19 to ER26, is interpreted as "most of the objects that comply with the filters R1 and R2 have the summarizers S1 and S2". In this case, the expected response of the indicators is low, because the quantifier "most" does not correspond to the number of objects that meet the summary conditions; since in this range of test classes, the objects also take low or medium values of certainty in the fuzzy sets of the variables of the filters and the summarizers. In this sense, Te_{1a} also reports better performance in these test classes.

Following the previous reasoning, in the test classes ER9, ER11, ER12, ER14, ER15, ER18, ER28 and ER29, the traditional indicator and the extension have a similar behavior and in correspondence with the expected result; while in the ER13 class the Te_{1a} report a better behavior. On the other hand, in the ER10, ER16, ER17 and ER27 classes for the data set used, the traditional T_1 reports a better performance compared to the Te_{1a} extension proposed (Fig. 4).

For the case of the summaries without filter represented in, it is identified that: before the test classes ER1, ER4 and ER8, the response of the indicators T_1 and Te_{1a} must reach medium or high values. This is inferred because the semantics of the summation in these test classes is interpreted as "Few objects comply with having values of summarizers S1 and S2". This expression is true because in these test classes the summarizers of the objects take low or medium values of certainty. Therefore, it could safely be said that it is true that few objects meet the expression. In this context, note that Te_{1a} has a better behavior with respect to T_1 in ER4 and ER8, while in ER1 T_1 had a better behavior.

In the test classes ER11, ER14 and ER17, the response of indicators T_1 and Te_{1a} should reach medium or high values. This is inferred because the semantics

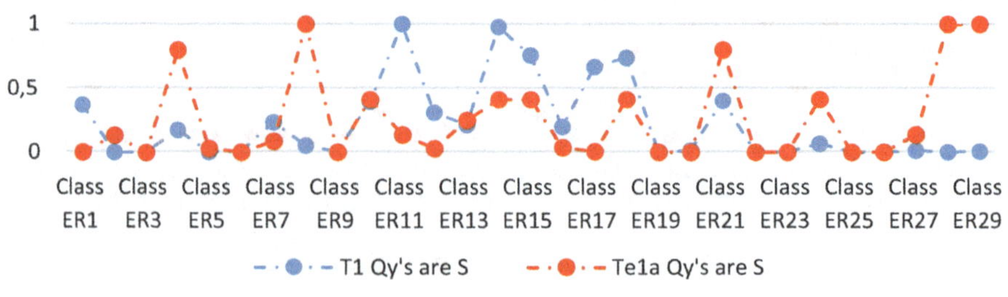

Fig. 4 Behavior of T_1 and Te_{1a} in summaries without filters

of the resume in these test classes is interpreted as “Approximately 50% of the objects comply with having values of summarizers S1 and S2”. This expression is true because in these test classes the object summarizers take mean values of certainty. Therefore, it could safely be said that it is true that approximately 50% of the objects comply with the expression. In this sense, note that in classes ER11 and ER14, Te_{1a} shows better behavior, since it represents values of the linguistic group “approximately 50%” and T_1 represents high values; while in the ER17 class it is the T_1 that shows the best behavior.

In the test class ER24 the indicators T_1 and Te_{1a} should reach high values. This is inferred because the summary semantics for this case is interpreted as “Most objects comply with summarizers S1 and S2.” This expression is true because in this kind of test, the summarizers take high values of certainty. Therefore, it could safely be said that it is true that most objects comply with the expression. In this context, note that Te_{1a} has a better performance than T_1.

In test classes ER28 and ER29 indicators T_1 and Te_{1a} should reach low values. This is inferred because summary semantics in these test classes are interpreted as “Most objects comply with summarizers S1 and S2.” As these classes are porous, when their objects are evaluated in the summary, the values of the summarizers are equal to zero. Therefore, it could be said that it is not true that most objects comply with the expression. In this context, note how T_1 has a better performance than Te_{1a}.

In test classes ER2, ER3, ER5, ER6, ER7, ER9, ER10, ER12, ER13, ER15, ER16, ER18, ER19, ER20, ER21, ER22, ER23, ER25, ER26 and ER27 the traditional indicator and the extension they have a similar behavior and in correspondence with the expected result.

It is identified that the summaries with the QRy’s are S structure present greater distances between the responses of T_1 and Te_{1a} (mean of the distances: 0.42) than in the summaries with the Qy’s are S structure (mean of the distances: 0.29).

4.2 Analysis of the Behavior of the Degree of Support Indicator and Its Extension

This subsection characterizes the behavior of the traditional T_3 indicators and the Te_3 extension. From the analysis of Fig. 5, it can be said that the T_3 indicator practically does not reflect variations, which is not a good behavior.

As can be seen, Te_3 reports better performance in most test classes. In classes where the data generated is below the threshold, the results of this indicator are expected to be low. Furthermore, this indicator should reach high values in the test classes ER9, ER18 and ER27 because it is where there are high values in both filters and summarizers, which guarantees that there is a high support for the summaries represented in these classes. The behavior of these test classes is clearly represented by the extended flag Te_3.

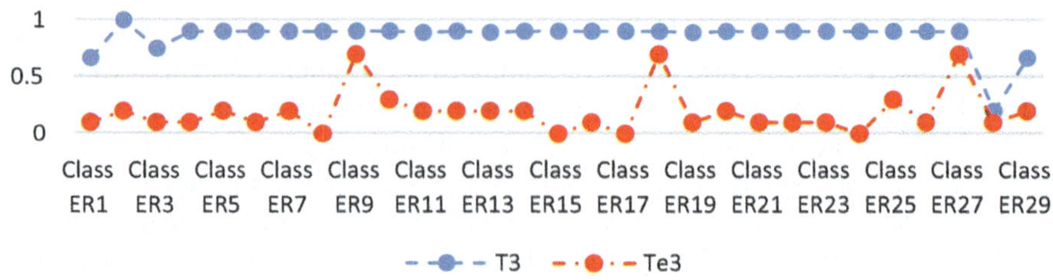

Fig. 5 Behavior of T_3 and Te_3

In general, it can be said that the extended Te_3 is stricter than the traditional one, and manages to better differentiate the scenarios represented by the test classes.

4.3 Analysis of the Behavior of the Degree of Appropriateness Indicator and Its Extension

In Fig. 6 it can be observed that in most of the test classes Te_4 has a greater discriminatory power of the different situations represented in the test classes with respect to T_4. In classes ER9, ER18 and ER27, the Te_4 indicator shows high peaks because the filters and the summarizers have high certainty values. In classes ER28 and ER29, Te_4 also presents high peaks because they are porous classes, where there is high certainty in few cases and low support.

In particular, the T_4 in the test classes ER1 and ER3, has an unexpected behavior, because in these test scenarios it is validated with objects that have a low certainty regarding the evaluation of the filters, which implies that there is a low certainty on the object's membership in the summary. However, in these cases a value of T_4 greater than the value of Te_4 is obtained, when the expected behavior is a very low

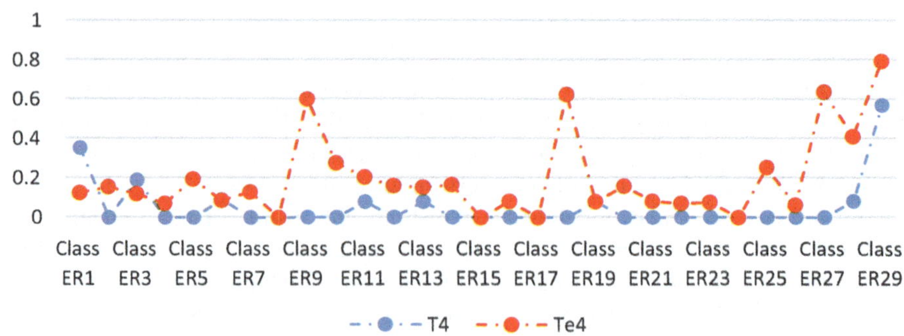

Fig. 6 Behavior of T_4 and Te_4

value of T_4. In these test classes the Te_4 indicator has a better performance than the T_4.

4.4 Analysis of the Behavior of the Indicator Length of a Summary and Its Extension

The T_5 analysis is best appreciated in the curves represented in Fig. 7. Note that the curve corresponding to the traditional T_5, the blue one, falls steeply. Also, it assigns a very high score (the maximum value) for summaries of size 1; while for summaries of size 2 and 3, the most common, it assigns a low score. On the other hand, the extended Te_5 assigns high values to the summaries that involve up to 4 variables; and the fall of the curve is smoother, with a behavior similar to the bell curve. Therefore, it is concluded that the extended Te_5 has a better performance than the traditional T_5, see Table 3.

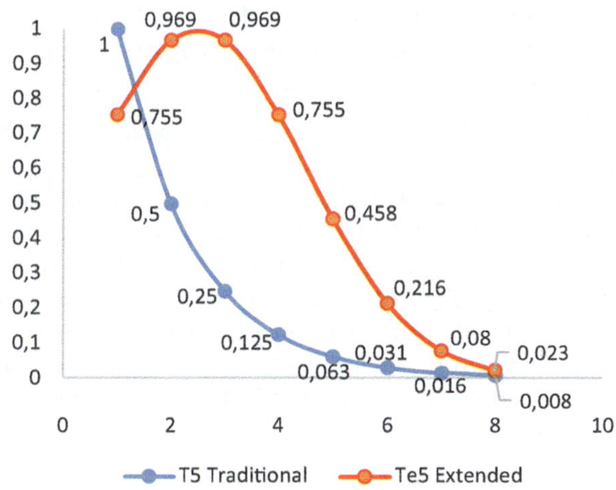

Fig. 7 Comparisons between T_5 and Te_5

Table 3 Comparison between T_5 and Te_5

Length of summary	T_5 Traditional	Te_5 Extended
1	1	0.755
2	0.5	0.969
3	0.25	0.969
4	0.125	0.755
5	0.063	0.458
6	0.031	0.216
7	0.016	0.08
8	0.008	0.023

Table 4 Comparación de indicadores respecto a tratamiento de incertidumbre

	Certainty	Indeterminacy	False	Parameterizable threshold
T_1	√			
T_2	√			
T_3	√			
T_4	√			
Te_{1a}	√			√
Te_{1b}	√			√
Te_{2a}	√	√	√	√
Te_{2b}	√	√	√	√
Te_{2c}	√	√		√
Te_{2d}	√		√	√
Te_3	√			√
Te_4	√	√	√	√

4.5 Summary of Comparison of Indicators Regarding the Treatment of Uncertainty

Table 4 shows a comparison of traditional and extended indicators regarding the treatment of uncertainty.

Table 4 shows that traditional indicators only deal with certainty and they do so use a fixed threshold (zero); while the extended indicators use a parameterizable threshold, which makes their use more flexible in different application domains. On the other hand, extended indicators manage to measure uncertainty with greater precision by combining the degrees of certainty, indeterminacy and falsehood, unlike traditional indicators. In addition, indicators such as Te_{2c} and Te_{2d} are incorporated that allow indeterminacy and falsehood to be measured in a particular way, respectively.

5 Conclusions

In this chapter, new indicators are proposed for the evaluation of linguistic summaries, from which it is concluded that:

- The indicators proposed in this chapter extend to the traditional indicators proposed in the bibliography, combining elements of the neutrosophic theory that allow the treatment of certainty, indeterminacy and falsehood in the evaluation of summaries.
- The introduction of concepts such as: the α-cut parameter, the upper approximation and the lower approximation, allows the level of rigor in the evaluation of linguistic summaries to be handled with greater flexibility.

- In measuring the degree of truth of the linguistic summaries, the Te_{1a} extension reported a significantly better behavior than the traditional T_1 variant in the test classes analyzed.
- The T_2 indicator is complemented by four other indicators that help to measure the imprecision of the summary, since the degrees of indeterminacy and falsehood are measured with greater precision through the Te_{2c} and Te_{2d} indicators respectively. On the other hand, the indicators Te_{2a} and Te_{2b} provide two ways of measuring uncertainty by combining the degrees of certainty, indeterminacy and falsehood in a single measure.
- Regarding the degree of coverage, it can be said that the extended Te_3 is stricter than the traditional T_3, it allows to better differentiate the scenarios represented by the test classes.
- Likewise, in the degree of appropriateness, Te_4 has a greater differentiation of the scenarios represented by the test classes with respect to T_4, because it promotes the identification of peculiar relationships between variables, such as relationships that are not present in the majority of database objects, but that are strong on those objects where they appear. The new indicator complements the original by providing a measure of appropriateness that combines the degree of certainty with indeterminacy and falsehood.
- Regarding the indicators to measure the length of a summary, the traditional T_5 has an abrupt behavior, so that it falls steeply from summaries of length 1 to summaries of length 3. While its replacement proposal, the extended Te_5, shows a smoother curve, favoring summaries of length 2 and 3 with more desired behavior.

References

1. Naamani-Dery, L., Kalech, M., Rokach, L., Shapira, B.: Reducing preference elicitation in group decision making. Expert Syst. Appl. **61**, 246–261 (2016)
2. Zyoud, D., Shaher, H., et al.: A framework for water loss management in developing countries under fuzzy environment: integration of fuzzy AHP with fuzzy TOPSIS. Expert Syst. Appl. 86–105 (2016)
3. Gomes, J., Romão, M.: Improving project success: a case study using benefits and project management. Procedia Comput. Sci. **100**, 489–497 (2016)
4. Boran, F.E., Akay, D., Yager, R.R.: An overview of methods for linguistic summarization with fuzzy sets. Expert Syst. Appl. **61**, 356–377 (2016)
5. Kacprzyk, J., Zadrożny, S.: On a fuzzy querying and data mining interface. Kybernetika **36**(6), 657–670 (2000)
6. García, J.A., Peña, A., Piñero, P.Y., Pérez, R.B.: Project control and computational intelligence: trends and challenges. Int. J. Comput. Intell. Syst. **10**(1), 320–335 (2017). ISSN 1875-6883. http://doi.org/10.2991/ijcis.2017.10.1.22
7. Yager, R.R.: A new approach to the summarization of data. Inf. Sci. **28**(1), 69–86 (1982). https://doi.org/10.1016/0020-0255(82)90033-0
8. Zadeh, L.A.: A computational approach to fuzzy quantifiers in natural languages. Comput. Math. Appl. **9**(1), 149–184 (1983)

9. Kacprzyk, J., Zadrożny, S.: Linguistic data summarization: a high scalability through the use of natural language? In: Scalable Fuzzy Algorithms for Data Management and Analysis: Methods and Design, pp. 214–237. IGI Global (2010)
10. Kacprzyk, J., Yager, R.R.: Linguistic summaries of data using fuzzy logic. Int. J. Gen. Syst. **30**(2), 133–154 (2001). https://doi.org/10.1080/03081070108960702
11. Kacprzyk, J., Wilbik, A.: Linguistic summaries of time series: on some extended aggregation techniques. Studia i Materiały Polskiego Stowarzyszenia Zarzdzania Wiedza **2010**(31), 326–337 (2001)
12. Smarandache, F., Pramanik, S.: New Trends in Neutrosophic Theory and Applications, vol. II. Infinite Study, Brussels (2018). ISBN 978-1-59973-559-7
13. Dubois, D., Prade, H.: Rough fuzzy sets and fuzzy rough sets. Int. J. Gen. Syst. **17**(2–3), 191–209 (1990). https://doi.org/10.1080/03081079008935107
14. Bello, R., Verdegay, J.L.: Rough sets in the soft computing environment. Inf. Sci. **212**, 1–14 (2012). ISSN 0020-0255. http://doi.org/10.1016/j.ins.2012.04.041

Part II
Planning and Sustainability of Projects Assisted by Artificial Intelligence

Chapter 7
Constraints Learning Univariate Estimation of Distribution Algorithm on the Multi-mode Project Scheduling Problem

Gaafar Sadeq S. Mahdi, Pedro Y. Piñero Pérez, Julio Madera Quintana, Salah H. Al-subhi, and Roberto García Vacacela

Abstract Project Scheduling Problems (PSP) constitute a family of problems that includes different variants, which range from simple task planning without taking into account the resources they consume, to more sophisticated variants that consider several modes of processing of projects tasks, generalization of precedence relationships, multiple projects simultaneously and projects with variable resources. In this sense, various algorithms, both exact and heuristic, have been used to find optimal or quasi-optimal project schedules. This research aims to propose a new Constraints Learning Univariate Estimation of Distribution Algorithm (CL_UMDA) as an extension of the Univariate Marginal Distribution Algorithm (UMDA). The new algorithm incorporates the constraints handling inside the probabilistic model, for the solution of the PSP problem in its multimode variant (MMRCPSP). For this purpose, a group of experiments was developed on five databases of the PSPLib library, comparing the proposed algorithm with others reported in the literature. The experimental results

G. S. S. Mahdi (✉)
Business Intelligence and Consulting Group, BusinessRedmine SA, Yemen Office, Sana'a, Yemen

P. Y. Piñero Pérez
Grupo de Investigaciones en Gestión de Proyectos, Universidad de las Ciencias Informáticas, Carretera a San Antonio km 2½, CP 54830 La Habana, Cuba
e-mail: ppp@uci.cu

J. Madera Quintana
Vicerrectoría de Informatización, Universidad de Camagüey, Camagüey, Cuba
e-mail: julio.madera@reduc.edu.cu

S. H. Al-subhi
Solar Plex Electrical and Electronics Contracting Co. Combined Security Group Co., Kuwait City, Kuwait

R. García Vacacela
Facultad de especialidades empresariales, Universidad Católica Santiago de Guayaquil, Guayaquil, Ecuador
e-mail: roberto.garcia@cu.ucsg.edu.ec

P. Y. Piñero Pérez et al. (eds.), *Artificial Intelligence in Project Management and Making Decisions*, Studies in Computational Intelligence 1035,
https://doi.org/10.1007/978-3-030-97269-1_7

show the superiority of the CL_UMDA performance over other algorithms used in the experimentation.

Keywords Multi-mode project scheduling problem · Project management · Univariate marginal estimation of distribution algorithm

1 Introduction

Nowadays, competitiveness in project development imposes the need for algorithms and tools for the rapid construction of schedules that optimize, as much as possible, the time and costs of project execution considering that the tasks can be executed in multiple modes and the existence of restrictions on the use of renewable and non-renewable resources.

Among the root causes of project failure can be insufficient training of human resources and frequent errors in planning processes, whether of scope, time, or logistics [1, 2] and poor management in the control and monitoring processes [3, 4]. This research focuses on contributing to the solution of the scheduling problem, understanding this as one of the fundamental problems of project management.

Scheduling problems generally involve several constraints, numerous variables and factors that make it a complex problem of the NP-hard type [5–7]. Considering the complexity involved in solving the Multi-mode Resource Constrained Project Scheduling Problem (MMRCPSP), several authors use metaheuristics [7–12]. One of the challenges in this regard lies in identifying the metaheuristics that solve this problem as efficiently as possible.

Lova et al. in [13] propose a hybrid genetic algorithm called MM-HGA, where the solution is represented by two lists, the first with the tasks and the other with the execution modes. Van Peteghem and Vanhoucke in [14] present a genetic algorithm, where, similar to other approaches, a preprocessing is performed before the start of the main algorithm. They use a version called GA Bi-Population in which the idea is to build two different populations of the same size where they use the crossover and mutation operators to obtain the possible solutions. Coelho and Vanhoucke [15] propose a new approach that divides the MMRCPSP into two interrelated sub-problems: the mode assignment and the project scheduling problem in a single mode [16]. Extend the scheduling problem with two new kinds of logical constraints that model the relationship between each pair of tasks by using the OR and BI (bidirectional) operators.

Damak et al. [17] propose a differential evolution algorithm in which a solution is represented by a mode assignment list and a task list that does not have to be ordered by precedence. Neighbor solutions are generated using two operators: mutation and crossover.

Also [18] develop a hybrid algorithm called (Adapted TS-NSGA-II), based on Tabu Search with the NSGA-II (Non-dominated Sorting Genetic Algorithm Two). Then Soliman and Elgendi [19] present a hybrid EDA with random local search techniques. The local search technique is based on the "delete" and "insert" operators to allow better exploitation of the search space. First, they use a list of tasks and modes, then they evaluate all the individuals by a task generation scheme called the Multi-Mode Serial Schedule Generation Scheme (MSSGS) and finally, they apply local search techniques to evaluate the selected individuals. However, checking the execution mode of each task in the two previous proposals in an iterative way implies a significant increase in the computational cost to evaluate the solutions obtained [20].

In [21, 22] Mayowa proposes Random Key based Estimation of Distribution Algorithm (RK-EDA) which is an EDA designed to solve problems based on permutations and apply it in solving the problem of resource allocation in scheduling tasks. This algorithm is similar to the UMDA and it is based on calculating the probability of occurrence of the values in the prominent solutions.

Another alternative was the combination of GA with the PBIL (Population-Based Incremental Learning) univariate EDA algorithms [23], presented in [24] as the PBGA-EDA algorithm, where the new children in the genetic algorithm are generated from the analysis of the distribution functions of the tasks with respect to the modes. In this algorithm, they work with two lists of solutions like [14] POP_L and POP_R. Solutions are generated by crossing an individual (possible solution) from one list with another tournament-selected solution from the same list; the resulting children will replace the individuals from the other list. That is, the POP_L is used to generate the POP_R solutions and vice versa. The sampling of the probabilities for the assignment of the modes of execution of the tasks may be interesting in the framework of this paper.

Ayodele in [20] also proposes the BPEDA (Bi-Population Estimation of Distribution Algorithm) as a result of the combination of the RK-EDA [22] for the generation of the tasks with the mechanism of generating the modes of execution proposed in [24]. As others mentioned above, this algorithm is based on a bi-population approach (POP_L and POP_R). POP_L is randomly initialized but introducing procedures to improve the execution modes of this initial population similar to the one used in [14] to reduce non-feasible solutions regarding the consumption of non-renewable resources. The solution is then evaluated using the SGS (Serial Schedule Generation Scheme). The best POP_L solutions are selected, later the probabilistic model is built regarding the tasks and modes. Through the probabilistic model built with respect to the tasks, the solutions are generated in POP_R. Each of these lists is then updated based on the probabilistic model built from the best solutions from the other list.

Furthermore, Tian et al. in [25] propose the Robust hGMEDA algorithm, Multi-phased Robust Scheduling method as an extension of the hGMEDA [26] and It is based on the hybridization between EDA and GA, for the solution of the MMRCPSP with uncertainty. Authors validate their proposal with databases of PSPLib. Results obtained by these algorithms will be compared with the proposal of this paper.

From the above analysis, the following limitations were identified:

- Despite the fact that the MMRCPSP are presented as optimization problems with the presence of correlated variables. However, in the searches

carried out, few works were found that apply EDAs in solving these problems [19–21, 24, 25, 27, 28].
- The absence of related works with EDAs that use information from the restrictions in the learning process of the probabilistic model [29].

This paper aims at presenting a solution to the MMRCPSP, based on the application of the Constraints Learning Univariate Estimation of Distribution Algorithm (CL_UMDA) which incorporates the constraints handling inside the probabilistic model.

The paper is organized as follows: In Sect. 2 the optimization problem is conceptualized. The third section presents the CL_UMDA algorithm. In Sect. 4 the results of the experimentation are analyzed and finally, the conclusions of the work are presented.

2 Modeling the MMRCPSP Optimization Problem

In this paper, we use the following notation:

- A project contains a set of tasks $j = \{1, \ldots, n\}$. Value $j = 1$ is the initial task and has no predecessor tasks and all tasks depend on it to start. Value $j = n$ is the final task, all other tasks precede it directly or indirectly.
- Each task has a set of predecessor tasks H_j that represent precedence's constraints such that in a feasible solution no task can start until has finished its preceding tasks.
- Each task j must be executed in an m_j mode, $M_j = \{1, \ldots m_j\}$ where M_j denotes the number of possible execution modes of task j. In the context of scheduling projects, the execution modes make it possible to represent, for example, the different competencies of human resources or the use of different techniques or technologies for the execution of tasks.
- t denotes the moment of execution of the task j, $t = 1, \ldots, T$, T is an upper bound of the project duration time, it is equal to the longest possible duration according to the task processing modes.
- d_{jm} denotes the duration time of task j if it is done in mode m (they are assumed to be whole quantities of the unit "day"). The duration of tasks 1 and $n = 0$.
- There are restrictions associated with the availability of resources such that $k \in K$ types of resources whose quantity in the project T is R_k. The term K such that $K = K_v \cup K_{nv}$ represents the set of all the resources available for the development of the tasks.
- The set K_v is associated with the renewable resource availability constraints. These resources can be used more than once and in different tasks in the context of the project (personnel and equipment).
- The set K_{nv} is associated with the non-renewable resources availability constraints that can only be used once during the life of the project. Examples of this are expendable materials such as paper, fuel, etc.

- Each task j according to the execution mode m_j can consume r_{jmk} units of the resource of type k with a duration of time d_{jm}.

2.1 Formalization of the Optimization Problem

The objective is to minimize the time and total cost of project execution associated with the objective functions f_1 and f_2 respectively, see Eq. (1).

$$\min F = \left(f_1, f_2\right) \tag{1}$$

f_1 Eq. (2) represents the start date of task n and f_2 Eq. (3) represents the cost of project execution calculated as the sum of the costs of the resources associated with all the tasks involved.

Continuous variables:

$$x_{jmt} = \begin{cases} 1, & \text{task } j \text{ starts on day } t \text{ and is done in mode } m, \\ 0, & \text{in other case} \end{cases}$$

Then, $\sum_{m=1}^{m_j} \sum_{t=1}^{T} t x_{jmt}$ represents the start day for j task and $\sum_{m=1}^{m_j} \sum_{t=1}^{T} d_{jm} x_{jmt}$, j task duration. The equation $\sum_{m=1}^{m_j} \sum_{t=1}^{T} t x_{jmt} + \sum_{m=1}^{m_j} \sum_{t=1}^{T} d_{jm} x_{jmt}$ represents the first due day of j task, that's mean the possible initial day for the successors of j task.

$$f_1 = \sum_{t=1}^{T} t x_{n1t} \tag{2}$$

$$f_2 = \sum_{j=1}^{n} \sum_{m=1}^{m_j} \sum_{t=1}^{T} \left[\sum_{k \in K_v} r_{jmk} \cdot C_k \cdot d_{jm} + \sum_{k \in K_{nv}} r_{jmk} \cdot C_k \right] x_{jmt} \tag{3}$$

Considering the complexity of the multi-objective's problem, to combine the objectives authors propose $\alpha f_1 + (1 - \alpha) f_2$, $\alpha \in [0, 1]$.

Each solution is evaluated independently with respect to each of the objectives. Then the evaluation result is added. Each solution must be subject to the following constrains:

$$\sum_{m=1}^{m_j} \sum_{t=1}^{T} x_{jmt} = 1, \quad \forall j = [1, \ldots, n] \tag{4}$$

$$\sum_{m=1}^{m_j} \sum_{t=1}^{T} t x_{jmt} \geq \sum_{m=1}^{m_i} \sum_{t=1}^{T} t x_{imt} + \sum_{m=1}^{m_i} \sum_{t=1}^{T} d_{im} x_{imt}, \quad i \in H_j \tag{5}$$

Equation (4) represents the restriction that guarantees that each task can be executed in only one mode on the same day t.

Equation (5) represents the constraint that guarantees that the precedence relationships among tasks are not violated.

$$\sum_{j=1}^{n}\sum_{m=1}^{m_j}\sum_{t=1}^{T} r_{jmk}x_{jmt} \leq R_k, \quad k \in K_{nv} \tag{6}$$

Equation (6) represents the restriction that guarantees that the use of non-renewable resources as expendable material resources is not exceeded, considering the mode of execution of the *j-th* tasks $j \in J$ of the project.

$$\sum_{j=1}^{n}\sum_{m=1}^{m_j}\sum_{t=max\{\bar{t}-d_{jm}+1,0\}}^{\bar{t}} r_{jmk}x_{jmt} \leq R_{k\bar{t}}, \quad \forall \bar{t} = 1, \ldots, T, k \in K_v \tag{7}$$

Equation (7) represents the restriction that guarantees the availability of renewable resources for each day of the project, such as human resources and equipment.

2.2 Constraints Learning Univariate Marginal Distribution Algorithm (CL_UMDA)

In this section, a new algorithm is proposed that incorporates the constraints of the optimization problem inside the probabilistic model. The first subsection presents the design of the solution to the optimization problem. The second subsection explains the algorithm CL_UMDA algorithm in detail.

2.3 Solution Design

Taking into account that the proposed approach is population-based, in which each population is made up of a set of individuals, which constitute possible solutions to the scheduling problem and where it is sought to simultaneously minimize time and cost objectives of the project while maintaining a balanced relationship among them.

The individual's design corresponds to an integer representation and is made up of a sequence of tasks. Each task has two relevant attributes that are the start date of task j (τ_j) and the mode in which $m \in M_j$ was performed. In the conceptualization of the scheduling problem, the mode of execution of a task defines the amount of renewable and non-renewable resources that are used and the duration of the task. In addition, there is a set of complementary attributes for each task that are calculated from the mode and the start date, see Fig. 1.

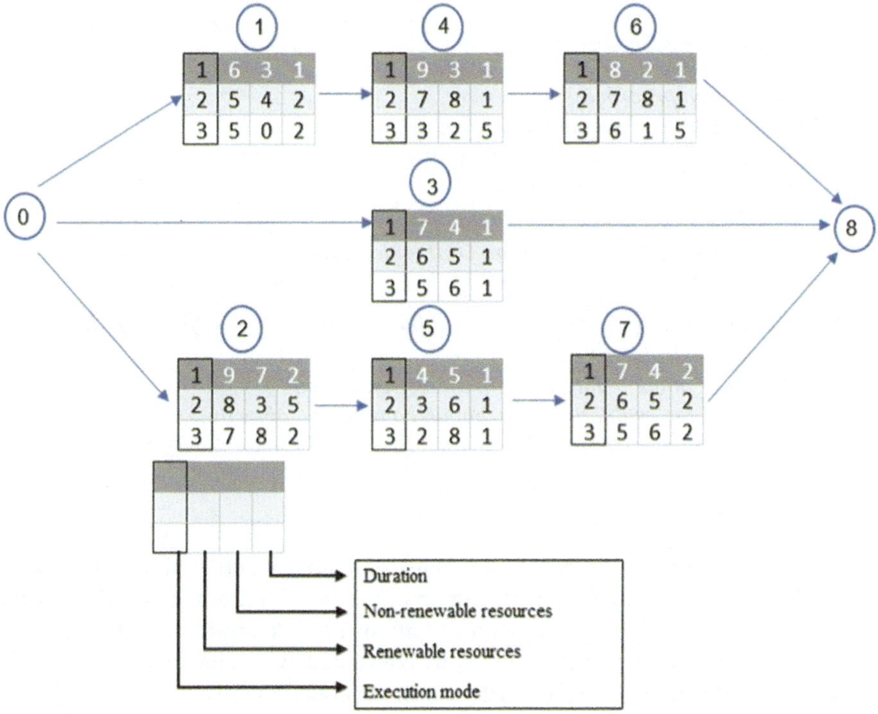

Fig. 1 Sample solution, representing precedence relationships and multiple modes

2.4 Detailed Formalization of the CL_UMDA

The Univariate Marginal Distribution Algorithm (UMDA) was proposed in [30], where in each iteration, the marginal distribution of each variable in the selected solution is estimated from their relative frequencies. This algorithm does not manage the dependency restrictions existing among the variables in the optimization problem. As an element to improve this algorithm, we proposed Algorithm 1, which handled the constraints of the problem inside the probabilistic model.

Algorithm 1 *CL_UMDA* pseudocode

```
Inputs
   constraints: problem constrains

1.  Information preprocessing
2.  popul = initial_population(constraints)
3.  popul = evaluation(popul, constraints)
4.  while not stop_condition(popul)
5.        D^S_{g-1} = schedule_selection(popul)
6.        p(J|D^S_{g-1}) = distribution_UMDA_learn(popul, D^S_{g-1}, constraints)
7.        ne                                   D^S_{g-1})
8.        new_popul = schedule_new_solutions(popul, new_popul, p(J|D^S_{g-1}),
            constraints)
9.        new_popul = Improve_solutions(new_popul)
10.       popul = evaluation(new_popul, constraints, objectives)
11.       indicators[g] += 1
12. end
13. return popul
```

In step 1, information preprocessing, which includes identification of the objectives of the optimization problem, definition of the stop condition and learning from the constraints the task-mode-resources relationships (determines for the problem for each task-mode how many resources are consumed and an indicator is calculated that expresses the relative cost of each task-mode combination).

In step 2, the initial population is generated randomly, but applying the following strategies that mitigate the possibility of generating non-feasible solutions:

In the generation of the initial population, explicit knowledge about the precedence of the tasks is used and the random generation of the start date of a task is conditioned so that no task has a start date before its predecessors.

In the initial generation, a pre-order is established for the assignment of the start date to each task, taking into account the graph with the precedence restrictions. For each day, the possible tasks that can be executed and that compete for renewable resources are identified. It is randomly selected which tasks should start that day and their execution mode in such a way that the number of renewable resources established as a restriction for each day is not exceeded.

Notwithstanding the proposed strategies, it is possible that in this step non-feasible solutions will be generated.

In steps 3 and 10, the population of individuals generated is evaluated; for this, individuals are evaluated in each of the proposed objectives (time–cost). In the MMRCPSP problem, the definition of the execution mode of a task determines the number of resources that will be used in this task, an element that directly influences the costs of the solution. During the evaluation process, those individuals who have been evaluated before and have not received modifications do not need to be reevaluated, but the population and the relative evaluation of each individual are reevaluated.

In step 5 the best individuals are selected to be considered in the learning of the probabilistic model and as part of the elitism process. Different alternatives such as roulette or establishing a rank and truncation can be considered. In the experimentation of this work, the ranking method is used and 30% of the population size is selected.

In step 6 the learning of the probabilistic model is carried out, which is formed by the conjunction of different probabilistic distribution functions as shown by Algorithm 2 and that respond to the following elements:

- Learning the relationships between tasks and their start dates. For example, the marginal probability of the occurrence of tasks in days is determined, given the start dates.
- Learning the best combinations of tasks-modes that are executed on a given day. Those combinations are learned that maximize the number of tasks that can be started on the same day, reducing the presence of waiting slack.
- Learning strategies to improve solutions, which influences the best use of resources. This learning presupposes the existence of a step to improve the solutions from the partial reconstruction of these solutions, following an approach of selecting modes that minimize the reserves of renewable and non-renewable resources present in the solution. This learning process is inspired by the heuristic that assumes that the maximum use of resources, by minimizing unexploited reserves, implies a reduction in the duration of the tasks and the project in general.

Algorithm 2 Distribution_CL_UMDA_learn algorithm

Inputs
popul: current population
sel_popul: selected individuals
constraints: optimization problem's constrains

1. $p_g(J) \Leftarrow \{\}$
2. $p_g(J) \Leftarrow$ *Learning the probabilistic model of relationships between tasks-start dates* $p_g^f(j_1, j_{2,}, \ldots, j_n)$ *and task-modes* $p_g^m(j_1, j_{2,}, \ldots, j_n)$
3. $p_g^f(j_1, j_{2,}, \ldots, j_n) = \prod_{i=1}^n p^f(j_i)$, $p_g^m(j_1, j_{2,}, \ldots, j_n) = \prod_{i=1}^n p^m(j_i)$
4. $p_g^e(J) \Leftarrow$ *Assign zero to the probabilities of the relationships that violate the precedence constraints in* $p_g(J)$ *given* $p_g^f(j_1, j_{2,}, \ldots, j_n)$
5. $p_g^e(J) \Leftarrow$ *Learning the best combinations of task-mode sets that are executed in the same day*
6. $p_g^e(J) \Leftarrow$ *Learning strategies to improve solutions, which influences the best use of resources.*
7. $p_g^e(J)$

In step 8 new solutions are generated from the probabilistic model and other information learned previously, see Algorithm 3.

Algorithm 3 Generation of the population from the learned distribution

Inputs
popul: current population
new_popul: new population
method, $\mathbf{p}_g^e(J)$*: strategy for new solutions generations*
constraints: optimization problem's constrains

```
1. renewable = constraints[:resource_renewable]
2. nonrenewable = constraints[:resource_nonrenewable]
3. amount = population_size - new_popul.length + 1
4. amount.times do
5.     solution = generate_feasible_solution(popul, method, p_g^e(J), constraints)
6.     new_popul.store(solution)
7. end
8. return new_popul
```

In this proposal, we work with populations of fixed size and the use of elitism is established in step 7, in such a way that the best individuals go directly to the new population and as many new individuals are generated as necessary to complete the population size. The process of generating new individuals is carried out considering the information learned in the probabilistic model.

In Algorithm 3, step 5 is essential, in which a feasible solution is generated. The essence of this step is that the generation of new individuals is carried out day by day and considering the set of tasks that can execute each day respecting the precedence constraints. At this time, the learned probabilistic model is applied and a task and its execution mode are selected. For this day, the process is repeated as long as tasks can be included without violating the restrictions associated with renewable resources.

In step 9, the solutions generated in the previous step are adjusted using the information from the reserves and the improvement strategies learned, for this the following steps are carried out:

- Identification of opportunities for improvement based on the analysis of its reserves of renewable and non-renewable resources. The reserves for a solution are made up of the set of unplanned resources for the project in this solution. An opportunity for improvement is constituted by a combination $activity_{\alpha} - mode_{\rho}$ in such a way that this $activity_{\alpha}$ could be executed in a different mode than $mode_{\rho}$ without violating the restrictions on the use of resources renewable.
- Generate a new task-mode combination for the identified opportunities using the information learned from the improvement strategies and respecting the constraints on the use of resources. Among the strategies for generating new combinations, the following can be used:
  - Generating a new task-mode combination stochastically.
  - The generation using the information learned $p_g^e(J)$, promoting those modes that minimize the consumption of resources.

- The generation using the information learned $p_g^e(J)$, promoting those modes that maximize the consumption of resources.
- The generation from the knowledge in $p_g^e(J)$ associated with the learning of the best combinations of sets of task-modes.

3 Experimental Results and Discussion

For the experimentation, data are taken from the PSPLib repository [31] in particular we use the databases *m4_12.mm*, *m4_32.mm* (16 tasks, two categories of renewable resources, two categories of non-renewable resources, four execution modes). In addition, we use *m5_12*, *m5_32* (16 tasks, two categories of renewable resources, two categories of non-renewable resources, five execution modes), *n0.mm* (16 tasks, two categories of renewable resources, two execution modes). These databases are selected because they have the maximum number of execution modes for each task and to be able to compare results with the literature. Each database contains 10 instances. Fixed-size populations of 150 individuals is used and selection based on the ranking of 30% of the population are used. Elitism is applied in the generation of the new population and 70% of the population is generated from the learned distributions. As a stop condition, a maximum number of iterations equal to 100 was set. To mitigate bias in experimentation, 20 runs of each algorithm are performed in each instance of the databases.

The proposed algorithm CL_UMDA has four variants: CL_UMDA/T.NR, CL_UMDA/T.R, CL_UMDA/MO.NR and CL_UMDA/MO.R. The configuration regarding the optimization objectives is represented using the /T notation for time, while /MO for the cost and time objectives simultaneously. The comparison is made with respect to the following variables:

"Mean Makespan" average of the project execution time (to minimize).

"Number of times the optimum founded" Number of times the optimum is founded (to maximize).

In the analysis of variables and in each database, the normality test is applied using the Shapiro–Wilk [32] and Kolmogorov–Smirnov [33] tests. Parametric and non-parametric tests are applied to compare the results, depending on the result of normality test. In all cases, we work with a significance of 0.05 and confidence intervals of 99%. The result of the comparisons in each database, the algorithms are organized in groups.

The algorithms in the same group do not present significant differences between them. While the algorithms in different groups present significantly different results. The algorithms in "Group A" report better results than the algorithms in "Group B" such that "Group A results" > "Group B results" > "Group C results" and so on. Bold numbers in tables indicate significant differences between algorithms.

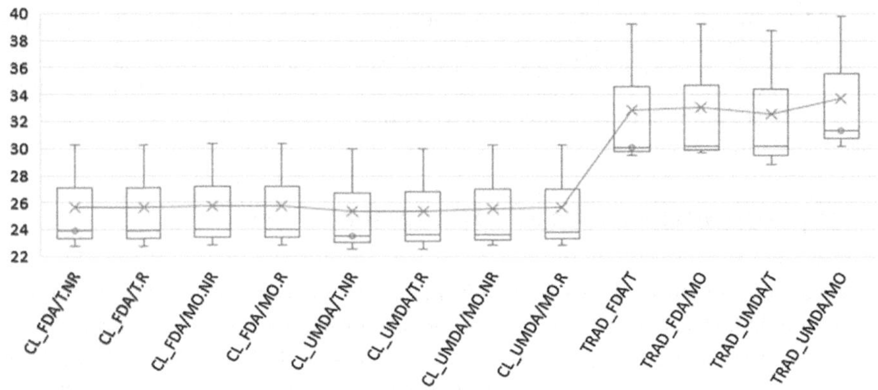

Fig. 2 Comparison of proposed algorithm with results reported in the literature, variable "Mean Makespan", database *n0*

3.1 "Mean Makespan" Variable

Regarding this variable, it is identified that the results in the databases *m4_12*, *m4_32* and *n0* do not follow a normal distribution, unlike *m5_12*, *m5_32* that do fit this distribution. Experimental results are presented in Fig. 2, Tables 1, 2 and 3, the algorithms *CL_UMDA/T.NR* and *CL_UMDA/T.R* report the best results.

It is also shown that the algorithms proposed in this work, with their different variants, report significantly better results than the traditional UMDA implementation (TRAD_UMDA). There are no statistically significant differences between the variants of the learning strategies. Nor are statistically significant differences found in the results obtained following the optimization approaches [just_time (T) and multi-objective (MO)].

In the consulted literature, results of the application of the Robust hGMEDA, hGMEDA and SPEA2 are reported on the database *n0*, in the instances: *n041_1*, *n042_1*, *n043_1*, *n044_1* and *n045_1*. It is shown in Fig. 2, that the performance of *CL_UMDA*, with its different variants, is better than traditional algorithm UMDA and also than Robust hGMEDA, hGMEDA and SPEA2 reported in [25, 26].

3.2 "Number of Times the Optimum Founded" Variable

Regarding the variable "Number of times the optimum founded", it is identified that in the five databases analyzed, the results do not follow a normal distribution. See Fig. 3, Tables 4, 5 and 6. These tables show that the *CL_UMDA* variants report better results than the rest of the algorithms, except in the *m5_12* database, where no statistically significant differences are found among the different *CL_UMDA* variants.

Table 1 Comparison results, variable *Mean_Makespan* for *n0 [38_1, 47_1]* dataset

n0 "Mean Makespan"	*sig.*
Wilcoxon test (Group A) CL_UMDA/T.R—CL_UMDA/T.NR	0.798
CL_UMDA/T.R—CL_FDA/T.R	**0.007**
Friedman test (Group B) CL_FDA/T.R—CL_UMDA/MO.R—CL_FDA/T.NR—CL_FDA/MO.R—CL_UMDA/MO.NR—CL_FDA/MO.NR	0.309
Wilcoxon test (Group C) TRAD_UMDA/T—TRAD_FDA/T	**0.032**
Wilcoxon test (Group D) TRAD_FDA/T—TRAD_FDA/MO	0.122
Wilcoxon test (Group E) TRAD_FDA/T—TRAD_UMDA/MO	**0.008**

Table 2 Comparison results, variable *Mean_Makespan* for *m4_12* and *m4_32* dataset

m4_12 "Mean Makespan"/Algorithms	*sig.*	*m4_32 "Mean Makespan"/Algorithms*	*sig.*
Friedman test (Group A) CL_UMDA/T.NR—CL_UMDA/T.R – CL_UMDA/MO.NR— CL_UMDA/MO.R	0.061	Wilcoxon test (Group A) CL_UMDA/T.NR—CL_UMDA/T.R	0.400
Wilcoxon test CL_UMDA/T.NR—CL_FDA/MO.R	**0.011**	Wilcoxon test CL_UMDA/T.NR—CL_FDA/T.R	**0.008**
Friedman test (Group B) CL_FDA/MO.R—CL_FDA/T.NR – CL_FDA/T.R—CL_FDA/MO.NR	0.709	Friedman test (Group B) CL_FDA/T.R—CL_UMDA/MO.NR – CL_UMDA/MO.R— CL_FDA/T.NR – CL_FDA/MO.R	0.069
Wilcoxon test		Wilcoxon test	
CL_FDA/MO.R—TRAD_UMDA/T (Group C)	**0.005**	CL_FDA/T.R—CL_FDA/T.NR	**0.038**
TRAD_UMDA/T—TRAD_FDA/T (Group D)	**0.025**	CL_FDA/T.NR—TRAD_FDA/T (Group C)	**0.005**
TRAD_FDA/T—TRAD_FDA/MO (Group E)	**0.027**	(Group D)	
		TRAD_FDA/T—TRAD_UMDA/T	0.202
TRAD_FDA/MO—TRAD_UMDA/MO (Group F)	**0.005**	TRAD_UMDA/T—TRAD_FDA/MO	0.086

Table 3 Comparison results, variable *Mean_Makespan* for *m5_12* and *m5_32* dataset

m5_12 "Mean Makespan"/Algorithms (t-test)	*sig.*	*m5_32 "Mean Makespan"/Algorithms (t-test)*	*sig.*
(Group A)		(Group A)	
CL_UMDA/T.NR—CL_UMDA/MO.R	0.735	CL_UMDA/T.NR—CL_UMDA/T.R	0.809
CL_UMDA/T.NR—CL_UMDA/T.R	0.309	CL_UMDA/T.NR—CL_UMDA/MO.R	**0.026**
CL_UMDA/T.NR—CL_UMDA/MO.NR	0.453	CL_UMDA/MO.R—CL_UMDA/MO.NR (Group B)	0.853
CL_UMDA/T.NR—CL_FDA/T.R	0.354	CL_UMDA/MO.R—CL_FDA/T.NR	**0.031**
CL_UMDA/T.NR—CL_FDA/T.NR	0.152	(Group C)	
CL_UMDA/T.NR—CL_FDA/MO.R	0.126	CL_FDA/T.NR—CL_FDA/T.R	0.867
CL_UMDA/T.NR—CL_FDA/MO.NR	0.063	CL_FDA/T.NR—CL_FDA/MO.NR	0.134
CL_UMDA/T.NR—TRAD_UMDA/T	**0.000**	CL_FDA/T.NR—CL_FDA/MO.R	0.104
(Group B)		CL_FDA/T.NR—TRAD_FDA/T	**0.000**
TRAD_UMDA/T—TRAD_FDA/T	0.079	TRAD_FDA/T—TRAD_UMDA/T (Group D)	0.778
TRAD_UMDA/T—TRAD_FDA/MO (Group C)	**0.037**	TRAD_FDA/T—TRAD_FDA/MO (Group E)	**0.000**
TRAD_FDA/MO—TRAD_UMDA/MO (Group D)	**0.000**	TRAD_FDA/MO—TRAD_UMDA/MO (Group F)	**0.000**

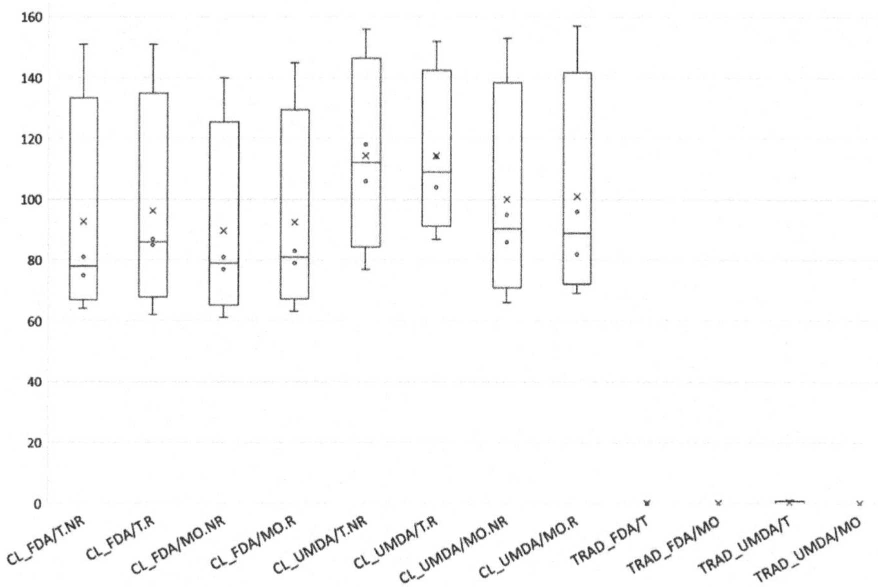

Fig. 3 Comparison results with respect to the variable "Number of times the optimum founded" for *m4_12, m4_32, m5_12, m5_32* and *n0 [38_1, 47_1]* dataset

Table 4 Comparison, number of times the optimum for *m4_12* and *m4_32* dataset

m4_12/Algorithms	*sig.*	*m4_32/Algorithms*	*sig.*
Friedman test (Group A) CL_UMDA/T.NR—CL_UMDA/T.R - CL_UMDA/MO.R	0.113	Wilcoxon test (Group A) CL_UMDA/T.NR—CL_UMDA/T.R	0.102
Wilcoxon test CL_UMDA/T.NR—CL_UMDA/MO.NR	**0.042**	Wilcoxon test CL_UMDA/T.NR—CL_FDA/T.R	**0.018**
Friedman test (Group B) CL_UMDA/MO.NR—CL_FDA/T.R - CL_FDA/MO.R	0.062	Friedman test (Group B) CL_FDA/T.R—CL_UMDA/MO.NR - CL_FDA/MO.R—CL_UMDA/MO.R - CL_FDA/MO.NR—CL_FDA/T.NR	0.901
Wilcoxon test CL_UMDA/MO.NR—CL_FDA/MO.NR	**0.042**		
Wilcoxon test (Group C) CL_FDA/MO.NR—CL_FDA/T.NR	0.833		
Wilcoxon test CL_FDA/MO.NR—TRAD_UMDA/T	**0.027**	Friedman test TRAD_FDA/T—TRAD_FDA/MO - TRAD_UMDA/T—TRAD_UMDA/MO	1.000
Friedman test (Group D) TRAD_UMDA/T—TRAD_UMDA/MO - TRAD_FDA/MO—TRAD_FDA/T	0.392		

Table 5 Comparison, number of times the optimum founded for *m5_12* and *m5_32* dataset

<table>
<tr><th>m5_12/Algorithms</th><th>sig.</th><th>m5_32/Algorithms</th><th>sig.</th></tr>
<tr><td rowspan="2">Friedman test (Group A)
CL_UMDA/MO.R
- CL_UMDA/T.NR
- CL_UMDA/MO.NR
- CL_UMDA/T.R
- CL_FDA/T.R
- CL_FDA/T.NR</td><td rowspan="2">0.788</td><td>Friedman test (Group A)
CL_UMDA/T.R
- CL_UMDA/T.NR
- CL_UMDA/MO.R
- CL_UMDA/MO.NR</td><td>0.124</td></tr>
<tr><td>Wilcoxon test
CL_UMDA/T.R—CL_FDA/T.NR</td><td>0.027</td></tr>
<tr><td>Wilcoxon test
CL_UMDA/MO.R—CL_FDA/MO.R</td><td>0.039</td><td>Friedman test (Group B)
CL_FDA/T.NR—CL_FDA/MO.R
- CL_FDA/T.R—CL_FDA/MO.NR</td><td>0.955</td></tr>
<tr><td>Wilcoxon test (Group B)
CL_FDA/MO.R—CL_FDA/MO.NR</td><td>0.275</td><td>Wilcoxon test
CL_FDA/T.NR—TRAD_UMDA/MO</td><td>0.028</td></tr>
<tr><td>Friedman test
TRAD_FDA/T
- TRAD_FDA/MO
- TRAD_UMDA/T
- Trad_UMDA/MO</td><td>1.000</td><td>Friedman test (Group C)
TRAD_UMDA/MO
- TRAD_UMDA/T
- TRAD_FDA/MO
- TRAD_FDA/T</td><td>1.000</td></tr>
</table>

Table 6 Comparison results, variable number of times the optimum founded for *m5_12* and *m5_32* dataset

<table>
<tr><th>m5_12/Algorithms</th><th>sig.</th><th>m5_32/Algorithms</th><th>sig.</th></tr>
<tr><td rowspan="2">Friedman test (Group A)
CL_UMDA/MO.R
- CL_UMDA/T.NR
- CL_UMDA/MO.NR
- CL_UMDA/T.R
- CL_FDA/T.R
- CL_FDA/T.NR</td><td rowspan="2">0.788</td><td>Friedman test (Group A)
CL_UMDA/T.R
- CL_UMDA/T.NR
- CL_UMDA/MO.R
- CL_UMDA/MO.NR</td><td>0.124</td></tr>
<tr><td>Wilcoxon test
CL_UMDA/T.R—CL_FDA/T.NR</td><td>0.027</td></tr>
<tr><td>Wilcoxon test
CL_UMDA/MO.R—CL_FDA/MO.R</td><td>0.039</td><td>Friedman test (Group B)
CL_FDA/T.NR—CL_FDA/MO.R
- CL_FDA/T.R—CL_FDA/MO.NR</td><td>0.955</td></tr>
<tr><td>Wilcoxon test (Group B)
CL_FDA/MO.R—CL_FDA/MO.NR</td><td>0.275</td><td>Wilcoxon test
CL_FDA/T.NR—TRAD_UMDA/MO</td><td>0.028</td></tr>
<tr><td>Friedman test
TRAD_FDA/T
- TRAD_FDA/MO
- TRAD_UMDA/T
- TRAD_UMDA/MO</td><td>1.000</td><td>Friedman test (Group C)
TRAD_UMDA/MO
- TRAD_UMDA/T
- TRAD_FDA/MO
- TRAD_FDA/T</td><td>1.000</td></tr>
</table>

The proposed algorithm reports better results than the traditional UMDA, the latter never found the optimal schedule in any of the 50 analyzed instances.

4 Conclusions

In this paper, a new approach on constraints handling inside the probabilistic model for the Univariate Marginal Estimation of Distribution Algorithm has been developed. In addition, the proposal was applied to solve the Multi-mode Resource-Constrained Project Scheduling Problem. Specifically, it was tested with several instances of the PSPLib database providing superior results than traditional UMDA and also than other algorithms proposed in the literature. With respect to variable "*Mean Makespan*" and "Number of times the optimum founded", the variants based on CL_UMDA/NR obtained significantly better results than the rest of the algorithms. Regarding the optimization approach, the *just_time* variants reported the best results. Furthermore, the proposed algorithms improve or equal the best results reported in the bibliography, by finding the global optimum in all instances of the experimental databases.

References

1. Delgado, R.: La Dirección Integrada de Proyecto como Centro del Sistema de Control en el Ministerio del Poder Popular para la Comunicación y la Información. La Habana, Cuba (2013)
2. Mahdi, G., Piñero, P.Y., García, R., Madera, J., Al-subhi, S., Peréz, I.: Nuevo algoritmo de estimación de distribuciones con diferentes métodos de aprendizaje aplicado a planificación de proyectos. Investigacion Operacional **41**(7), 945–960 (2020)
3. Bermúdez, N.V., Abreu, M.P., Valareso, S.B., Pérez, I.: Experiencias en la integración de procesos en las organizaciones orientadas a proyectos de software. Revista Cubana de Ciencias Informáticas **10**, 171–185 (2016)
4. Pacelli, L.: Grandes Errores en la Gestión de Proyectos. Resumen autorizado de: The Project Management Advisor: 18 Mayor Projetc Screw-Ups and Cut Them off the Pass. Financial Times Prentice Hall (2004)
5. Balouka, N., Cohen, I.: A robust optimization approach for the multi-mode resource-constrained project scheduling problem. Eur. J. Oper. Res. (2019). http://doi.org/10.1016/j.ejor.2019.09.052
6. Carter, M., Price, C.C., Rabadi, G.: Operations Research: A Practical Introduction, 2nd edn. CRC Press, New York (2019)
7. Tirkolaee, E.B., Goli, A., Hematian, M., Sangaiah, A.K., Han, T.: Multi-objective multi-mode resource constrained project scheduling problem using Pareto-based algorithms. Computing **101**(6), 547–570 (2019). https://doi.org/10.1007/s00607-018-00693-1
8. Feng, M., Kong, J., Liu, L., Liu, G., Zhao, S., Zhang, Y.: Modified Estimation of Distribution Algorithm for Solving Flow-shop Scheduling Problem with Setup Times, pp. 4041–4046. Chinese Automation Congress (CAC) (2019). http://doi.org/10.1109/CAC48633.2019.8997384
9. Kannimuthu, M., Raphael, B., Palaneeswaran, E., Kuppuswamy, A.: Optimizing time, cost and quality in multi-mode resource-constrained project scheduling. Built Environ. Proj. Asset Manag. **9**(1), 44–63 (2019). https://doi.org/10.1108/BEPAM-04-(2018)-0075
10. Ratajczak-Ropel, E.: Experimental evaluation of agent-based approaches to solving multi-mode resource-constrained project scheduling problem. Cybern. Syst. **49**(5–6), 296–316 (2018). https://doi.org/10.1080/01969722.2017.1418269

11. Rosłon, J.H., Kulejewski, J.E.: A hybrid approach for solving multi-mode resource-constrained project scheduling problem in construction. Open Eng. **9**(1), 7–13 (2019). https://doi.org/10.1515/eng-2019-0006
12. Zamani, R.: An effective mirror-based genetic algorithm for scheduling multi-mode resource constrained projects. Comput. Ind. Eng. **127**, 914–924 (2019). https://doi.org/10.1016/j.cie.2018.11.031
13. Lova, A., Tormos, P., Cervantes, M., Barber, F.: An efficient hybrid genetic algorithm for scheduling projects with resource constraints and multiple execution modes. Int. J. Prod. Econ. **117**(2), 302–316 (2009). https://doi.org/10.1016/j.ijpe.2008.11.002
14. Van Peteghem, V., Vanhoucke, M.: A genetic algorithm for the preemptive and non-preemptive multi-mode resource-constrained project scheduling problem. Eur. J. Oper. Res. **201**(2), 409–418 (2010). https://doi.org/10.1016/j.ejor.2009.03.034
15. Coelho, J., Vanhoucke, M.: Multi-mode resource-constrained project scheduling using RCPSP and SAT solvers. Eur. J. Oper. Res. **213**(1), 73–82 (2011). https://doi.org/10.1016/j.ejor.2011.03.019
16. Vanhoucke, M., Coelho, J.: An approach using SAT solvers for the RCPSP with logical constraints. Eur. J. Oper. Res. **249**(2), 577–591 (2016). https://doi.org/10.1016/j.ejor.2015.08.044
17. Damak, N., Jarboui, B., Siarry, P., Loukil, T.: Differential evolution for solving multi-mode resource-constrained project scheduling problems. Comput. Oper. Res. **36**(9), 2653–2659 (2009)
18. Tao, S., Dong, Z.S.: Multi-mode resource-constrained project scheduling problem with alternative project structures. Comput. Ind. Eng. **125**, 333–347 (2018). https://doi.org/10.1016/j.cie.2018.08.027
19. Soliman, O.S., Elgendi, E.A.: A hybrid estimation of distribution algorithm with random walk local search for multi-mode resource-constrained project scheduling problems. IJCTT **8**(2), 57–64 (2014). https://doi.org/10.14445/22312803/IJCTT-V8P111
20. Ayodele, M.: Effective and efficient estimation of distribution algorithms for permutation and scheduling problems. Ph.D. Thesis, Robert Gordon University (2018)
21. Ayodele, M., McCall, J., Regnier-Coudert, O.: Estimation of Distribution Algorithms for the Multi-mode Resource Constrained Project Scheduling Problem, pp. 1579–1586. IEEE Congress on Evolutionary Computation (CEC), Donostia, San Sebastián, Spain (2017). http://doi.org/10.1109/CEC.2017.7969491
22. Ayodele, M., McCall, J., Regnier-Coudert, O.: RK-EDA: A Novel Random Key Based Estimation of Distribution Algorithm, pp. 849–858 (2016)
23. Rao, M.S., Prasad, K.L., Anusha, P.: Automatic timetable generation using PBIL algorithm. i-Manager's J. Inf. Technol. **8**(2), 31 (2019)
24. Ayodele, M., McCall, J., Regnier-Coudert, O.: BPGA-EDA for the multi-mode resource constrained project scheduling problem. In: Evolutionary Computation (CEC) IEEE Congress, pp. 3417–3424 (2016)
25. Tian, J., Hao, X., Gen, M.: A hybrid multi-objective EDA for robust resource constraint project scheduling with uncertainty. Comput. Ind. Eng. **130**, 317–326 (2019). https://doi.org/10.1016/j.cie.2019.02.039
26. Tian, J., Hao, X., Murata, T.: Markov network based EDA and robust project scheduling with multi-mode resource constraint and rework uncertainty. In: Third Asian Conference on Information Systems, vol. 370 (2014)
27. Gao, S., Silva, C.W.: Estimation distribution algorithms on constrained optimization problems. Appl. Math. Comput. **339**, 323–345 (2018). https://doi.org/10.1016/j.amc.2018.07.037
28. Wang, L., Fang, C.: A hybrid estimation of distribution algorithm for solving the resource-constrained project scheduling problem. Expert Syst. Appl. **39**(3), 2451–2460 (2012). https://doi.org/10.1016/j.eswa.2011.08.095
29. Mahdi, G., Quintana, J., Pérez, P., Al-subhi, S.: Estimation of distribution algorithm for solving the multi-mode resource constrained project scheduling problem. EAI Endorsed Trans. Energy Web **7**(27) (2020). http://doi.org/10.4108/eai.13-7-2018.164111

30. Mühlenbein, H., Mahnig, T., Ochoa-Rodríguez, A.: Schemata, distributions and graphical models in evolutionary optimization. J. Heuristics **5**(2), 215–247 (1999)
31. Kolisch, R., Sprecher, A.: PSPLIB—a project scheduling problem library: OR software—ORSEP operations research software exchange program. Eur. J. Oper. Res. **96**(1), 205–216 (1997). https://doi.org/10.1016/S0377-2217(96)00170-1
32. Shapiro, S.S., Wilk, M.B.: An analysis of variance test for normality (complete samples). Biometrika **52**(3/4), 591–611 (1965). https://doi.org/10.2307/2333709
33. Lilliefors, H.W.: On the Kolmogorov-Smirnov test for normality with mean and variance unknown. J. Am. Stat. Assoc. **62**(318), 399–402 (1967)

Chapter 8
New Methods for Feasibility Analysis of Investment Projects in Uncertain Environments

Marieta Peña Abreu, Carlos R. Rodríguez Rodríguez, Pedro Y. Piñero Pérez, and Yadira García García

Abstract Project feasibility analysis is a key process in organizations to predict the success of projects in the future. This process becomes complex because it involves multiple criteria, which are evaluated in uncertain environments. This paper presents a model to perform the feasibility analysis of investment projects in uncertain environments using soft computing techniques. The model includes technical, economic, social and commercial criteria and the methods for its calculation. Traditional economic criteria are extended by introducing triangular fuzzy numbers, which provides greater flexibility and certainty in predicting economic feasibility. The 2-tuple linguistic representation model is used in the qualitative evaluation of the criteria and the calculation of the final results. The model output is the feasibility analysis of the projects, integrating the results of all the evaluated criteria without loss of information, which provides greater interpretability for decision making. The case study and experiments conducted demonstrate that the model is applicable in real environments and that its ability to predict feasibility is greater than that of other proposals.

Keywords Project feasibility · Project management · Soft computing

M. Peña Abreu (✉) · C. R. Rodríguez Rodríguez · Y. García García
Vicedecanato de Investigación y Posgrado, Facultad 3, Universidad de las Ciencias Informáticas, La Habana, Cuba
e-mail: mpabreu@uci.cu

C. R. Rodríguez Rodríguez
e-mail: crodriguezr@uci.cu

Y. García García
e-mail: yggarcia@uci.cu

C. R. Rodríguez Rodríguez
Kazan Federal University, Kazan, Tatarstan, Russia

P. Y. Piñero Pérez
Grupo de Investigaciones en Gestión de Proyectos, Universidad de las Ciencias Informáticas, La Habana, Cuba
e-mail: ppp@uci.cu

P. Y. Piñero Pérez et al. (eds.), *Artificial Intelligence in Project Management and Making Decisions*, Studies in Computational Intelligence 1035,
https://doi.org/10.1007/978-3-030-97269-1_8

1 Introduction

The feasibility analysis of investment projects is the basic step for the success of projects to be executed in an organization. In recent years there has been an increase in the number of projects evaluated for execution, among which software projects and Building Information Modeling (BIM) technologies [1] stand out. The analysis becomes complex given the number of criteria considered and the uncertainty of the environment [2]. This makes it possible to formulate the problem with a basic structure of decision analysis, taking into account: environment, number of criteria, number of experts, changes over time and domains of expression [3]. This analysis allows decision-making at the strategic, tactical and operational levels and reducing biases and uncertainty. This provides the project with greater sustainability over time [4].

Feasibility analyses are indispensable for better use of the available resources. Approximately 34% of the resources are sub used and 95% of the information generated is not used as lessons learned in new projects [5] Although there are numerous schools of project management: IPMA [6], PMI [7], CMMI [8], standard ISO 21500 [9] which offer good practices for project management, they do not detail how to perform feasibility analyses. In addition, there is a lack of solutions that integrate quantitative and qualitative analyses in a single model for evaluating investment projects, which affects the interpretability of the results. Generally, the solutions consider only economic aspects with traditional methods and do not deal with uncertainty. For uncertainty management in similar problems, researchers have used different soft computing techniques depending on the context [2, 10, 11]. Among the most prominent paradigms are fuzzy mathematics [12] and computing with words (CWW) [11, 13].

2 Background

Around 1920, there began to be a more comprehensive approach to the company as a system, which led to a greater impact in the evolution of computational models for the analysis of project feasibility [14], which at the beginning only assessed economic rentability. At this stage, the investment theory proposed by Schneider in 1956 was the main reference for the models developed, although it only was applied in private businesses. During the period 1950–1973, the need to deal with the uncertainty in financial analysis began to be perceived [15] and was handled using probabilistic techniques. In 1973, social criteria were included in the feasibility analyses [16]. For the first time, in 1978, different aspects were considered in the same evaluation process [17], although no operators were used to aggregate the information. At the end of the twentieth century, there was a boom in the use of the economic criteria Net Present Value (NPV), Internal Rate of Return (IRR) and Payback Period (PRI), recognized as traditional criteria [18]. These methods operate with classical mathematics and have limitations associated with a lack of flexibility since they do not handle uncertainty [19].

There are several classifications in the literature for qualitative criteria [20], but technical, commercial, environmental and social criteria stand out [21–23]. Their main shortcomings lie in the absence of mechanisms for using existing information that is already useful for the analysis.

This paper considers only methods that deal with uncertainty in their computations. The 2-tuple linguistic representation model is used due to its multiple advantages. This model allows the continuous representation of linguistic information in a given domain and its computational model based on symbolic translation attempts to deal with the loss of information. Moreover, fuzzy triangular numbers are used to modify the economic criteria since they overcome the principle of the excluded third party and offer a confidence interval to the evaluator [3, 11].

This paper proposes a model for the feasibility analysis of investment projects in environments with uncertainty. Traditional economic criteria are extended using triangular fuzzy numbers. Besides, several qualitative criteria (technical, commercial, social) are evaluated using the 2-tuple model. The model output is a list of evaluated projects, ordered according to the values of feasibility. This feasibility value consists of a 2-tuples that indicate the degree of feasibility and the certainty of that evaluation. This allows for greater flexibility and predictability of project success.

3 Model for the Feasibility Analysis of Investment Projects in Environments with Uncertainty

This section proposes MFac-PS, a model for the feasibility analysis during project conceptualization, to know its probability of success. MFac-PS has an adaptive capacity since it allows the inclusion of new evaluation criteria and algorithms over time. Among its main contributions are the definition of economic, technical, commercial and social criteria and the methods for their calculation, dealing with uncertainty and loss of information. The model consists of four components, which are briefly described below.

The model is proposed from the study of traditional models [24] because they use deterministic methods, the major novelty of the proposal is the treatment of uncertainty in all its conception, as well as the integration of multiple criteria to evaluate the feasibility analysis.

Component I. Defining evaluation criteria

Overview: Criteria for evaluating the feasibility of the project are defined. In each evaluation, these criteria may change according to the characteristics of the project. In this component authors apply the following steps:

- I-1 Define requirements for the selection of experts, see Algorithm 8.1
- I-2 Select set of experts E: $e_i \in E : i \in [1..n]$ based on his competences applying **curriculum synthesis analysis techniques** [25]
- I-3 Define set of feasibility criteria C by combine Delphi and 2-tuple methods. Reuse Algorithm 8.1. C consists of technical, economic, commercial and social criteria: $c_j \in C : j \in [1..m]$
- I-4 Conceptualize criteria, detailing each criterion and the domain in which it will be evaluated (linguistic, numerical, or interval).

Algorithm 8.1 *Defining requirements.*

Inputs:

R - Candidate requirements for the selection of experts, $r_q \in R : q \in [1..z]$; *V - Set of evaluators,* $v_l \in V : l \in [1..n]$

Outputs: R_e *List of requirements selected for evaluation*

Begin

1. *Define the basic set of linguistic terms for expert evaluation (BSLT)* $S_T = \{VLC, NVC, MC, C, VC\}$[1].
2. *For each requirement* $r_q \in R$ *do*
3. *For each evaluator* $v_l \in V$ *do*
4. *Collect the preferences of the evaluator* v_l *on the requirement* r_q *using BSLT. The evaluation matrix is built, where each element* M_{ql} *represents the preference of the evaluator* l *on the requirement* q.
5. *End step 3*
6. *Calculate the q-th element* R_q *that represents the aggregate value of the evaluators' preferences for the requirement* r_q *once the 2-tuple method has been applied.*
7. *End step 2*
8. *Order the requirements according to their evaluation and select the* R_e *requirements that will finally be considered to evaluate the experts.*
9. *Return the list of selected requirements* R_e.

End

Component II. Data collection for evaluation

Overview: All the essential information on the projects is detailed in order to carry out the evaluation. In this component authors apply the following steps:

- II-1 Select the set of projects P: $p_y \in P : y \in [1..s]$
- II-2 Collect project data by detailing technical information of projects
- II-3 Select the evaluation criteria (by using CWW 2-tuple method) from the set C defined in component I
- II-4 Determine the criteria weight vector W^c: $w^{c_j} \in W^c : j \in [1..m]$, by using AHP method, see Algorithm 8.2.

Algorithm 8.2 *Determine the weight vector of the criteria,* W^c.

```
Inputs:
  C- Criteria list, c_j ∈ C: j ∈ [1..m];
  E - Set of experts, e_i ∈ E: i ∈ [1..n]
Begin
 1. For each expert e_i ∈ E do
 2.    For each criterion c_j ∈ C do
 3.       Define the importance of criterion c_j, comparing it with the others and apply expert judgment. //An n x
          m matrix is gradually built up.
 4.    End step 2
 5. End step 1
 6. For each criterion c_j ∈ C do
 7.    Calculate w^{c_j} for c_j, as the geometric mean of the scores given by the experts.
 8.    Constraint Σ_{j=1}^{m} w^{c_j} = 1 is satisfied.
 9. End step 6
End
```

Component III. Evaluation of feasibility criteria

Overview: The criteria are evaluated considering the uncertainty. Two consecutive sub-processes are defined for the instrumentation:

- Sub-process: calculation of economic criteria.
- Sub-process: Qualitative evaluation of all criteria.

Sub-process: Calculate economic criteria. It receives baseline information about the project and its environment. The output of this sub-process consists on values NPV, IRR and PRI expressed in triangular fuzzy numbers $\widetilde{VAN(p,c,o)}$, $\widetilde{TIR(p,c,o)}$, $\widetilde{PRI^d(o,c,p)}$ see [26]. This sub-process consist on the following steps:

III-1 Estimate income and expenses under uncertainty for each period t, using Expert judgment and considering three scenarios: pessimistic, most accurate, and optimistic; $\widetilde{I_n} = (i_p, i_c, i_o)$ and $\tilde{E} = (e_p, e_c, e_o)$.

III-2 Determine Fuzzy Cash Flow $\widetilde{CF(p, c, o)}$ according to Definition 8.1

III-3 Calculate $\widetilde{NPV(p,c,o)} = [VAN_{p\alpha}; NPV_{o\alpha}]$. Detailed in [26].

III-4 Calculate $\widetilde{IRR(p,c,o)} = [IRR_{p\alpha}; IIR_{o\alpha}]$. Detailed in [26].

III-5 Calculate $\widetilde{PRI^d(o,c,p)} = [PRI^d_{o\alpha}; PRI^d_{p\alpha}]$. Detailed in [26].

III-6 Calculate the unified value of each criterion using the parametric technique Estimation by three values, see Eq. (5).

Definition 8.1 The fuzzy cash flow $\widetilde{CF(p,c,o)}$ is the succession of t triangular fuzzy numbers expressing the difference between the collections and payments generated by the project in each one of its (t) periods of operation, from the first investment expense is made ($t = 0$) to its liquidation or replacement at the end of its useful life ($t = n$). $\widetilde{CF(p,c,o)}$ considers three possible scenarios (pessimistic, more accurate and optimistic) and is defined according to (1).

$$\widetilde{\mathrm{CF(p, c, o)}} = \left\{\widetilde{\mathrm{CF(p, c, o)}}_0, \widetilde{\mathrm{CF(p, c, o)}}_1, \ldots, \widetilde{CF(p, c, o)}_n\right\} \quad (1)$$

For a period of operation t the fuzzy cash flow $\widetilde{CF(p, c, o)}_t$ is obtained according to Eq. (2), from calculating the fuzzy balance $\tilde{S}$ (Eq. 3) and the fuzzy tax $\widetilde{P}$ (Eqs. 4, 5).

$$\widetilde{CF(p, c, o)}_t = -\widetilde{A} + \left(\tilde{S} - \tilde{P}\right) + \widetilde{V_r} + \tilde{D} \quad (2)$$

$$\tilde{\mathrm{S}} = \widetilde{\mathrm{I_n}} + \widetilde{\mathrm{B_e}} - \tilde{\mathrm{E}} - \tilde{\mathrm{D}} \quad (3)$$

$$\tilde{P} = \tilde{S} \times \widetilde{f_{imp}} \quad (4)$$

where

$\widetilde{A}$	Estimated investment expenses for period t. $\tilde{A} = (a_p, a_c, a_o)$.
$\tilde{S}$	Estimated balance for period t. $\tilde{S} = (s_p, s_c, s_o)$.
$\widetilde{I_n}$	Estimated income for period t. $\widetilde{I_n} = (i_p, i_c, i_o)$.
$\widetilde{B_e}$	Extraordinary benefits estimated for period t. $\widetilde{B_e} = (b_p, b_c, b_o)$.
$\tilde{E}$	Expenses estimated for period t. $\tilde{\mathrm{E}} = (\mathrm{e_p}, \mathrm{e_c}, \mathrm{e_o})$.
$\tilde{D}$	Estimated depreciation for period t. $\tilde{\mathrm{D}} = (\mathrm{d_p}, \mathrm{d_c}, \mathrm{d_o})$.
$\tilde{P}$	Estimated income tax for period t. $\tilde{\mathrm{P}} = (\mathrm{p_p}, \mathrm{p_c}, \mathrm{p_o})$.
$\widetilde{f_{imp}}$	Estimated factor for period t. $\widetilde{f_{imp}} = (f_p, f_c, f_o)$.
$\widetilde{V_r}$	Residual value of the investment estimated for period t. $\widetilde{V_r} = (v_p, v_c, v_o)$.

Generally, for the case of $t = 0$ only the initial investment is made $\widetilde{A}$ and $\tilde{S} = \widetilde{P} = \widetilde{V_r} = \tilde{D} = 0$, so that for $t = 0$, $\widetilde{FC(p, c, o)}_0 = -\widetilde{A}$.

Estimation by three values

$$FEC_e = \frac{C_o + 4C_c + C_p}{6} \quad (5)$$

where

FEC_e	Estimated value of a given fuzzy economic criterion.
C_o	Optimistic value of a given fuzzy economic criterion.
C_c	Most accurate value of a given fuzzy economic criterion.
C_p	Pessimistic value of a given fuzzy economic criterion.

Sub-process: Qualitative evaluation of all criteria. It receives: the set of projects, the set of experts, the values of $\widetilde{NPV(p, c, o)}$, $\widetilde{IRR(p, c, o)}$, $\widetilde{PRI^d(o, c, p)}$, and the set of technical, commercial and social criteria. The outputs of the sub-process are the homogeneous values of criteria expressed in 2-tuples (feasibility, accuracy). This sub-process consist on the following steps:

III-1 Collect the experts' preferences for all criteria, see Algorithm 8.3

III-2 Unify the experts' preferences in the linguistic domain using the transformations defined in [27].

Algorithm 8.3 *Evaluation of feasibility criteria.*

Inputs:

C: Criteria list, $c_j \in C: j \in [1 \dots m]$;

E: Set of experts. $e_i \in E: i \in [1..n]$; *P- Set of projects.* $p_y \in P: y \in [1 \dots s]$;

D: *domains "numeric", "interval-valued and "linguistic"*

- *"numeric"* $x_j^{yi} \in D: x_j^{yi} \in [1,5]$,
- *"interval-valued"* $x_j^{yi} \in D: x_j^{yi} \in I([0,1]) = [a_j^{yi}, b_j^{yi}] : a_j^{yi}, b_j^{yi} \in [0,1]\ ;\ a_j^{yi} \leq b_j^{yi}$
- *"linguistic"* $x_j^{ki} \in S = \{S_0, \dots, S_g\}$ *where g+1 represents the cardinality of S, the Set of Linguistic Terms (SLT). Each linguistic term* $S_i \in S$ *has an associated membership function* $\mu_{S_i}(y), y \in [0,1]$.

Begin

1. *For each project* p_y *do*
2. *For each expert* e_i *do*
3. *For each criterion* c_j *do*
4. *Express the preference* x_j^{yi} *using one of the defined domains*
5. *End step 3*
6. *End step 2*
7. *End step 1*

10. *Return the vector of preferences X X- vector of preferences,* $x_j^{yi} \in C: x_j^{yi} \in [1 \dots x_n^{sm}]$, *where* x_j^{yi} *represents the opinion of expert* e_i *on project* p_y *according to criterion* c_j.
11. *Verify expert concordance using Kendall's coefficient*

End

Component IV: Comprehensive response of the model

Overview: The values of the criteria obtained from the previous component are integrated using the aggregation operators for 2-tuples to calculate the integral evaluation of each project. The results allow the ranking of each project according to its overall feasibility or according to a specific criterion. In this component authors apply the following steps:

IV-1 Calculate the collective value of each criterion for each project using the Arithmetic mean extended for 2-tuple operator [28].

IV-2 Aggregate the value of the criteria for each project using the Weighted average extended for 2-tuple operator [28].

IV-3 Interpret the results using the comparison operators for 2-tuple [28]. With these tools, it is possible:

- Select the projects considering their overall feasibility, or make more detailed analyses by comparing the criteria individually.
- Decide on each project individually, analyzing the most relevant criteria for each one and considering the current status and plans of the executing organization. This allows evaluating at the same time projects with different characteristics in terms of the criteria analyzed and also considering other aspects such as duration, costs, risks, etc.

4 Experimentation

A data set of completed projects (Feasility_Dataset), published by the Research Laboratory in Project Management of the University of Informatics Sciences, was used for the experimentation. It consists of 30 software projects varying in scope and size.

A metrics to analyze the results [24] was proposed to calculate the distance between the results of each method and the actual feasibility of the projects. The predictive capacity of the feasibility is calculated using the Mean Squared Error (MSE). The Wilcoxon test is used for the statistical analysis, considering the differences as significant when the p-value < 0.05. In the Wilcoxon test, the Monte Carlo method is applied for a 99% confidence interval. Positive and negative cases, mean and standard deviation are also measured.

Quasi-experiment 1: Assessment of ability to predict the economic feasibility.

Objective: Determine whether there are statistically significant differences in the feasibility order of the projects between the results obtained with the fuzzy economic criteria and those obtained with the traditional economic criteria.

Procedure: The projects in the database are evaluated concerning the economic criteria using fuzzy and traditional methods. Two different lists of the projects are created, according to the feasibility order given by each method. Both lists are compared with the actual feasibility ranking (pattern). The distance between the actual ranking and the order obtained in each listing is calculated. These distances are represented as related samples and compared with the Wilcoxon statistical test.

Results: The results of fuzzy and traditional economic criteria showed that the former ones performed better in all cases. The values of MSE indicate that the fuzzy criteria better predicted the actual values, i.e., the values obtained with fuzzy criteria were closer to the actual feasibility ranking. Table 1 shows a statistical summary of the quasi-experiment, which shows that the results of the fuzzy criteria are equal to or better than those of the traditional criteria.

The results of the Wilcoxon test for NPV and IIR criteria reported significant differences. For NPV, the Z rate reached a value of − 3.491 and for IRR, the Z rate gave a value of− 3.907. The asymptotic significance and Monte Carlo method

Table 1 Statistical summary of the quasi-experiment experiment 1

	MSE	Mean	StdDev	False positives	False negatives
NPV trad	31.2	2.4	5.13003	3	3
NPV fuzzy	0.2	0.2	0.40684	0	0
IRR trad	3.73	1.4667	1.27937	3	1
IRR fuzzy	0.73	0.4667	0.7303	0	1
PRI trad	0.87	0.6	0.72397	0	1
PRI fuzzy	0.47	0.3333	0.60648	0	0

significance were equal to zero for both criteria. This value is smaller than the decision rule established ($p \leq 0.05$) and therefore, there are significant differences. For the PRI criterion, the Wilcoxon test also found significant differences. Although, as can be seen in the values of Z rate (− 2.309), asymptotic significance (0.021), and unilateral and bilateral significance with Monte Carlo method (0.018; 0.037), they are less strong than for the NPV and IRR criteria. These results confirm that the fuzzy criteria performed better in the sample analyzed.

Quasi-experiment 2: Assessment of ability to predict the overall feasibility

Objective: Determine whether there are statistically significant differences regarding project feasibility prediction between the results provided by the MFac-PS model and the FACTCOM method [29].

Procedure: The projects in the database are evaluated concerning 15 feasibility criteria using MFac-PS and FACTCOM. The projects are ranked according to the feasibility order given by each method, obtaining two lists. Each list is compared with the actual feasibility ranking (pattern) and the distance between the actual (desired) ranking and the value obtained with each method is calculated. These distances are represented as related samples and compared using the Wilcoxon test.

Results: The MFac-PS results show a better prediction of feasibility and integrate all the proposed criteria. Unlike the traditional method, MFac-PS also includes social criteria and takes into account the value of the fuzzy economic criteria, even if these have not yielded satisfactory results. This feature gives the decision-maker the possibility of analyzing the project, even if it is not economically feasible, since it could have a greater social impact and therefore be implemented.

When calculating the MSE of the prediction of both methods, FACTCOM shows a value of 223.96 while MFac-PS reaches only 5. The remarkable difference is because FACTCOM recommends that 11 of the 30 projects should not be developed after evaluating their technical and commercial feasibility. However, after a more comprehensive evaluation (with MFac-PS), including economic criteria, it is observed that these 11 projects generate income and are profitable. This confirms the need for a comprehensive evaluation of all available information.

Applying the nonparametric Wilcoxon test, significant differences were found in favor of MFac-PS, with a value of $Z = -3.145$. The asymptotic significance and Monte Carlo method significance were $\leq$0.002 for both criteria and therefore less than the defined threshold (0.05). Table 2 provides a statistical summary of the quasi-experiment, which shows that the results of MFac-PS are better than those of FACTCOM.

Table 2 Statistical summary of the quasi-experiment experiment 2

	ECM	Mean	StdDev	False positives	False negatives
MFac-PS	5	1.4667	1.7145	3	2
FACTCOM	223.96	9.9667	11.1955	5	11

4.1 Case Study

A case study was conducted to assess the applicability of the proposal. The study was practiced in a software development organization using a sample of three real projects $P = \{p_1, p_2, p_3\}$, for which the actual feasibility results are known. The information on the projects can be found in BusinessRedmine solution [30]. Considering the features of the projects, the curricular synthesis method was applied to select the experts $E = \{e_1, e_2, e_3, e_4, e_5\}$. The experts selected 13 criteria (technical 7, economic 3, commercial 1, and social 2), assigning each one a weight according to its relevance. Based on the estimated $\widetilde{CF(p, c, o)}$, the fuzzy economic criteria were calculated.

Table 3 shows the results of these criteria and those of the traditional methods.

Comparison of the actual values with the fuzzy criteria results shows that the former ones are within the confidence intervals offered by the fuzzy ones. This demonstrates the predictive capability of the proposal. In addition, having the confidence interval allows assessing the impact of potential risks.

Table 4 shows the feasibility value of the economic criteria and their order following the comparison criteria defined. With these data, the overall economic feasibility of each project was calculated. It turned out that {p_1 = (HF; −0.1) > p_2 = (HF; −0.27) > p_3 = (MF; −0.43)}. This means that project p_1 is the most economically feasible project even though p_2 has a better result for $\widetilde{VAN(p, c, o)}$.

Computation of technical feasibility of the projects yielded {p1 = (VHF; −0.42) > p2 = (HF; −0.04) > p3 = (MF; −0.04)}. For the social criteria, it was found that {p_2

Table 3 Results of fuzzy and traditional economic criteria

	NPV		PRI (months)		IRR (percent)	
	$\widetilde{NPV(p, c, o)}$	Real	$\widetilde{PRI^d(o, c, p)}$	Real	$\widetilde{IRR(p, c, o)}$	Real
P_1	[214.581,89; 263.796,31]	239.189,1	[9,40; 9,18]	9,29	[0,58; 0,61]	0,59
P_2	[627.120,0; 639.906,0]	633.513,0	[9,55; 9,54]	9,54	[0,53; 0,53]	0,53
P_3	[24.464,35; 25.656,25]	25.060,3	[10,25; 10,23]	10,24	[0,17; 0,17]	0,17

Table 4 Economic feasibility criteria and their ranking according to results

	$\widetilde{NPV(p, c, o)}$		$\widetilde{IRR(p, c, o)}$		$\widetilde{PRI^d(o, c, p)}$	
	Feasibility	Ranking	Feasibility	Ranking	Feasibility	Ranking
p_1	(VHF; – 0.5)	2	(MF; 0.3)	1	(HF; – 0.1)	1
p_2	(VHF; 0.18)	1	(MF; 0)	2	(MF; 0)	2
p_3	(HF; – 0.1)	3	(LF; 0.4)	3	(VLF; 0.42)	3

= (VHF; −0.18) > p_1 = (VHF; −0.31) > p_3 = (LF; 0.4)}. Thus, project p_2 is more feasible, which is in correspondence with the characteristics of the project. The evaluated commercial criterion indicated that p_3 (which is a service project) was the most feasible: {p_3 = (FA; −0.4) > p_1 = (FM; 0.3) > p_2 = (FB; 0.1)}. Finally, we note that the overall feasibility of the three projects taking into account the 13 evaluated criteria, had the following behavior: {p_1 = (FA; 0.31) > p_2 = (FA; −0.14) > p_3 = (FM; −0.15)}. As a summary of this study, the three projects were found to be feasible, which agrees with the analyses carried out in practice before their inception.

5 Conclusions

- Software project feasibility analysis can be modeled as a decision-making problem, with evaluation criteria handling heterogeneous information, evaluated by experts with different degrees of expertise.
- The proposed model for feasibility analysis following an integrated approach and considering uncertainty showed better results than traditional techniques by incorporating triangular fuzzy numbers and computing with words.
- Validation through quasi-experiments and case studies demonstrates the capacity of the model to predict the feasibility with a lower degree of uncertainty, which contributes to improved decision making.
- The use of confidence intervals helps assess the impact of risks on the project.

References

1. Mesároš, P., Mandičák, T., Behúnová, A.: Use of BIM technology and impact on productivity in construction project management. Wirel. Netw. 1–8 (2020)
2. Pérez, I., Piñero, P.Y., García-Vacacela, R., Bello, R., Santos, O., Vázquez, M.: Extensions to linguistic summaries indicators based on neutrosophic theory, applications in project management decisions. Infinite Study (2018)
3. Zulueta, Y., Rodríguez, R., Bello, R., Martínez, L.: A hesitant heterogeneous approach for environmental impact significance assessment. J. Environ. Inf. **29**(2), 74–87 (2017)
4. Soler, J.A., Delgado, F.M., Sanjuán, A.M., García, M.N.: Modelos para evaluar la sostenibilidad de las organizaciones. Estudios Gerenciales 63–73 (2018)
5. Dave, B., Buda, A., Nurminen, A., Främling, K.: A framework for integrating BIM and IoT through open standards. Autom. Constr. **95**, 35–45 (2018)
6. ICB.: Swiss individual competence baseline: Version 4.0: Beschreibung der individuellen Kompetenz im Management von Programmen gemäss IPMA (2016)
7. PMI.: The standard for Project Management and a guide to the Project Management Body of Knowledge, 7th Edition. Project Management Institute, ISBN: 978-1-62825-664-2 370 (2021)
8. SEI.: CMMI for Development v1.3, SEI CMMI Production, CMU/SEI-2010-TR-034, Technical report, Software Engineering Institute (2010)
9. Zandhuis, A., Stellingwerf, R.: ISO 21500: Guidance on Project Management; a Pocket Guide: Van Haren (2013)

10. Al–Subhi, S.H., Mahdi, G.S., Alava, M.V., Pérez, M.P., Vázquez, M.L.: Operador media potencia pesada lingüística y su aplicación en la toma de decisiones. Int. J. Innov. Appl. Stud. **22**(1), 38–43 (2017)
11. Zulueta-Veliz, Y., García-Cabrera, L.: A choquet integral–based approach to multiattribute decision–making with correlated periods. Granular Computing **3**(3), 245–256 (2018)
12. Kaufmann, A., Aluja, J.G., Pirla, J.M.: Introducción a la teoría de los subconjuntos borrosos a la gestión de las empresas: Milladoiro (1986)
13. Felix, G., Nápoles, G., Falcon, R., Froelich, W., Vanhoof, K., Bello, R.: A review on methods and software for fuzzy cognitive maps. Artif. Intell. Rev. **52**(3), 1707–1737 (2019)
14. Cruz, A.D., Hidalgo, K.M., Medina, M.Q.: Evaluación de factibilidad de proyectos de inversión. Revista Caribeña de Ciencias Sociales (2014)
15. Ochoa, E.A.: Investigación y Desarrollo, un Método Comparativo de su Valuación. Paper presented at the Sixth LACCEI International Latin American and Caribbean Conference for Engineering and Technology Honduras (2008)
16. Little, I., Mirrlees, J.A., Mirrlees, J.A.: Estudio social del costo beneficio en la industria de países en desarrollomanual de evaluación de proyectos (1973)
17. ONU.: Manual para la preparación de Estudios de Viabilidad Industrial. Naciones Unidas, Nueva York (1978)
18. Dubs de Moya, R.: El proyecto factible: una modalidad de investigación. Sapiens. Revista Universitaria de Investigación, **3**(2) (2002)
19. Milanesi, G.S.: Un modelo de opciones reales fuzzy y funciones de utilidad isoelásticas para valorar I&D en mercados incompletos. Estocástica: finanzas y riesgo, **8**(2), 205–232 (2018)
20. Miranda, J.J.: Gestión de proyectos: identificación, formulación, evaluación financiera–económica–social–ambiental: MMEditores (2005)
21. Pollack., J.F.: Back propagation is sensitive to initial conditions 860–867 (1991)
22. Rodríguez, T.C., Montibeller, G., Oliveira, M.D., Costa, C.A.: Modelling multicriteria value interactions with reasoning maps. Eur. J. Oper. Res. **258**(3), 1054–1071 (2017)
23. Kengpol, A., O'Brien, C.: The development of a decision support tool for the selection of advanced technology to achieve rapid product development. Int. J. Prod. Econ. **69**(2), 177–191 (2001)
24. Abreu, M.P.: Modelo para el análisis de factibilidad de proyectos de software en entornos de incertidumbre (Tesis doctoral), Universidad de las Ciencias Informáticas. La Habana, Cuba (2017)
25. Torres, S.: Modelo de evaluación de competencias a partir de evidencias durante la gestión de proyectos (Tesis doctoral), Universidad de las Ciencias Informáticas La Habana, Cuba (2015)
26. Abreu, M.P., Rodríguez, C.R., García-Vacacela, R., Piñero, P.Y.: Economic feasibility of projects using triangular fuzzy numbers. In IWAIPR 2018, pp. 288–298. Springer, Cham (2018)
27. Herrera F., Martínez L., Sánchez P.J.: Managing non-homogeneous information in group decision making. Eur. J. Oper. Res. **166**(1), 115–132 (2005)
28. Herrera, F., Martinez, L.: A 2–tuple fuzzy linguistic representation model for computing with words. IEEE Trans. Fuzzy Syst. **8**(6), 746–752 (2000)
29. Castro, M.D.: Método de evaluación de proyectos para decidir su aceptación. (Maestría) Grupo de Investigaciones en Gestión de Proyectos, Universidad de las Ciencias Informáticas La Habana (2010)
30. Piñero, P.Y., Pérez, I., Rivero, C., Rojas, C., Sosa, R., Torres, S.: Repositorio de datos para investigaciones en gestión de proyectos. Rev. Cuba. Cienc. Informáticas **13**(1), 176–191 (2019)

Chapter 9
Sustainability Risk Management for Project-Oriented Organizations

Yasmany Aguilera Sánchez, Lisardo Garcia Jane, Juan Antonio Plasencia Soler, and Fernando Marrero Delgado

Abstract Sustainability is used as a reference in the management of organizations, so that, in addition to increasing economic benefits, social and environmental strategies are applied. To this end, approaches and models are used to define the dimensions for measuring sustainability. The increase of Cuban companies makes necessary the implementation of tools that contribute to the sustainability of an organization. This research aims to develop a procedure to contribute to sustainable risk management in project-oriented organizations. The proposed procedure consists of five stages: Prior preparation, Organizational analysis, Risk assessment, Treatment and Follow-up and continuous improvement. It incorporates fundamental characteristics of sustainability, such as multidimensional assessment, process approach and ethical character. The validation of the proposed strategy was carried out using the following methods: expert criteria using the Likert scale, taking into account that it is a useful validation method to check the reliability of a research, obtaining a percentage index higher than 90.77 in all the questions. The case study as a method that allows verifying the applicability of the proposal made possible its application in a real environment, which shows a high level of efficiency and effectiveness of the procedure.

Keywords Project management · Risks management · Sustainability

Y. A. Sánchez (✉) · L. G. Jane · J. A. P. Soler
Grupo de Investigaciones en Gestión de Proyectos, Universidad de las Ciencias Informáticas, Havana, Cuba
e-mail: yasmanyas@uci.cu

L. G. Jane
e-mail: ljane@uci.cu

J. A. P. Soler
e-mail: juanps@uci.cu

F. M. Delgado
Facultad de Ingeniería Industrial, Universidad Central "Marta Abreu" de lasVillas, Santa Clara, Cuba

P. Y. Piñero Pérez et al. (eds.), *Artificial Intelligence in Project Management and Making Decisions*, Studies in Computational Intelligence 1035,
https://doi.org/10.1007/978-3-030-97269-1_9

1 Introduction

Today, the main challenges for organizations come from changes in the environment: globalization, competition, technology and knowledge, which require changes in their structures and development strategies.

In recent years and associated with the need to measure an organization's performance and under the auspices of international standards, the concept of sustainable development has evolved into corporate sustainability (CS). CS aims to increase or stabilize the organization's performance over time, reducing negative impacts on the environment, profitability and providing long-term value in financial, social, ecological and ethical terms.

In the "Pulse of the Profession" report of the study conducted by the Project Management Institute (PMI) in 2017, it is identified that 27% of the projects that fail present undefined opportunities and risk management; this indicator has been present among the five most affecting software development projects that fail in the last 5 years.

At the same time, the "project management fundamentals guide" [1] does not fully recognize or incorporate the elements of sustainability and the social element remains somewhat diffuse in various areas of knowledge.

The relationship between Sustainability and Project Management (PM) is an emerging field of study, which is why the literature is still considered poor [2].

A study on Sustainable Project Management (SPM) identifies as the most accurate definition the planning, monitoring and control of product delivery and support processes; taking into account the environmental, economic and social aspects of the life cycle of project resources, processes, deliverables and impacts aimed at obtaining benefits for stakeholders. This is done in a transparent, fair, ethical manner and includes proactive stakeholder participation [3].

An essential part that is directly related to the Quality of Sustainable Project Management is risk management. While some aspects of sustainability are observed in the various project management standards, there is a growing need to apply criteria and concepts to project management processes [4].

A study on risk management showed that 92% of the definitions studied (12) have a process approach, as well as 83% make explicit reference to the phases of risk management. In addition, 58% refer to risk monitoring strategies. The definition issued by [5] complies with all the above-mentioned characteristics; however, it is valid to highlight that the multidimensionality and ethical nature, fundamental characteristics of sustainability, are not evident.

Existing research highlights a new research trend: Sustainability Risk Management (SRM), which, according to [6], focuses fundamentally on the sustainability of supply chain risk. SRM is an enterprise risk management practice that addresses environmental and social responsibility risks. Companies are being pressured by society to face environmental and social responsibility in their performance, in addition to the traditional bottom line [7].

The above definitions show the incorporation of the fundamental characteristics of sustainability into the risk management process; mainly in the one issued by [8], which is considered the most accurate because it contains the dimensions of the 3BL approach combined with the main phases of classic risk management, although it lacks a strategic approach.

Based on the previously exposed, it is proposed as a definition of sustainable risk management for the pre-sent research: systemic process that starts with the strategic analysis of the organization for the development of a set of activities of prevention, diagnosis, mitigation, control, evaluation in a multidimensional way and monitoring; to minimize the risks and their adverse effects, fomenting the positive ones that materialize in the processes that affect the fulfillment of the objectives, as well as to enhance the ethical character of the organization and the decision making in the actions [4].

In Cuba, Resolution 60 of the Office of the Comptroller General of the Republic of Cuba has come into force since 2011. This resolution in its Article 9 defines that the Internal Control System is made up of five interrelated components, these are: Control Environment, Risk Management and Prevention, Control Activities, Information and Communication, in addition to Supervision and Monitoring.

The second component is closely related to the concept of risk management, taking into account that this resolution seeks in this component the way to mitigate the negative economic-financial, environmental, political, social and technological impacts referred to in the resolution.

Resolution 60 describes the activities in a very general manner and handles in a very compact way several areas that, even when they are related, make their interpretation and implementation difficult. In addition, it does not suggest the techniques or tools to be used to make an impact from risk management on the sustainability of project-oriented organizations, which would facilitate the work and reduce the time dedicated by process engineers to research in order to make the proposal.

Besides, the development of projects in Cuba has demonstrated the lack of formalized and applied risk management, which has minimized the possibility of generating savings, meeting delivery schedules and budgets and using resources more efficiently [9].

Just 34.43% of the Organizational Units (OU) of the Cuban Software Industry (ICSW) have completely de-fined a process to manage risks [9], leaving most of the project oriented organizations with many weaknesses in this aspect, taking into account that they execute isolated activities that do not guarantee the quality of the execution of the same, nor the result that is really expected.

To provide a solution for this problem, in 2016 the Software Quality Center (CALISOFT) and the Ministry of Informatics and Communications (MIC) jointly developed the Cuban Model for the Development of Computer Applications (MCDAI). This model describes the basic process to carry out a correct risk management aligned with international standards, among which CMMI level II and the fifth edition of PMBOK stand out.

However, this author considers that the basic process for risk management of the MCDAI has some limitations to contribute to the sustainability of project-oriented organizations. In first place, it does not perform a multidimensional risk assessment, which results in the organization's management not being able to develop intentional actions based on the dimensions of sustainability; it also lacks a strategic approach and does not identify the regulatory framework associated with risk assessment as suggested by the International Standard ISO 19600 (Compliance Management System). And last but not least, it does not take into account responsible and ethical behaviors of the members of the organization, which is evidenced by the fact that the values and ethical principles most affected by the occurrence of risks are not handled in the risk identification process.

For the reasons stated above, the objective of the research is to develop a procedure to contribute to the sustainability risk management on project-oriented organizations.

2 Procedure

In order to develop a procedure to contribute to the sustainability of project-oriented organizations from risk management, the authors suggest a methodology based on the standards, systems and indicators proposed by previous research, in addition to the following methods: virtual linguistic analysis and the Pareto technique.

The procedure as a contribution to the sustainability of project-oriented organizations takes into account the following principles [1, 10–18]:

Integration: it has the capacity to integrate all its interrelated components for the improvement of risk management as a contribution to the sustainability of project-oriented organizations.

Adaptability: it has the capacity to adapt to the conditions of the environment and the characteristics of the project-oriented organization in which it is applied.

Participation and learning: include the participation of managers and workers, as well as group work methods, surveys and expert methods for the selection, evaluation and prioritization of the elements, which requires training in the use of these techniques.

Flexibility: it allows the total or partial application to other organizations, regardless of their activity and size.

Contextuality: the general procedure responds to the characteristics of the Cuban environment and regulatory framework.

Continuous improvement: it contemplates the feedback from previous stages, which allows a gradual improvement of the organization under study.

The procedure is structured by five stages with a group of activities to be developed as shown below:

Stage 1 Previous preparation

- Step 1. Establishment of the organization's commitment
- Step 2. Creation and preparation of the work team

- Step 2.1. Selection of the group of experts
- Step 3. Verification of the Assumptions.

Stage 2 Organizational analysis

- Step 1. Strategic diagnosis of the organization
- Step 2. Analysis of the compliance with regulations
- Step 3. Identify the stakeholders
- Step 4. Identifying the organization's shared values.

Stage 3 Risk assessment

- Step 1. Identification of the legal and ethical context
- Step 2. Identification of risks
- Step 3. Risk evaluation
- Step 4. Confection of the MOROVAL map.

Stage 4 Risk treatment

- Step 1. Definition of actions to treat risks.
- Step 2. Elaboration of risk management programs
- Step 3. Execute actions to treat risks.

Stage 5 Monitoring and continuous improvement

- Step 1. Reevaluation of the organization's diagnosis
- Step 2. Improvements.

2.1 Stage 1. Previous Preparation

The first part of the research will prepare the conditions for the implementation of the general procedure. The verification and scope of the premises, the formation of the work team and the group of experts are the main results of the first phase. In the event that the evaluations issued by the management and the workers interviewed of any of the variables corresponding to the premises concerning step 3, obtain results of unacceptable or insufficient, actions must be implemented by the top management in such a way that, in order to continue with the application of the procedure, compliance with the premises can be achieved.

2.2 Stage 2. Organizational Analysis

The diagnosis is structured in two fronts of the organization: the internal analysis, where the main weaknesses and strengths are determined; and the external analysis through the identification of strengths and threats. In this case, it is proposed to focus the strategic diagnosis on the dimensions of sustainability according to the sustainability model selected to be applied in the organization.

Identifying stakeholders is a key process within the project, in which a periodic analysis will be conducted to document relevant information regarding their interests, participation, interdependencies, influence and potential impact on the success of the project.

In order to identify the shared values of the organization, it is necessary to return to the group of experts, which will identify the values that the members of the organization must comply with so that the organization becomes more and more socially responsible and sustainable.

2.3 Stage 3. Risk Evaluation

This phase is based on the analysis of the stakeholder register and the organization's core values, as well as the analysis of compliance with the regulations carried out in phase 2. In this step, the organization's obligations are identified and documented in a systematic manner.

These obligations come from norms and resolutions relevant to the organization and the sector to which it belongs; rules or guidelines issued by regulatory agencies; treaties, agreements and protocols; agreements with community groups; environmental commitments; laws and decrees, among others from the external point of view. From the internal point of view, the policies, codes, standards and other obligations that govern the organization's behaviour.

Identifying the organization's risks considers both the individual risks of its component projects and the sources of overall risk. Participants in risk identification activities may include: project manager, project team members, project risk management specialist (if designated), stakeholders and subject matter experts external to the organization [4].

This process should begin with the analysis of the results of the application of control actions, audits or the application of the "Self-Control Guide" proposed by the Office of the Comptroller General of the Republic; the internal and external control actions file should also be consulted, as well as the knowledge base; an element that will make it possible to identify the risks that in the history of the organization may have a high level of materialization at present. Subsequently, the strategic diagnosis and the register of stakeholders will be used as elements that may be sources of risks in the organization [4].

The classification of risks can be developed in different ways, by sources of risk, area or process of the affected organization or other useful category to determine the areas of the project that are more exposed to the effects of uncertainty. This research proposes to classify them taking into account the dimensions of sustainability, which will make it possible to increase the effectiveness of decision making.

Next, the causes that originate the risks must be identified, which guarantees that the corrective actions to be defined will have the required effectiveness, so that the risk mitigation actions also address the cause that gave rise to the risk.

While describing and recording risks, a consistent format for risk statements should be used to ensure that each risk is easily understood and unambiguous to support the effective development of risk analysis and response.

After applying the risk identification phase, the results of the strategic diagnosis of the organization and especially its objectives are taken into account, which together with the identified risks will allow defining the control objectives, which is a key element to be taken into account in the monitoring and control of risk management.

Identifying risks is an iterative process, because new individual risks may emerge from the projects, as well as from the organization, while the project progresses.

For risk assessment, the possible consequences and probability of occurrence are identified. It is proposed to use a Risk Map, which will provide greater organization in the work.

The main modification made, which constitutes a contribution to previous research, is the evaluation of the consequences in the dimensions of sustainability as shown in Eq. (1) according to the model selected for implementation in the organization [4, 19]:

$$Ic_i = P_i * Imax_i(Eco_i;\ Soc_i;\ Ecol_i) \tag{1}$$

where:

P_i Probability of risk occurrence (i).

$Imax_i$ Maximum value of the impact after assessing the risk (i) in the following dimensions: economic, social, eco-logical, social and environmental.

As a part of the assessment procedure, the technique analysis virtual linguistic of risk is a proposed, based on the virtual linguistic model proposed by [20] and extended by [21, 22]. This technique was applied in the risk management process of a project oriented organization by [23]. The steps for applying this technique are described below:

Define the set of people who are members of the risk management team $E = \{e_1, \ldots, e_k\}$.

Define the set of linguistic labels to be used.

From this step on, you will have a working team that will $E = \{e_1, \ldots, e_k\}$ that analyze the risk exposure of a project by issuing their criteria on the probability of occurrence (P) and the impact (I), where the set of labels $P = I$.

The participants in the analysis express their preferences on the risks identified.

Based on the operator defined by [20] for the virtual linguistic multiplication, which also makes use of the also makes use of [4, 19, 23] $S_\alpha \otimes S_\beta = S_\beta \otimes S_\alpha = S_{\alpha\beta}$.

The collective preference is calculated for each risk being analysed through the Equation of the arithmetic average proposed by $\overline{r_n} = \frac{1}{n}\sum_{i=1}^{n} S^n_{\alpha_i} \otimes S^n_{\beta_i}$.

where: $\overline{r_n}$ is the value of the collective preference for risk n.

n is the amount of risks that are analysed.

$S^n_{\alpha_i}$ is the value of the impact of risk n issued by person i.

$S^n_{\beta_i}$ is the value of the probability of risk n issued by person i.

Then, the MOROVAL map is created, a tool that allows to relate the legal and ethical context (M) to the organization's obligations (O), to these commitments its risks (R), to the risks its control objectives (O) and the possible values that can mitigate or reduce them (VAL).

2.4 Stage 4. Risk Treatment

In order to define the actions to address the risks, it is essential to follow the classification given to the risks previously in the evaluation phase, where they were defined as follows:

- Positive or opportunities.
- Negatives or threats.

Once the strategies to be followed for each of the risks have been defined, a cost–benefit analysis of the measures to be taken for each risk is carried out. This will allow to evaluate the level of effectiveness of these measures, which together with the correct planning of the organization will not allow unnecessary budget execution for the measures taken. Once the correct measures have been selected, the risk management program is updated, assigning one or more measures to be taken, the person responsible for their compliance and the date of compliance.

Based on the MOROVAL map, we propose the elaboration of risk management programs by processes and taking into account the criticality index. This allows integrating in a single program the mitigation and contingency actions, together with other elements of the ethics and compliance programs, such as: codes and values, actions and those responsible for compliance with the actions; in addition to having the risks ordered for a better treatment and monitoring of the most critical risks.

The risk management program is communicated to all employees and actions and behaviours are implemented to mitigate risks in each of the business processes, promoting a culture of ethics and compliance throughout the organization. Workers must be aware of their ethical and legal obligations, especially those related to the processes they perform.

As part of the communication process, training and education activities are carried out for workers on issues related to occupational safety, anticorruption practices, as well as the legal and ethical framework of the sector. The risk management program during this phase becomes a guide for the decision-making process of the organization, which does not replace the risk prevention plan, but becomes a support document that will be the basis for the process of mitigating the risks identified.

2.5 Stage 5. Monitoring and Continuous Improvement

In this phase, the execution of the risk management program is monitored and controlled, changes are made if necessary, the level of compliance with the actions is evaluated and the causes of deviations are analysed and documentary evidence is left of the meetings, where the causes and consequences of deviations in the implementation schedule resulting from the application of the measures of the risk management program are analysed, as well as the possible postponement of tasks as a result of breaches of the planned measures. In the latter case, a re-evaluation of the risks is recommended, taking into account that new risks may arise from this re-evaluation.

This phase allows the feedback of the process, with this objective in mind, improvement plans are drawn up taking into account the results of the previous phases. These plans will address the main problems detected during the implementation and control of the risk management program.

The author recommends applying the diagnostic guide proposed in the initial phase to check the effectiveness of the control actions, as well as their behaviour. This would be the starting point to evaluate how much progress has been made in the implementation of the risk management program in the organization (Phase 2. Organizational analysis).

In this phase also, improvement actions are drawn up based on the results of the risk management program evaluation. The fundamental sources of improvements are the results of the application of the checklist, the values of the quantitative indicators defined and the criticality indexes of risks according to [4].

For the evaluation of the impact of risk management on the sustainability of the project-oriented organization, the construction of a Comprehensive Sustainability Indicator (CSI) is recommended.

This phase concludes with the preparation of an improvement action plan, integrating the proposals of the three phases. This plan must contain, for example: name of the action, person responsible for the action, date of completion, expected impact of the improvement, also indicating to which dimension and process it corresponds.

3 Results

The application of the expert judgment method was carried out by performing the following steps:

- Identification of potential experts.
- Selection of experts.
- Carrying out the consultation with the experts.
- Processing and evaluation of the information obtained.

In order to identify potential experts, professional experience in relation to the object of research, participation in research related to this topic, theoretical mastery of the subject, academic and scientific preparation and experience were taken into account, so that they would be able to offer pertinent assessments and recommendations. For the application of the method, 17 possible experts were identified, of which 13 were selected, 8 of them are Doctors of Science, 4 are Masters and 1 is a software production specialist.

A five-point Likert scale was used to process the survey, where a score between 1 and 5 is given to each item, allowing the interviewees to give an evaluation and assess the degree of agreement with the ideas put forward. With this technique, the percentages of agreement of the experts with each of the possible answers to the questions asked are calculated. A percentage index (PI) is then calculated, which integrates into a single value the acceptance of each approach by the evaluators by means of Eq. (2):

$$\mathrm{IP} = \frac{5(\%\ deT) + 4(\%\ BM) + 3(\%\ AM) + 2(\%\ PM) + 1(\%\ NM)}{5} \tag{2}$$

where: *T* (strongly agree), *BM* (good measure), *AM* (some measure), *PM* (little measure) and *NM* (No measure).

Figure 1 shows a graph with the percentage index of the experts in each of the questions, which shows that the PI in all cases is greater than 90.77. This evidences the high valuation by the experts on the clarity and synthesis of the proposal; possibility of its use in project-oriented organizations and adherence to international consensus best practices.

The questions used in the instrument are listed below:

1. The proposed procedure defines activities that can be integrated into the phases of the development process of project-oriented organizations.
2. There is coherence and interrelation between the elements of the procedure.
3. The proposal is based on the existing theoretical references for risk management.
4. The proposed procedure guarantees the monitoring and control of risks.

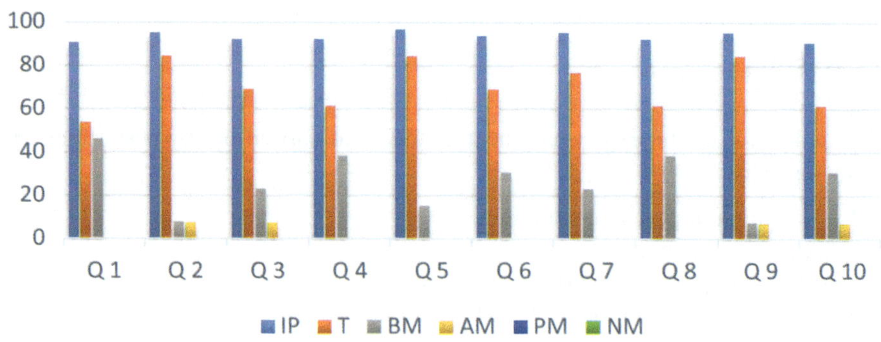

Fig. 1 Expert appraisal of the proposed procedure

5. The implementation of the phases proposed by the procedure will reduce the work of the project members, facilitating the monitoring of risks.
6. Applicability and flexibility of the procedure to the risk management models and to the continuous changes in the organizational environment.
7. The assessment of risks in the dimensions of sustainability made in the procedure is accurate.
8. The treatment given to the organization's shared values from the risk management process is considered pertinent.
9. The strategic approach given to the risk management process is correct.
10. The proposed procedure has a process approach.

3.1 User Satisfaction with the Proposed Procedure

In order to determine the personal and group satisfaction index of the specialists, the Iadov Technique was applied. The analysis of the answers was carried out by means of the "Iadov's logical table", which consists of five questions, three closed and two open; it expresses a list of the possible answers to the three closed questions.

The number resulting from the interrelation of the three questions indicates the position of each one on the following satisfaction scale:

1. Clear satisfaction
2. More satisfied than unsatisfied
3. Not defined
4. More unsatisfied than satisfied
5. Clearly unsatisfied
6. Contradictory.

The level of satisfaction is need to calculate the Group Satisfaction Index (GSI) by means of Eq. (3):

$$ISG = \frac{A(+1) + B(+0,5) + C(0) + D(-0,5) + E(-1)}{N} \tag{3}$$

To measure the level of satisfaction, a sample of 12 users of the CIDI center was taken, representing five of the organization's seven projects, which represents 71.43% of the organization (see Fig. 2).

The ISG according to the above formula was 0.91. Placing this result in the group categories: Dissatisfaction: from (– 1) to (– 0.5), Contradictory: from (– 0.49) to (+ 0.49) and Satisfaction: from (+ 0.5) to (1) shows that the specialists are satisfied with the proposed procedure.

The Iadov technique also contemplates two complementary open questions. These allow us to obtain an evaluation of the usefulness of the proposal and suggestions for the present and future research. The aspects identified according to the evaluation were as follows:

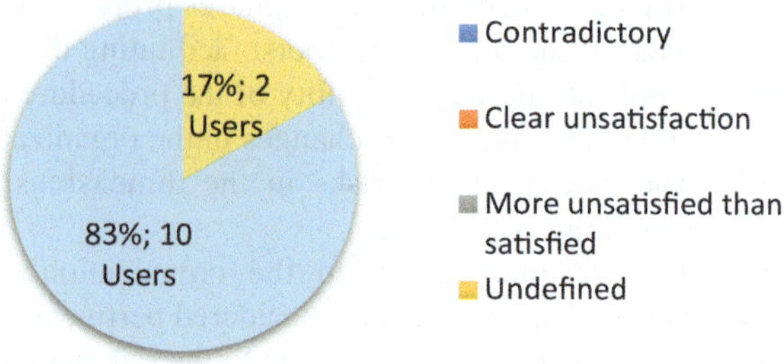

Fig. 2 Results of user satisfaction assessment

In general, the interviewees rated the proposal positively, highlighting its usefulness, flexibility and ease to understand. Also, the high level of traceability of risks is noted.

3.2 Case Study

Once the risks and their associated control objectives have been identified, the risks are evaluated using the virtual linguistic analysis technique. To do this, we start by forming a group of experts composed by 7 members of the center, who are asked for their criteria on the probability (P) and impact on the dimensions of sustainability, according to the selected model. Subsequently, the maximum probability value to be used to prioritize the risks (Imax) is selected; where the set of labels $P = I$.

Then, each expert expresses his or her preferences about P and I and the collective preferences are calculated. In the virtual linguistic analysis of risks, individual criteria are used as a starting point, a consensus is reached and a collective criterion is determined. Once the collective preferences have been obtained, the risks are ranked. The ordering of the previous example is as follows: R12; R04; R03; R13; R09; R21; R11; R14; R20; R01; R10; R22; R24; R16; R08; R15; R17; R18; R05; R23; R06; R07. Next, the Pareto technique was used see Table 1.

Pareto technique is a simple but powerful tool that allows us to visually identify in a single review the minorities of vital risks to which it is important to pay attention and thus use all the necessary resources to carry out an improvement action without wasting efforts, since the analysis discards the trivial majorities.

In identifying the shared values of the organization, it is evident that the University of Computer Sciences has a group of shared values, which will be used in this research; taking into account that the project-oriented organization in question belongs to it. Of these, when assessing the impact that the risks may have, it can be seen that the values most affected are: honesty, responsibility, hard work, modesty and creativity.

Then an analytical hierarchy was structured with three dimensions (economic, social and ecological), to which 10 control objectives were associated. Then, the

Table 1 List of vital risks with their possible manifestations

Code	Risk	Possible manifestations
R05	Loss of information	Lack of connectivity
R06	Loss of human lives	Fire occurrence
		Work accidents
R07	Deterioration of health due to infectious diseases	Affectation of workers' health
R08	Ineffective internal control system	Incorrect preparation of the annual control action plan
		Lack of knowledge of the workers and the board of directors of the results of the control actions
		Non-compliance with laws, regulations, Rector's resolutions and instructions from the Vice
R15	Shortages and surpluses of Tangible Fixed Assets (TFA)	Rector's resolutions and instructions from Vice Rector's Offices or indications from Central Management
R17	Shortages and surpluses of tools and equipment	Inaccuracy of the AFT account
R18	Insufficient research results	Inaccuracy of the records of norm 40
		Low rate of scientific publications
R23	Breakage of information storage devices	Low impact of research

24 identified risks were linked to the objectives, to which the previously defined indicators were associated.

Subsequently, Saaty's AHP method was applied to calculate the weights of each of the elements that make up the analytical hierarchy defined. The experts assigned the weights by consensus, through paired comparisons of each of the dimensions, control objectives and indicators of the hierarchy.

Then, the impact values are calculated for each risk, control objective, sustainability dimensions and finally the organization's sustainability coefficient, which for the closing of the month of September 2019 offers a value of 0.52. This value evaluated on the scale offers as a result a Medium sustainability level ($0.47 \leq Ne \leq 0.62$).

Next, a reassessment is made of the sustainability level of the organization under study at the end of September 2020 using the same indicators, to verify the contribution to sustainability of the GRICSOOP procedure. In this case it is found that the sustainability coefficient is 0.61; which results in a Medium level of sustainability ($0.47 \leq Ne \leq 0.62$).

4 Conclusions

A procedure was defined for risk management as a contribution to the sustainability of project-oriented organizations, consisting of five phases: Prior preparation, Organizational analysis, Risk assessment, Treatment and Monitoring and continuous improvement, based on the results obtained from the construction of the theoretical and referential framework of the research.

The procedure combined with the techniques and methods provide an instrument that allows a complex process to be carried out in a simple way and that can be fully or partially applied in project-oriented organizations as a contribution to sustainability from the risk management process.

From the results obtained after the application of the defined validation methods, it is concluded that the proposal was satisfactorily valued by experts and accepted by users.

The case study evidenced the possibility of applying the procedure, the results of which were favourable as they contributed to the sustainability of the project-oriented organization.

References

1. PMI.: The standard for Project Management and a guide to the Project Management Body of Knowledge Seventh Edition. Ed.: Project Management Institute, ISBN: 978-1-62825-664-2, pp. 370 (2021)
2. Porras, N.: A look at sustainability in project management. Int. J. Good Conscience, **12**(3) (2017)
3. Chawla, V., Chanda, A., Angra, S., Chawla, G.: The sustainable project management: A review and future possibilities. J. Project Manage. **3**(3), 157–170 (2018). https://doi.org/10.5267/j.jpm.2018.2.001
4. Sánchez, Y.A., Soler, J.A., Delgado, F.M.: Procedimiento para determinar el impacto de la gestión de riesgos en la sostenibilidad de las organizaciones. Dirección y Organización **73**, 39–49 (2021). https://doi.org/10.37610/dyo.v0i73.591
5. Bolaño, Y.: Modelo de dirección estratégica basado en la administración de riesgos para la integración del sistema de dirección de la empresa. Universidad Tecnológica de La Habana José Antonio Echeverría, La Habana, Cuba, Tesis doctoral (2014)
6. Giannakis, P.: Supply chain sustainability: A risk management approach. Int. J. Prod. Econ. **171**(4), 455–470 (2016). https://doi.org/10.1016/j.ijpe.2015.06.032
7. Wijethilake, C., Lama, T.: Sustainability core values and sustainability risk management: Moderating effects of top management commitment and stakeholder pressure. Bus. Strateg. Environ. **28**(1), 143–154 (2019). https://doi.org/10.1002/bse.2245
8. Schulte, J., Hallstedt, S.I.: Company risk management in light of the sustainability transition. Sustainability **10**(11), 4137 (2018). https://doi.org/10.3390/su10114137
9. Gutiérrez, L.M.: Proceso base de gestión de riesgos para las pequeñas y medianas empresas de desarrollo de software. Universidad de las Ciencias Informáticas, La Habana, Cuba, Tesis de maestría (2016)
10. Contraloría General de la República, Resolución 60. Normas del Control Interno, Gaceta Oficial de la República, vol. 109 (2011)

11. ISO.: ISO 10006 Quality management systems—Guidelines for quality management in projects. (2009) [Online] Available: https://www.iso.org/obp/ui/#iso:std:iso:10006:ed-2:v1:es [Accessed: 12 May 2021]
12. ISO.: ISO 31000 AENOR: Standard UNE–ISO (2018). [Online] Available: http://www.aenor.es/aenor [Accessed: 22 Apr 2021]
13. Chatzitheodorou, K., Tsalis, T.A., Tsagarakis, K.P., Evangelos, G., Ioannis, N.: A new practical methodology for the banking sector to assess corporate sustainability risks with an application in the energy sector. Sustain. Prod. Consumption **27**, 1473–1487 (2021). https://doi.org/10.1016/j.spc.2021.03.005
14. Trubetskaya, A., Horan, W., Conheady, P., Stockil, K., Merritt, S., Moore, S.: A methodology for assessing and monitoring risk in the industrial wastewater sector. Water Resour. Ind. **25**. https://doi.org/10.1016/j.wri.2021.100146 (2021)
15. Hegde, J., Rokseth, B.: Applications of machine learning methods for engineering risk assessment–A review. Safety Sci. **122**. https://doi.org/10.1016/j.ssci.2019.09.015 (2020)
16. Chowdhury, M.M., Quaddus, M.A.: Supply chain sustainability practices and governance for mitigating sustainability risk and improving market performance: A dynamic capability perspective. J. Cleaner Prod. 278. https://doi.org/10.1016/j.jclepro.2020.123521 (2021)
17. Fan, D., Lo, C.K., Zhou, Y.: Sustainability risk in supply bases: The role of complexity and coupling. Transp. Res. Part E: Logistics Transp. Rev. **145**. https://doi.org/10.1016/j.tre.2020.102175 (2021)
18. Modrak, V., Soltysova, Z., Onofrejova, D.: Complexity assessment of assembly supply chains from the sustainability viewpoint. Sustainability **11**(24). https://doi.org/10.3390/su11247156 (2019)
19. Aguilera, Y., Plasencia, J., Marrero, F.: Procedimiento para evaluar los riesgos de cumplimiento normativo en organizaciones orientadas a proyectos. Foll. Gerenciales **XXIII**(2), 93–104 (2019)
20. Xu, Z.: Consistency of interval fuzzy preference relations in group decision making. Appl. Soft Comput. **11**(5), 3898–3909 (2011). https://doi.org/10.1016/j.asoc.2011.01.019
21. Xu, Z., Xia, M.: Distance and similarity measures for hesitant fuzzy sets. Inf. Sci. **181**(11), 2128–2138 (2011). https://doi.org/10.1016/j.ins.2011.01.028
22. Xu, Z., Xia, M.: Identifying and eliminating dominated alternatives in multi-attribute decision making with intuitionistic fuzzy information. Appl. Soft Comput. **12**(4), 1451–1456 (2012). https://doi.org/10.1016/j.asoc.2011.08.046
23. Pérez, O.: Proceso para gestionar riesgos en los proyectos de desarrollo de software del Centro de Informatización Universitaria. Universidad de las Ciencias Informáticas, La Habana, Cuba, Tesis de maestría (2013)

Chapter 10
New Extensions of Fuzzy Cognitive Maps for Sequential Multistage Decision-Making Problems: Application in Project Management

S. Salah Hasan Al-subhi, Iliana Pérez Pupo, Pedro Y. Piñero Pérez, Gaafar Sadeq S. Mahdi, and Nicolas Villavicencio Bermúdez

Abstract From a systematic review on the use of FCMs and their extensions, it is identified that there are shortcomings in the works reported in the consulted bibliography regarding the treatment of indeterminacy and the solution of multi-stage sequential decision-making problems. In this paper, two new extensions of Fuzzy Cognitive Maps (FCMs) for multistage sequential decision-making problems are proposed. The Multistage Sequential Triangular Neutrosophic Cognitive Map (MSTrNCM) combines neutrosophic theory with computer with words techniques to represent the map's relationships and the inference process. This extension improves the modeling of indeterminacy and the interpretability of results. The second map, which is called Neutrosophic Cognitive Map based on linguistic Data Summarization (NCM-LDS), uses linguistic summaries to represent the map's relations and to carry out the inference process. One of the main advantages of this extension is that it facilitates the maps construction and interpretability. Furthermore, the suggested extensions are applied as a decision-making support tool for projects evaluation using a dataset with 1011 projects records. In experimental analysis, the two proposed extensions MSTrNCM and NCM_LDS report better results than the traditional FCM and NCM_Indeterminacy reported in bibliography.

S. S. H. Al-subhi
Solarplex Electyrical and Electronics Contracting Co. Combined Security Group Co, Kuwait, Kuwait

I. Pérez Pupo (✉) · P. Y. Piñero Pérez
Grupo de Investigaciones en Gestión de Proyectos, Universidad de las Ciencias Informáticas, Carretera a San Antonio km 2 ½, 54830 La Habana, Cuba
e-mail: iperez@uci.cu

G. S. S. Mahdi
Business Intelligence and Consulting Group, Business Redmine SA, Aden, Yemen

N. Villavicencio Bermúdez
Facultad de Especialidades Empresariales, Universidad Católica Santiago de Guayaquil, Guayaquil, Ecuador
e-mail: nicolas.villavicencio@cu.ucsg.edu.ec

P. Y. Piñero Pérez et al. (eds.), *Artificial Intelligence in Project Management and Making Decisions*, Studies in Computational Intelligence 1035,
https://doi.org/10.1007/978-3-030-97269-1_10

Keywords Fuzzy cognitive maps · Linguistic data summarization · Project management

1 Introduction

Nowadays, the number of failed and renegotiated projects worldwide is still high [1]. This situation has negative social and economic impacts. One of the fundamental causes of the project failures are the decision-making errors during the evaluation process [2]. In this work, authors present the project evaluation as a multi-stage sequential decision-making problem. This problem constitutes a non-structured decision-making problem, such that the solution D is the aggregation of the non-structured decision-making subproblem k solutions. There is a sequential dependence between the sub-problems k and k−1, such that the sub-problem S_i solution represented by decision D_i depends on the sub-problem S_i-1 solution represented by decision D_i-1 [3]. In this context, each subproblem is modeled as a stage of the general decision-making problem, this is the reason for the use of the term multistage.

Projects evaluation process consist on a set of weekly cuts that occurs during the project cycle life. In each cut, experts carry out sequentially, the processes of: project evaluation (diagnosis), definition of measures to correct project deviations (decisions) and prediction of what will happen according to the decisions made in the previous stage. During this process, projects managers consider qualitative and quantitative indicators. Besides, this process is characterized by the presence of uncertainty, vagueness, indeterminacy and imprecision in data [4]. For this reason, many authors employ different artificial intelligence techniques to be used as decision-making support tools. One of the most suitable artificial intelligence tools used in this context are Fuzzy Cognitive Maps (FCMs) because of the interpretability of its results [5].

In this investigation, we carried out a systematic review about the FCMs evolution during the period 1986–2021 and its impact in decision-making problems. In the systematic review authors selected 150 publications and were analyzed in depth. The result of this analysis is presented below.

In order to increase the map interpretability, in the literature there are some efforts focused on the use of computing with words (CWW) in FCMs [6–8]. In general, linguistic FCMs do not consider the indeterminacy relationship.

Fuzzy gray cognitive maps (FGCMs) [9] combine FCMs and gray systems theory. They improve the treatment of uncertainty associated with the existence of different ways of representing the map's relationships. These maps represent the indeterminacy as a black relationship (relationships with unknown intensity), but they are not able to handle in a differentiated way the indeterminacy and falsity levels.

Evidential cognitive maps (ECMs) are the combination of FCMs and evidential theory [10]. They represent the indeterminacy as a calculable value from the degrees of positive relationships, negative relationships and the non-existence of relationship, beside indeterminacy cannot be expressed directly by experts. In the inference process, these maps combine the degrees represented in the experts' preferences and obtain an interval; as a consequence, it is not possible to know the indeterminacy degree in the map's outputs.

The belief degree-distributed fuzzy cognitive maps (BDD-FCMs) [11] allow experts to express their preferences about the relationships among concepts through numerical values, linguistic terms, intervals or belief structures composed of more of a linguistic term. The inference process is carried out after transforming belief structures to numerical values. In this process, the degree of indeterminacy is lost, which is not reflected in the final decisions.

The intuitionistic fuzzy cognitive maps (iFCMs) [12] extend the traditional FCMs considering the of hesitation degree $\pi_A(x)$ of experts at the time of determining the weights of the map's relationships. In the consulted bibliography, only 8% of the identified works use this type of maps which have been gradually replaced by the neutrosophic cognitive maps NCMs_Indeterminacy.

Neutrosophic cognitive maps (NCM_Indeterminacy) [13, 14] combine the neutrosophic theory [15] with the FCMs. In these maps the relationships among concepts are represented by values in the interval $[-1, 1]$ or by symbol I [16]. However, the use of the symbol I during the inference does not allow to quantify the indeterminacy levels and causes loss of information, affecting map efficacy regarding prediction capability and results interpretability. This situation denotes the need to develop new extensions that modify the traditional inference mechanisms using mathematical models established in the neutrosophic theory.

Summarizing the foregoing, it is identified that, in the development of the FCMs and their extensions for the decision-making problems in project management, the following limitations persist:

There are not FCMs applications in project management that solve the diagnosis, decision and prediction processes in an integrated way. In this sense, authors such as Amirkhani, Papageorgiou propose the need to develop new extensions of FCMs that integrate these three processes as a single decision-making problem [17, 18].

Insufficiencies of FCMs topologies to analyze and model multistage sequential decision-making problems [18] that appear during projects cut evaluation.

There are insufficiencies in the FCMs extensions in the treatment of indeterminacy [3, 19]. Besides, there is a loss of information in the neutrosophic cognitive maps reported in the consulted literature, affecting the maps prediction capacity [20].

This paper aims at developing new extensions of FCMs improving their efficacy in solving multistage sequential decision-making problems by the combination of neutrosophic theory and soft computing techniques. Section 2 and 3 describe the two new extensions of FCMs for multistage sequential decision-making problems. Section 4 stands for experimental results. Section 5 is devoted to the paper conclusions.

2 Multistage Sequential Triangular Neutrosophic Cognitive Map (MSTrNCM)

The first extension is the Multistage Sequential Triangular Neutrosophic Cognitive Map (MSTrNCM). This extension overcomes the limitations of neutrosophic cognitive maps reported in literature (NCM_Indeterminacy) regarding the treatment of indeterminacy. MSTrNCM is based on triangular neutrosophic numbers (TrNNs), see Definition 1

Definition 1 [21]: a TrNNs $\tilde{A} = \langle(a_1, a_2, a_3); \propto_{\tilde{A}}, \theta_{\tilde{A}}, \beta_{\tilde{A}}\rangle$represents the set of real values in the interval $[a_1, a_3]$ whose most probable value is a2 and it is characterized by $T_{\tilde{A}}(x), I_{\tilde{A}}(x)\text{y}F_{\tilde{A}}(x)$ that represent the certainty, indeterminacy and falsity degrees, respectively, of each specific value. The values of $T_{\tilde{A}}$: IR $\in \left[0, \propto_{\tilde{A}}\right]$, $I_{\tilde{A}}$: IR $\in \left[\theta_{\tilde{A}}, 1\right]$ and$F_{\tilde{A}}$: IR $\in \left[\beta_{\tilde{A}}, 1\right]$ can be calculated by the triangular functions represented by (a_1,a_2,a_3), where $a_1 \leq a_2 \leq a_3$ and the values of$\propto_{\tilde{A}}$,$\theta_{\tilde{A}}$, $\beta_{\tilde{A}} \in [0, 1]$ represent the maximum degree of membership (certainty), the minimum degree of indeterminacy and the minimum degree of non-membership (falsity), see Fig. 1.

In the definition of triangular neutrosophic numbers which are assigned to the map's relationships, the values of $\propto_{\tilde{A}}$, $\theta_{\tilde{A}}$and$\beta_{\tilde{A}}$ are determined by experts.

2.1 *Representation of the Relationships Among Concepts and Map Construction*

The representation of map's relationships combines triangular neutrosophic numbers (TrNNs) [21] and the semantic model of CWW. The proposed approach differs

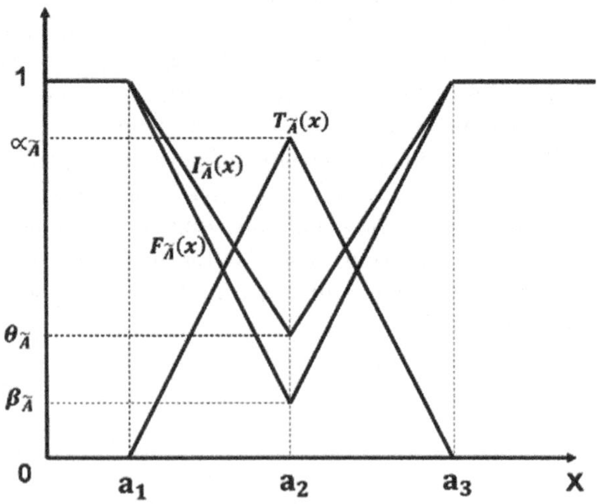

Fig. 1 Example of triangular functions associated with a triangular neutrosophic number $\tilde{A}$

significantly from the NCM_Indeterminacy identified in literature, which use the symbol I to represent indeterminacy.

For the MSTrNCM construction, the participation of several experts in the domain is recommended to reduce the bias that occurs when a single expert participates. Algorithm 1 explains the steps for the map construction.

Algorithm 1 MSTrNCM construction process.

1. *Definition of the set of stages S such that* $| S | = k$*, which represent the multistage sequential decision-making problem.*
2. *Selection of the set of experts with knowledge in the application domain E such that* $|E| = m$ *and* e_j *represents the j-th expert.*
3. *Experts propose the set of concepts C organized by stages such that* $C^{(Si)}$ *brings together the set of objects associated with the* S_i *stage.*
4. *Definition of the set of linguistic terms L to describe the relationships among concepts.*
5. *Definition of the triangular neutrosophic numbers associated with the linguistic terms.*
6. *Each expert* e_j *builds a map* $M_{e_j} = \{M_{S1}^{e_j}, M_{S2}^{e_j}, \ldots, M_{Sk}^{e_j}\}$ *expressing the relationships among concepts in C in each of the S stages of the problem in question, using the set of linguistic terms L defined above.*
7. *Construction of the unified* M_{agg} *map from the aggregation of experts' preferences expressed in the individual maps* M_{e_j} *where* $e_j \in E$.
8. *Return the* M_{agg} *map made up of the unified criteria of all the experts with the structure* $M_{agg} = \{M_{S_1}, M_{S_2}, \ldots, M_{S_k}\}$ *such that* M_{S_i} *represents the aggregated map of the E experts with respect to the stage* S_i.

In step 1, the different stages of the problem in question are identified. In step 2, domain's experts are selected for the map construction. In step 3, experts propose the set of concepts C that will be represented in the map. In step 4, the set of linguistic terms *L* is defined Table 1.

In step 5, a triangular neutrosophic number (TrNNs) is associated to each linguistic term, see Definition 1 and Table 1, which allows aggregating the experts' preferences using computing with word techniques. In step 6 of the algorithm, each expert builds his map based on his individual beliefs, expressing the relationships among concepts, using the linguistic terms *L* established in the previous steps.

Subsequently, in step 7, we proceed to the construction of the unified map Magg by the aggregation of experts' preferences, expressed in the individual maps M_{e_j} where ej ∈ E. For the preferences aggregation, neutrosophic aggregation operators [21] are applied in combination with a semantic model of computation with words based on the extension principle. The operations among single-valued neutrosophic numbers are defined as shown in [21], which constitute the basis of both the construction and inference processes of the MSTrNCM. Finally, the aggregated map Magg is returned as a single map that will be used in the inference process.

Table 1 Assignation example of TrNNs to linguistic terms

Linguistic terms set L	Triangular neutrosophic numbers
Negatively very high (−VH)	$\langle(-1, -1, -0.8); 0.99, 0.02, 0.01\rangle$
Negatively high (−H)	$\langle(-1, -0.8, -0.6); 0.99, 0.03, 0.02\rangle$
Negatively medium (−M)	$\langle(-0.8, -0.6, -0.4); 0.99, 0.03, 0.02\rangle$
Negatively low (−L)	$\langle(-0.6, -0.4, -0.2); 0.99, 0.04, 0.03\rangle$
Negatively very low (−VL)	$\langle(-0.4, -0.2, 0); 0.99; 0.04, 0.03\rangle$
Indeterminacy (IND)	$\langle(-0.2, 0, 0.2); 0.5, 0.99, 0.5\rangle$
Very low (VL)	$\langle(0, 0.2, 0.4); 0.99, 0.04, 0.03\rangle$
Low (L)	$\langle(0.2, 0.4, 0.6); 0.99, 0.04, 0.03\rangle$
Medium (M)	$\langle(0.4, 0.6, 0.8); 0.99, 0.03, 0.02\rangle$
High (H)	$\langle(0.6, 0.8, 1); 0.99, 0.03, 0.02\rangle$
Very high (VH)	$\langle(0.8, 1, 1); 0.99, 0.02, 0.01\rangle$

2.2 Map Inference and Activation Function

The inference process as similar as traditional FCMs. Having an initial vector, which represents a set of concepts with input values, an iterative process of mutual activation among concepts occurs until a certain stop condition, see Algorithm 2.

Algorithm 2. MSTrNCM inference process

Inputs

I_0*: initial vector to be used as inputs to the inference process.*

M_{agg}*: aggregated map in its different S stages* $M_{agg} = \{M_{S_1}, M_{S_2}, \ldots M_{S_k}\}$.

Initial vector loading $I_0 = (b_1, b_2, \ldots, b_m)$

Transformation of each real value of the input vector I_0 *into a triangular neutrosophic number*

$R_{S_0} = I_0$

From $j = 1$ *until n do // for each* S_j *stage*

$t = 0$

$B^t_{S_j} = do\_initial\_activation(R_{S_{j-1}}, M_{S_j})$

While stop_condition is not met(t*,* $B^t_{S_j}$*)*

  For each $\tilde{b}^t_i \in B^t_{S_j}$ *do*

8. $\tilde{b}_i^{t+1} = f\left(\tilde{b}_i^t \oplus \sum_{i=1}^{n} \tilde{w}_{ji} \otimes \tilde{b}_j^t\right): \tilde{w}_{ji} \in M_{S_j}, \tilde{b}_i^t, \tilde{b}_j^t \in B_{S_j}^t, \tilde{b}_i^{t+1} \in B_{S_j}^{t+1}$
9. *End for*
10. $t = t + 1$
11. *End while*
12. *R_{S_j} = order_concepts($B_{S_j}^t$)*
13. *End from*
14. *Return ($R_{S_1}, R_{S_2}, \ldots, R_{S_k}$)*

In step 1, a set of input values to the map is determined with the form $I_0 = (b_1, b_2, \ldots, b_m)$, the nature of these values depends on the application domain and they represent the initial activation values of concepts. Then, in step 2, the input values vector is loaded to the inference process ($b_1, b_2, \ldots, b_m$). Each of the numerical values are transformed into triangular neutrosophic numbers TrNNs, see Definition 1 in such a way that the value b_i is transformed into the number $\langle(b_i - u, b_i, b_i + u); T_{b_i}, I_{b_i}, F_{b_i}\rangle$, symmetric with respect to b_i. The parameters $u, T_{b_i}, I_{b_i}, F_{b_i}$ are established in a prescriptive way taking into account the application domain and the certainty of the input data. The parameters $u, T_{b_i}, I_{b_i}, F_{b_i}$ represent a threshold value and the values of certainty, indeterminacy and falsity, respectively.

In steps 4 to 12, each of the stages are evaluated sequentially, the evaluation process of each stage is divided into two moments. At first, based on the concepts with a higher degree of activation of the previous stage, an input vector is determined and the initial values of the concepts of the stage in question are calculated. In a second moment, an iterative process for calculating the concepts activation levels is carried out, which constitutes a competitive process in which there are exciting relationships and inhibitory relationships among the concepts. The competitive process runs continuously until a predetermined stop condition is met. In this process, the activation function used by traditional FCMs is extended to triangular neutrosophic numbers.

In step 9, the activation level of each concept is calculated. The normalization function used is the hyperbolic tangent [22]. In step 13, the outputs of each stage are identified; the concepts are ordered using the Score function defined in [21] and the concept that reach the highest level of activation are considered as part of the stage outputs. These outputs are inputs to the next stage. In the output nodes, the certainty, indeterminacy and falsity degrees appear together with each decision. Decisions with higher values of falsity or indeterminacy are considered to be of greater risk.

The computational complexity of Algorithm 2 is O(ku) where k is the number of stages and u is the maximum number of iterations established for the inference process. The main advantages of the MSTrNCM extension are focused on improving the interpretability of the map's results and the treatment of indeterminacy, offering greater degrees of freedom to represent experts' preferences. Nevertheless, offering higher degrees of freedom for experts introduces additional complexity in the map construction process. This limitation constituted the main motivation for the development of the second extension NCM_LDS proposed in this paper.

3 Neutrosophic Cognitive Map Based on Linguistic Data Summarization

The extension NCM_LDS is the neutrosophic cognitive map based on linguistic data summaries, whose relationships are expressed by linguistic summaries created automatically or with the help of experts through an active learning process.

3.1 Representation of the Relationships Among Concepts and Map Construction

In general, relationships among the map's concepts are defined by linguistic summaries based on the protoform 5 defined in [23], see 1.13.

Example 1 *32: Some records indicate that in 100.0% of the times, projects with "high quality of data" they have "defects in tasks control".*

Algorithm 3 explains in detail how the NCM_LDS is constructed tic summaries.

Algorithm 3 NCM_LDS construction

Inputs

D: database formed by $Y = \{y_1, \ldots, y_n\}$, such that each record y_i represents an object.

1. *Definition of the set of stages S, such that $|S| = k$, which make up the multistage sequential decision-making problem.*
2. *Selection of the set of experts with knowledge in the application domain E, such that $|E| = m$ y e_i represents the i_{th} expert.*
3. *Experts propose the set of concepts C organized by stages such that C^{S_i} brings together the set of objects associated with the stage S_j.*
4. *Generating of linguistic summaries using records in D and linguistic data summarization algorithms.*
5. *Refinement of linguistic summaries with active learning techniques.*
6. *Construction of the map M_{LDS} by organizing the generated linguistic summaries in the S stages.*
7. *Return the map M_{LDS} made up of the generated linguistic summaries.*

In step 1, the different stages of the problem in question are identified, then in step 2, experts in the domain under study are selected in order to participate in the NCM_LDS construction and in step 3 experts propose the map's concepts. Steps from 4 is associated to the generation of linguistic summaries and we recommend the use of the algorithms proposed in [23] for the construction of linguistic summaries since these algorithms registered better results than other algorithms reported in the consulted literature. Later, in step 5, an active learning process occurs where the generated linguistic summaries are validated and improved. Experts participate in this process and by consensus, a unique set of linguistic summaries is formed that represent the relationships among the concepts. Finally, in step 6, the NCM_LDS map is constructed by organizing linguistic summaries generated in the previous steps in the S stages. The constructed map has a node topology, where summaries are included within each node. Each node is made up of the following elements:

- Concept that represents the node.
- The set of linguistic summaries (SUMM) used for the evaluation of the concept; each node includes all those summaries that contain in the summarizer the concept that this node represents.
- The activation function of a node g(SUMM) that evaluates the linguistic summaries included in the node and returns the activation level of node $\widetilde{a_p} = (a_i, svnn_{a_i})$, where a_i represents the activation level of the i-th concept and $svnn_{a_i}$ represents a single-valued neutrosophic number (SVNN), see Definition 2.

Definition 2 [24]: Let X be a universe of discourse with a generic element x. A single-valued neutrosophic set (SVNS) A in X is $A = \{x, T_A(x), I_A(x), F_A(x) \mid x \in X\}$, where T_A is the truth-membership function, I_A is the indeterminacy-membership function and F_A is the falsity-membership function. $T_A(x), I_A(x)$ and $F_A(x) \in [0, 1]$ and $0 \leq T_A(x) + I_A(x) + F_A(x) \leq 3$. For convenience, the triple $(T_A(x), I_A(x), F_A(x))$ is called single-valued neutrosophic number SVNN. The operations among SVNNs are defined as shown below [24]. Let $a = (T_1, I_1, F_1)$ and $b = (T_2, I_2, F_2)$ be two SVNNs.

$$a \oplus b = (T_1 + T_2 - T_1.T_2, I_1 \cdot I_2, F_1 \cdot F_2) \tag{1}$$

$$a \otimes b = (T_1 \cdot T_2, I_1 + I_2 - I_1 \cdot I_2, F_1 + F_2 - F_1 \cdot F_2) \tag{2}$$

$$k \cdot a = \left(1 - (1 - T_1)^k, (I_1)^k, (F_1)^k\right) \text{ such that } k > 0, kIR \tag{3}$$

$$a^k = \left((T_1)^k, 1 - (1 - I_1)^k, 1 - (1 - F_1)^k\right), \text{ such that } k > 0, kIR \tag{4}$$

Definition 3 Let $a = (T, I, F)$ be a single-valued neutrosophic number SVNN, then the degree of cosine similarity $S(a)$ of a is defined as follows:

$$S(a) = \frac{T}{\sqrt{T^2 + I^2 + F^2}} \tag{5}$$

This equation measures the cosine similarity between $a = (T, I, F)$ and the ideal solution (1, 0, 0) for the comparison of SVNNs.

Definition 4 : Let $a = (T_1, I_1, F_1)$ and $b = (T_2, I_2, F_2)$ be two SVNNs, if $S(a) \leq S(b)$, then $a \leq b$.

3.2 Inference Process of NCM_LDS

The essence of NCM_LDS inference is the linguistic summaries evaluation process. Linguistic summaries have the form: "*QRy's* are *S*", where *R* is a set of filters, *S* is a set of summarizers, *Q* is a quantifier and each summary contain a set of quality values *T*. The inference process is described in Algorithm 4.

Algorithm 4 Inference process of NCM_LDS

u: maximum number of iterations
n: stages
M_{LDS}*: constructed map with the help of experts*
list_t_values: set of quality indicators of linguistic summaries T

1- Load the map with the different stages $M_{LDS} = \{M_{S_1}, M_{S_2}, \ldots M_{S_k}\}$.
2- Initial vector loading $R_{S_0} = (b_1, b_2, \ldots, b_m)$.
3- From $j = 1$ to k do // for each stage S_j
4- $\mathrm{B}_{\mathrm{S_j}}^{0}$ = do_evaluate_layer($R_{S_{j-1}}$, $M_{S_j}[: input\_layer]$, t, list_t_values)
5- If connection_method = = :low_connectivity
6- Then Layer = $M_{S_j}[: competence\_layer]$
7- If not Layer = $M_{S_j}[: competence\_layer] + M_{S_j}[: merge\_layer]$
8- End if
9- $t = 0$
10- While not stop_condition(t, u, $\mathrm{B}_{\mathrm{S_j}}^{\mathrm{t}}$)
11- $B_{S_j}^{t+1}$ = do_evaluate_layer($B_{S_j}^{t}$, Layer, t)
12- $t = t + 1$
13- End while
14- R_{S_j} = order_concepts($B_{S_j}^{t}$)
15- End from
16- Return (R_{S_1}, R_{S_2}, ..., R_{S_k})

This algorithm allows the evaluation of each stage of the defined multistage sequential problem. In step 2, the input values vector is loaded to the inference process $(b_1, b_2, \ldots, b_{\mathrm{m}})$. The input numerical values are transformed into SVNNs, see Definition 2, in such a way that the value b_i is transformed into the number $\langle b_i; T_{b_i}, I_{b_i}, F_{b_i} \rangle$. The parameters T_{b_i}, I_{b_i}, F_{b_i} are set in a prescriptive way taking into account the application domain and the certainty of the input data. The inference process occurs between steps 3 to 15 sequentially considering the order of the different stages, where the outputs of one stage constitute the input for the next.

The inference process at each stage consists of two fundamental moments. The first moment occurs in step 4, where a preliminary evaluation of the stage is performed using the summaries located in the initialization layer. In a second moment, represented by steps 5 to 13, a competitive process occurs among the stage's concepts until a certain stop condition is met.

An input to the second moment of inference is the map loading type for the inference process, this occurs between steps 5 and 8 with two possible options.

In the first option that we call "low_connectivity", the map is formed considering only the summaries that were initially located in the internal competition layer. In the second alternative called "high_connectivity", the map is formed by summaries whose filters combine concepts from both inputs and competitive layers.

In steps 10 to 13, occurs an iterative competence between concepts until a certain stop condition is met. In the last step of Algorithm 4, concepts activation values (R_{S_1}, $R_{S_2}, \ldots, R_{S_k}$) in the different stages of the multistage sequential problem are returned.

The computational complexity of Algorithm 4 is O(kupL) where k is the number of stages, u is the number of maximum iterations established in the inference process, p is the number of linguistic summaries and L is maximum linguistic summaries length.

4 Validation and Results Analysis

In the comparison process, authors used the database "Decision making in project management" (200821_gp_toma_decisiones) from Project Management Research repository [25] made up of 1011 records of project cuts. For the validation of this investigation, authors build the maps MSTrNCM and NCM_LDS. Then, they are compared with traditional mFCMs [18] and NCM_Indeterminacy [14]. The following experiments are carried out:

- Experiment 1: performance analysis of the algorithms MSTrNCM, mFCM and NCM_Indeterminacy with respect to the parameter lambda λ, considering the indicator "Arithmetic mean of the error in prediction" defined in Eq. (6).

$$ME = \frac{1}{N}\sum_{i=1}^{N} E_i \tag{6}$$

- Experiment 2: comparison of the extensions MSTrNCM and NCM_LDS with the implementations of mFCM and NCM_Indeterminacy, considering the indicators: "Arithmetic mean of the error in prediction" defined in Eq. (6), precision (P1) declared in Eq. (7) and global precision (PG) defined in Eq. (8).

$$P_1 = 100 \cdot \frac{\text{Number of cases with first prediction as response}}{\text{Total of processed cases}} \tag{7}$$

$$P_G = 100 \cdot \frac{\text{Number of cases with first three predictions as responses}}{\text{Total of processed cases}} \tag{8}$$

- Experiment 3: algorithms applicability analysis which consist in the comparison of the LDS_NCM and MSTrNCM with mFCM and NCM_Indeterminacy approaches regarding the ease of decision-making process using a multi-criteria evaluation method based on the 2-tuples computing with words technique [26].

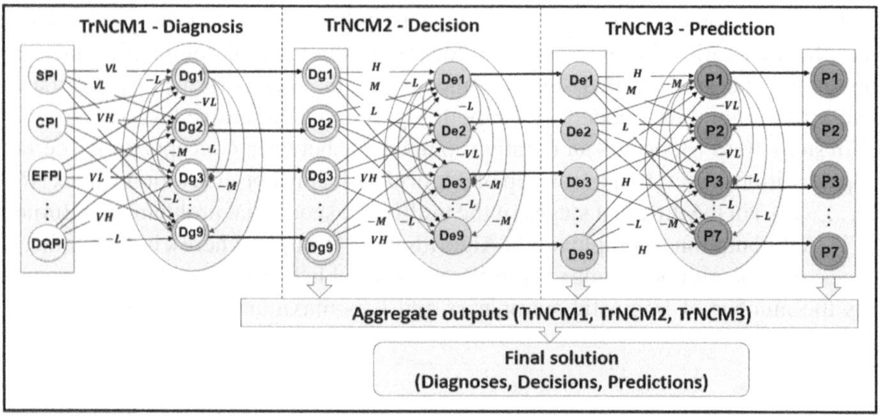

Fig. 2 Partial view of the aggregated MS_TrNCM for project evaluation

- Experiment 4: evaluation of the efficiency of algorithms considering the indicator "execution time".

The MSTrNCM construction was carried out consulting three experts and in general 33 concepts were identified. During the construction process, each expert builds an individual map describing the relationships among concepts using linguistic terms to express their preferences. Subsequently, individual maps are aggregated using the operator (OMPNNT) [21] and the aggregated MSTrNCM is obtained, see Fig. 2.

The NCM_LDS was constructed by using Algorithm 4 and for this experimentation authors applied the algorithm proposed by Pérez et al. [23] in combination with active learning techniques. A total of 490 summaries were obtained, distributed as follows: 164 summaries correspond to the diagnosis stage, 225 to the decision stage and 101 to the prediction stage.

4.1 Experiment 1: Analysis of the Algorithms Regarding the Parameter Lambda Λ

The algorithms are compared regarding the indicator "Arithmetic mean of the error in the prediction" with different values of lambda $\lambda \in \{0.5, 1, 2.5, 5\}$ [27], Table 2.

Table 2 shows the results in the stages of diagnosis, decision and prognosis, respectively. In the three stages, similar behavior of the algorithms is evidenced regardless of the value of λ, being MSTrNCM the one with the best results, followed by m-mFCM and finally NCM_Indeterminacy. Regarding the influence of the value of λ on the performance of the three algorithms, it is identified that the best results are found with $\lambda = 0.5$. This value of λ is used in the rest of experiments.

Table 2 Results of the arithmetic mean of the error in the tree stages with variations of λ

Lambda λ	Algorithms	Diagnosis	Decision	Prediction
0.5	NCM_Indeterminacy	0.335	0.436	0.334
	mFCM	0.231	0.218	0.185
	MSTrNCM	0.048	0.043	0.071
1	NCM_Indeterminacy	0.44	0.488	0.334
	mFCM	0.225	0.218	0.193
	MSTrNCM	0.06	0.047	0.087
2.5	NCM_Indeterminacy	0.461	0.488	0.461
	mFCM	0.255	0.437	0.287
	MSTrNCM	0.076	0.057	0.113
5	NCM_Indeterminacy	0.461	0.488	0.461
	mFCM	0.394	0.462	0.318
	MSTrNCM	0.082	0.088	0.115

4.2 *Experiment 2: Comparison Regarding the Error in Prediction and Precision*

For the comparison of the algorithms with respect to the indicator "Arithmetic mean of the error in the prediction", a sample is formed for each algorithm with the size $N = 1\,011$ made up of the error in the prediction in each case of the database. The results in the three stages are independently evaluated. The Shapiro–Wilk test is applied to the samples, results show that these samples do not have a normal distribution. In the comparison of the algorithms, a descriptive analysis is carried out, see Tables 3 and Table 4, where the Friedman and Wilcoxon tests are conducted.

Table 4 shows the comparison results of the error in the diagnostic, decision and prediction stages, respectively. It is shown that there are significant differences between the algorithms, being the NCM_LDS algorithm the one with the best results, followed by MS-TrNCM, then m-FCM and the one with the worst results was NCM_Indeterminacy.

Table 3 Results of "Arithmetic mean of the prediction error" in the tree stages

$N = 1\,011$		NCM_Indeterminacy	mFCM	MSTrNCM	NCM_LDS
Diagnosis	Mean	0.3346	0.2315	0.0478	0.0418
	Std dev	0.12958	0.14645	0.04157	0.11032
Decision	Mean	0.4356	0.2184	0.0432	0.0365
	Std dev	0.04796	0.17064	0.03347	0.09776
Prediction	Mean	0.3340	0.1851	0.0709	0.0086
	Std dev	0.06242	0.16045	0.06037	0.05188

Table 4 Comparative analysis of the algorithms regarding the indicator "Arithmetic mean of the error in prediction" in diagnosis, decision and prediction stages

Diagnosis stage	Decision stage	Prediction stage
NCM_LDS – MSTrNCM – mFCM – NCM_Indeterminacy (Friedman p-value = 0.000)	NCM_LDS–MSTrNCM – mFCM – NCM_Indeterminacy (Friedman p-value = 0.000)	NCM_LDS–MSTrNCM – mFCM – NCM_Indeterminacy (Friedman p-value = 0.000)
NCM_LDS [Group A] – MSTrNCM (Wilcoxon p-value = 0.000)	NCM_LDS[Group A] – MSTrNCM (Wilcoxon p-value = 0.000)	NCM_LDS[Group A] – MSTrNCM (Wilcoxon p-value = 0.000)
MSTrNCM [Group B] – mFCM (Wilcoxon p-value = 0.000)	MSTrNCM[Group B] – mFCM (Wilcoxon p-value = 0.000)	MSTrNCM[Group B] – mFCM (Wilcoxon p-value = 0.000)
mFCM[Group C] – NCM_Indeterminacy[Group D] (Wilcoxon p-value = 0.000)	mFCM[Group C] – NCM_Indeterminacy[Group D] (Wilcoxon p-value = 0.000)	mFCM[Group C] – NCM_Indeterminacy[Group D] (Wilcoxon p-value = 0.000)

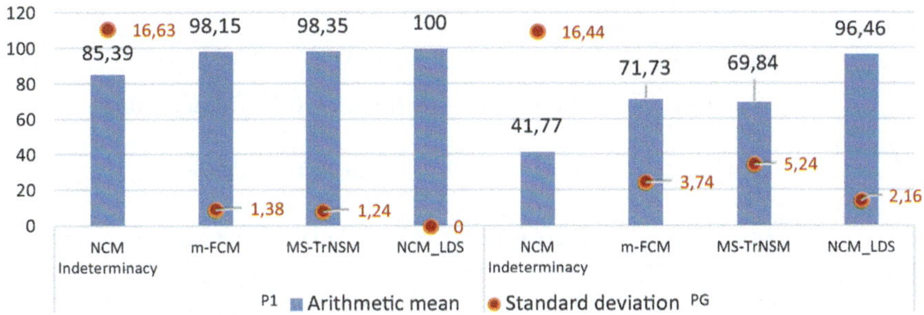

Fig. 3 Performance of algorithms with respect to the P_I and P_G indicators

In addition, the results are analyzed considering the precision indicators P_I and P_G, the arithmetic mean and standard deviation, as shown in Fig. 3.

Figure 3 shows that the best results regarding both indicators P_I and P_G are reported by the NCM_LDS algorithm, while the worst results are reported by the NCM_Indeterminacy algorithm. In the case of the P_I indicator, the MSTrNCM algorithm reports better results than the mFCM algorithm; however, regarding the P_G indicator, the mFCM algorithm outperformed the MSTrNCM algorithm.

4.3 Experiment 3: Algorithms Applicability Analysis

Table 5 presents the aggregated preferences of 10 project management's experts, for each criterion and concordance for each algorithm. The evaluation process is based on nine criteria: C1 map interpretability, C2 ease of map construction, C3 flexibility in expressing elicitations during the map construction. C4 time saved in the construction process. C5 influence of using phrases or sentences in facilitating the map construction, C6 satisfaction with responses in diagnosis stage, C7 satisfaction with responses in decision stage, C8 satisfaction with responses in prediction stage and C9 treatment of indeterminacy.

It can be seen that NCM_LDS obtained the experts highest evaluation (Very high, − 0.5), followed by MSTrNCM with the evaluation (High, − 0.5) and then mFCM with the evaluation (Medium, − 0.2), while the lowest evaluation (Average, − 0.4) was given to NCM_Indeterminacy. It is also evidenced that there is agreement in the experts' preferences, in the four algorithms an agreement index $R^*_{WG} > 0.89$ is reached.

Concerning the maps and prediction results interpretability C1, the NCM_LDS algorithm obtained the highest evaluation, followed by MSTrNCM and then NCM_Indeterminacy, while mFCM obtained the lowest evaluation.

Regarding the ease of the map construction, the NCM_LDS algorithm obtained the highest evaluation with respect to the criteria C2, C4 and C5, whereas MSTrNCM algorithm reported the lowest evaluation. The MSTrNCM low evaluation is due to

Table 5 Aggregated experts' elicitations

Criteria	NCM_LDS	MSTrNCM	mFCM	NCM_Indeterminacy
C1	(Very high, – 0.1)	(High, – 0.1)	(Mean, – 0.4)	(Mean, – 0.1)
C2	(Very high, – 0.1)	(Low, 0.3)	(Mean, 0.4)	(Mean, 0.1)
C3	(High, 0.0)	(Very high, – 0.4)	(Low, 0.1)	(Mean, 0.1)
C4	(Very high, – 0.2)	(Low, – 0.2)	(Mean, 0.4)	(Mean, – 0.1)
C5	(Very high, – 0.4)	(High, 0.1)	(High, – 0.4)	(Low, 0.4)
C6	(High, 0.3)	(High, – 0.5)	(High, – 0.3)	(Low, 0.2)
C7	(High, 0.5)	(High, – 0.5)	(High, – 0.3)	(Low, 0.4)
C8	(High, – 0.4)	(Very high, – 0.3)	(Very high, 0.0)	(Mean, – 0.3)
C9	(Very high, – 0.2)	(Mean, 0.2)	(Low, 0.1)	(Low, 0.1)
Aggregated preference	(Very high, – 0.5)	(High, – 0.5)	(Mean, – 0.2)	(Mean, – 0.4)
Concordance	0.940	0.893	0.939	0.915

the fact that experts must express the degrees of certainty, indeterminacy and falsity of their preferences.

In relation to the criteria C6, C7 and C8 referring to the satisfaction of the experts with the algorithms' responses in the three stages, NCM_LDS registered the highest evaluation. Regarding the treatment of indeterminacy C9, the best evaluation was reported by MSTrNCM, followed by NCM_LDS and then NCM_Indeterminacy, while mFCM obtained the lowest evaluation.

4.4 Experiment 4: Evaluation of the Efficiency of Algorithms Considering the Indicator "Execution Time"

The efficiency of algorithms is measured by the "execution time" indicator in seconds. The most efficient algorithm was mFCM with 11.4187 s, followed by NCM_indeterminacy with 13.36 s, then MSTrNCM with 79.139 s and finally NCM_LDS with 89.8 s.

5 Conclusion and Future Work

Based on the systematic review performed in this work, it is identified that traditional FCMs and most of their extensions still have insufficiencies in the representation of indeterminacy and neutrosophic cognitive maps (NCM_Indeterminacy) represent indeterminacy through the symbol I, affecting the predictability of these maps and the interpretability of its results. It is appreciated also that FCMs and their extensions cannot be applied naturally for the of multistage sequential decision-making problems.

In this research, authors propose two new FCMs extension for solving multistage sequential decision-making problems, the multistage sequential triangular neutrosophic cognitive map (MSTrNCM) and the neutrosophic cognitive map based on linguistic summaries (NCM_LDS). The MSTrNCM extension improves the treatment of indeterminacy and interpretability by combining triangular neutrosophic numbers with computing with word techniques. The MSTrNCM map provides greater degrees of freedom for the representation of map relationships. The NCM_LDS extension makes easy the map construction process and results interpretability. This map allows modeling the indeterminacy and enables the representation of multiple relationships between two concepts.

In the validation process, the two new extensions MSTrNCM and NCM_LDS are applied for diagnosis, decision and prediction during projects evaluation. MSTrNCM and NCM_LDS are compared with the traditional implementations of fuzzy cognitive maps (mFCM) and of neutrosophic cognitive maps (NCM_Indeterminacy) and it is concluded that:

The MSTrNCM, mFCM and NCM_Indeterminacy algorithms obtain the best results with the parameter lambda $\lambda = 0.5$.

In relation to the indicator "Arithmetic mean of the error in the prediction", it is shown that the NCM_LDS algorithm obtains better results. The NCM_Indeterminacy algorithm reports the worst results.

Considering the precision indicators P_1 and P_G, the best results are reported by NCM_LDS, while NCM_Indeterminacy registers the worst results.

Regarding the applicability of the algorithms in the application scenario, in general, the proposed algorithms NCM_LDS and MSTrNCM obtained the highest evaluation of experts. A concordance $R^*_{WG} > 89$ is achieved.

In relation to execution time, the most efficient algorithm was mFCM, while the NCM_LDS algorithm was the most time consuming.

For future works, authors propose to continue this research by introducing machine learning techniques to adjust automatically the MSTrNCM's weights. It is also suggested to extend the use of the proposed extensions in other application scenarios and investigate new forms of visual presentation of the inference results.

References

1. Johnson, J.: CHAOS report: decision latency theory: it is all about the interval. The standish group (2018)
2. Pérez, I., García, R., Piñero, P.Y., Mahdi, G.S., Peña, M.: Experiencias en el uso de tecnicas de softcomputing en la evaluacion de proyectos de software. Investigación Oper. **41**, 108–120 (2020)
3. Al-subhi, S.H., Papageorgiou, E.I., Pérez, P.P., Mahdi, G.S., Acuña, L.A.: Triangular neutrosophic cognitive map for multistage sequential decision-making problems. Int. J. Fuzzy Syst., pp. 1–23 doi: https://doi.org/10.1007/s40815-020-01014-5 (2021).
4. Dursun, M., Goker, N., Mutlu, H.: A cognitive map integrated intuitionistic fuzzy decision-making procedure for provider selection in project management. J. Intell. Fuzzy Syst **39**, 6645–6655 (2020). https://doi.org/10.3233/JIFS-189125
5. Kosko, B.: Fuzzy cognitive maps. Int. J. Man Mach. Stud. **24**, 65–75 (1986). https://doi.org/10.1016/S0020-7373(86)80040-2
6. Rickard, J.T., Aisbett, J., Yager, R.R.: Computing with words in fuzzy cognitive maps. Annual conference of the North American fuzzy information processing society (NAFIPS) held jointly with 5th world conference on soft computing (WConSC), pp. 1–6 (2015). doi: https://doi.org/10.1109/NAFIPS-WConSC.2015.7284135
7. Frías, M., Filiberto, Y., Nápoles, G., Vanhoof, K., Bello, R.: Fuzzy cognitive maps reasoning with words: the ordinal case. 2nd International Symposium on Fuzzy and Rough Sets, Cuba (2017)
8. González, M.P., De La Rosa, C.G., Moran, F.J.: Fuzzy cognitive maps and computing with words for modeling project portfolio risks interdependencies. Int. J. Innov. Appl. Stud. **15**, 737–742 (2016)
9. Salmeron, J.L.: Modelling grey uncertainty with fuzzy grey cognitive maps. Expert Syst. Appl. **37**, 7581–7588 (2010). https://doi.org/10.1016/j.eswa.2010.04.085
10. Kang, B., Deng, Y., Sadiq, R., Mahadevan, S.: Evidential cognitive maps. Knowl.-Based Syst. **35**, 77–86 (2012). https://doi.org/10.1016/j.knosys.2012.04.007
11. Mkrtchyan, L., Ruan, D.: Belief degree-distributed fuzzy cognitive maps. IEEE Int. Conf. Intell. Syst. Knowl. Eng., pp. 159–165 (2010). doi: https://doi.org/10.1109/ISKE.2010.5680815
12. Jia, Z., Zhang, Y., Dong, X.: An extended intuitionistic fuzzy cognitive map via Dempster-Shafer theory. IEEE Access **8**, 23186–23196 (2020). https://doi.org/10.1109/ACCESS.2020.2970159
13. Vasantha, W.B., Kandasamy, I., Devvrat, V., Ghildiyal, Sh.: Study of imaginative play in children using neutrosophic cognitive maps model. Neutrosophic Sets Syst. **30** (2019). doi: https://doi.org/10.5281/zenodo.3569702
14. Chithra, B., Nedunchezhian, R.: Dynamic neutrosophic cognitive map with improved cuckoo search algorithm (DNCM-ICSA) and ensemble classifier for rheumatoid arthritis (RA) disease. J. King Saud Univ. Comput. Inform. Sci. (2020). https://doi.org/10.1016/j.jksuci.2020.06.011
15. Smarandache, F.: A Unifying Field in Logics: Neutrosophic Logic, Neutrosophy, Neutrosophic Set, Neutrosophic Probability. American Research Press, Rehoboth N.M (1999)
16. Kandasamy, W.B., Smarandache, F.: Fuzzy Cognitive Maps and Neutrosophic Cognitive Maps. Xiquan, New Mexico, USA (2003)
17. Amirkhani, A., Papageorgiou, E.I., Mohseni, A., Mosavi, M.R.: A review of fuzzy cognitive maps in medicine: Taxonomy, methods and applications. Comput. Methods Prog. Biomed **142**, 129–145 (2017). https://doi.org/10.1016/j.cmpb.2017.02.021
18. Rezaee, J., Yousefi, M., Valipour, S., Dehdar, M.: Risk analysis of sequential processes in food industry integrating multi-stage fuzzy cognitive map and process failure mode and effects analysis. Comput. Ind. Eng. **123**, 325–337 (2018). https://doi.org/10.1016/j.cie.2018.07.012
19. Martin, N., Aleeswari, A., Merline, W.: Risk factors of lifestyle diseases – analysis by decagonal linguistic neutrosophic fuzzy cognitive map. Mater. Today: Proc **24**, 1939–1943 (2020). https://doi.org/10.1016/j.matpr.2020.03.621

20. Bhutani, K., Kumar, M., Garg, G., Aggarwal, S.: Assessing it projects success with extended fuzzy cognitive maps & neutrosophic cognitive maps in comparison to fuzzy cognitive maps. NSS **12**, 9–19 (2016)
21. Liu, P., Wang, Y.: Multiple attribute decision-making method based on single-valued neutrosophic normalized weighted Bonferroni mean. Neural Comput. Appl. **25**, 2001–2010 (2014). https://doi.org/10.1007/s00521-014-1688-8
22. Bueno, S., Salmeron, J.L.: Benchmarking main activation functions in fuzzy cognitive maps. Expert Syst. Appl. **36**, 5221–5229 (2009). https://doi.org/10.1016/j.eswa.2008.06.072
23. Pérez, I., Piñero, P.Y., Bello, R., Acuña, L.A., García, R.: Linguistic Summaries Generation with Hybridization Method Based on Rough and Fuzzy Sets. Rough Sets, pp. 385–397. Springer International Publishing, Havana, Cuba. doi: https://doi.org/10.1007/978-3-030-52705-1_29 (2020)
24. Wang, H., Smarandache, F., Zhang, Y., Sunderraman, R.: Single valued neutrosophic sets. Multisp. Multistruct. **4**, 410–413 (2010)
25. Piñero, P.Y., Pérez, I., Hechavarría, C.C., Rojas, C., González, R., Torres, S.: Repositorio de datos para investigaciones en gestión de proyectos. Revista Cubana de Ciencias Inform **13**, 176–191 (2019)
26. Martínez, L., Rodriguez, R.M., Herrera, F.: The 2-tuple Linguistic Model: Computing with Words in Decision Making. Springer International Publishing, Switzerland, doi: https://doi.org/10.1007/978-3-319-24714-4 (2015).
27. Nápoles, G., Grau, I., Papageorgiou, E., Bello, R., Vanhoof, K.: rough cognitive networks. Knowl.-Based Syst. **91**, 46–61 (2016). https://doi.org/10.1016/j.knosys.2015.10.015

Chapter 11
A Software Ecosystem for Project Management in BIM Environments Assisted by Artificial Intelligent Techniques

Pedro E. Piñero Ramírez, Iliana Pérez Pupo, Pedro Y. Piñero Pérez, Yosvani Marquez Ruiz, and Yulia Fustiel Alvarez

Abstract This paper presents a new platform called BusinessRedmine for project management in BIM environments. The platform constitutes a software ecosystem that combines traditional project management techniques with soft computing techniques for decision-making in projects. The proposed platform is business intelligence tools and consist of a set of software actives that cover the different processes of project management. The platform allows the management of multiple processes such as: scope management, planning management, schedule construction, risk management, stakeholder management, communications and quality. In addition, it facilitates integration with tools for 2D and 3D design using the IFC format. The platform includes facilities for managing project files aligned with the ISO 19650 series of standards. One of the fundamental elements it provides is the application of soft computing techniques to artificial intelligence for decision-making in project cuts. We compare the algorithm FIS_{BR} proposed in BusinessRedmine systems for project evaluation with other algorithms reported in bibliography based on neural networks and genetic algorithms techniques. The algorithms selected learn from a database that contains projects already evaluated and after that classify other projects. In experiments, we applied cross validation techniques combined with Friedman test and Wilcoxon test. Finally, a comparison with different project management tools is presented.

Keywords Artificial intelligence · BIM · Project management · Soft computing

P. E. Piñero Ramírez (✉)
Intelligent Software Ecosystems Group, BusinessRedmine Corporation, Mississauga, Canada

I. Pérez Pupo · P. Y. Piñero Pérez · Y. Marquez Ruiz · Y. Fustiel Alvarez
Grupo de Investigaciones en Gestión de Proyectos, Universidad de las Ciencias Informáticas, CP 54830 La Habana, Cuba
e-mail: iperez@uci.cu

P. Y. Piñero Pérez
e-mail: ppp@uci.cu

P. Y. Piñero Pérez et al. (eds.), *Artificial Intelligence in Project Management and Making Decisions*, Studies in Computational Intelligence 1035,
https://doi.org/10.1007/978-3-030-97269-1_11

1 Introduction

There is a growing demand for tools with high added value for the management of project-oriented organizations. Within this context there is a group of elements that must be carefully planned to achieve the project objectives with an adequate balance of cost, time and quality. Two spheres of knowledge that exemplify this situation are the growing development of Building Information Modeling (BIM) technologies [1–3] for construction projects and the development of software projects [4].

In particular, in the construction sector, BIM technologies are aimed to simulate the complete execution of projects in virtual environments. The use of BIM methodologies and tools is still not very standardized and not generalized at a global level [5]. They constitute an open field for the development of new research [6–8]. At the current moment of BIM development, progress has been made fundamentally in the integration and parameterization of the designs or plans of the works with the planning activities. In this context, it is proposed that around 34% of the resources invested in a project are wasted and that approximately 95% of the information that is generated later is underused and is not used as lessons learned in new projects [9]. In these same sources, it is stated that the main factor that affects the inappropriate use of resources is found in inefficient planning. Considering this context, the development of new algorithms and techniques that allow the optimal or quasi-optimal construction of project schedules and take into account the limitations of resources and the variability in the ways of executing the tasks, remains a necessity.

Similarly, it happens in the development of software projects. In this area of knowledge, it is identified by several sources that, in medium-sized projects, approximately 26% need to be renegotiated and 31% are canceled [10–12]. This scenario is characterized by high dependence on human resources and their skills, accompanied by tasks that can be performed in various ways. Among the fundamental causes of project failure are: insufficient training of human resources and frequent errors in planning processes, whether in scope, time or logistics, and poor management and inadequacies in control and monitoring processes [13–15]. In addition, it is identified that there is a group of factors that influence this problem related to the inadequacies of the available computer tools [13].

In this sense, a study of the tools reported in the bibliography was carried out and around 125 solutions associated with project management were identified [13]. Authors analyze 125 solutions, they identified that only 25 are based on free software, which represents 20% of the total. A characterization of these tools regarding the coverage of project management processes is shown in Table 1.

From the detailed analysis of the tools reported in the bibliography, the following shortcomings could be identified:

- Need to use numerous tools with different interfaces and low integration to manage the data they generate.
- Only 44% of the tools reported in the bibliography include facilities for the management of indicators for data analysis.

Table 1 Analysis of project management tools based on open source

Features that include	Percent (%)
Portfolio management	44
Planning management	60
Human and non-human resource management	56
Document management	60
Control and monitoring reports	32

- Among the most frequent indicators in the tools are those associated with cost management described in the PMBOK [16], such as: earned value (EV), cost performance index Cost Performance Index (CPI), the Schedule Performance Index (SPI), Cost Variance (CV), Planned Cost (PV), and Actual Cost (AC).
- The few tools that allow the management of project portfolios are notable, in this case only 11 tools with this functionality.
- Although some tools include indicators for evaluating costs and time, they do not propose specific mechanisms for managing uncertainty in their calculation.
- In general, the existing tools do not allow treatment of the imprecision, vagueness or uncertainty contained in the information.
- High cost of solutions and low level of customization for different scenarios.
- Little availability of software ecosystems with solid architectures that take advantage of free software for reuse and rapid development of low-cost solutions.
- Among the open source or free solutions, only 8 sufficiently include the control and monitoring reports that they represent. In this way, a significant insufficiency is identified in the decision-making support processes in project management.

This paper presents a software ecosystem, based on free software technologies called BusinessRedmine for project management in different production environments. The second section presents a brief study of the state of the art associated with software ecosystems. The third section presents the architecture proposal of the proposed software ecosystem. Then in the fourth section is the analysis of results in the application of the ecosystem. Finally, the last section presents the conclusions of the work.

2 Brief Analysis of Software Ecosystems

Traditional software development models manage projects independently, an element that causes each project to be presented as a custom development and production costs rise. These traditional models include cascade development, prototype-based development, original component-based model, and spiral development. As an alternative to these traditional models, others such as software product lines (LPS) and software ecosystems have emerged. These new models with a focus on architecture and reuse enhance cost reduction and increase quality in productions.

The software product lines constitute industrial models of software development, inspired by models of chain production used in other spheres of society. They are created from the evolution of development models based on components and software factories, improving the organization of processes and the division of responsibilities between the departments of the organization. This development model encourages the management of project programs, over and above the management of independent projects.

The elements that define an LPS are: domain, product families, the set of reusable assets, the mode of production and the architecture, as a central element that guarantees the integration of solutions and reuse [17, 18].

Software ecosystems, for their part, are aimed at creating alliances between different organizations, focusing on outsourcing the relationships of organizations, enhancing productivity and sharing market sectors [19–23]. According to Bosh, a software ecosystem constitutes a set of computer systems that enable, automate and support a certain business need [20]. Other authors, for their part, consider that the basis of software ecosystems is in the establishment of solid business architectures that are supported by the integration of their components [24].

Among the important elements to guarantee the sustainability of ecosystems is the need to define the income models that regulate the relationships between the different entities of the ecosystem. Let's see below different entry strategies that can be used and the trends in this regard within free software [20]:

- Custom development: in this scheme, users pay for the development of customized software for their specific requirements. It is used by 1.8% of those surveyed.
- Services and support: generally supported by phone calls, training, and consulting contracts. Employed by 7.9% of those surveyed.
- Business licensing: a licensing under a business license in the traditional way. Used by 24.6% of those surveyed.
- Software as a service (SaaS): users pay to access and use the software via the internet. Used by 5.3% of those surveyed.
- Embedded Software: Open source is embedded in commercial end-to-end solutions packages. Used by 5.3% of those surveyed.
- Embedded hardware: Open source is embedded in hardware that is generally distributed under commercial licenses. Used by 9.7% of those surveyed.
- Advertising: the software is free to be used under advertising conditions. Employed by 2.6% of those surveyed.
- Subscriptions: where support services and updates are contracted, for a period of time that is generally annual. Employed by 34.2% of those surveyed.
- Other products and services: this model refer to the fact that open source is not used directly to generate income, but other products and services that are generally proprietary are those that generate the income and contain parts based on open source. This model is used by 8.8% of those surveyed.

It is also important that the licensing strategies for the ecosystem products are defined, promoting good relations between the entities and guaranteeing the legal protection of the products. In this sense, the following strategies are presented [25]:

- Open Source—The solution is released under an open source license only. Employed by approximately 24.6%.
- Assembled Open Source: The product or service includes code from multiple open source projects using multiple licenses. Scheme used by 16.7% of those surveyed.
- Closed: The product is based on open source, but it is not available under open source licensing. Scheme used by 5.2% of those surveyed.
- Open and closed licensing: open source projects are developed from the union of independent parties under commercial licensing, developed as closed source. Method used by 14.9% of those surveyed.
- Open-core licensing: the core of the solution is open under an open source license. But the professional versions of the products include open and closed source developments and are commercially licensed. Scheme used by 23.7% of those surveyed.
- Dual licensing: the base code is licensed to different users as open source or as commercial licensing. Method used by 14.9% of those surveyed.

Note that the open-core-based licensing model is one of the most widely used, taking advantage of the advantages of free software, but managing the professional versions of the products under commercial licenses, which include closed source.

Another important element to consider is the development strategies that are the basis for the licensing of the solutions. Furthermore, these strategies have a direct influence on the level of code reuse and the reduction of development costs [25]:

- Open source community development: the software is distributed using open source licenses and is publicly developed by a community of supplying individuals or organizations. Method used by 9.7% of those surveyed.
- Open source development as a single provider: the software is distributed using a free software license and all contributions and extensions are published but development and is dominated by a single provider organization. Employed by approximately 21.1% of respondents.
- Mixed open source development: the product or service is provided by a single provider and is based on the combination of projects developed in free software, publicly developed by communities or other organizations. Employed by approximately 19.3% of respondents.
- Completely closed development: all software components and functionalities are closed by the producing entity and are not based on free software.
- Hybrid development: refers to development models where the software is distributed using open source licenses, but some functionalities are developed behind closed doors, the following three categories stand out in this regard:
  - Hybrid community: most of the software is developed publicly by a community of individuals or providers, while some functionalities are developed behind closed doors by providers of the product or service. Scheme used by 7.9% of those surveyed.

– Hybrid providers: development is dominated by employees of a single organization, and some functionalities are developed behind closed doors. Scheme used by 33.3% of those surveyed.
– Mixed hybrid: the product or service is based on the combination of projects that use open source developments developed publicly by communities or multiple providers and some functionalities are developed behind closed doors. Scheme used by 8.8% of respondents.

Based on the analysis of the models and the different strategies considered, a software ecosystem is proposed for the development of solutions and its use is proposed to help decision-making in project management.

An example of software ecosystems is the Microsoft Office suite, made up of a set of IT solutions for office management. Which includes facilities for project management among others. This ecosystem is also made up of all the relational knowledge of the Microsoft company that enhances the integration of its products with different technologies and solutions. This ecosystem is based on a commercial licensing model.

Another example of an ecosystem is the development of the Eclipse package. This is a set of solutions specialized in the development of IDEs for programming computer solutions [26]. It is an ecosystem based on open source solutions.

3 Architecture of the BusinessRedmine Software Ecosystem

The proposal constitutes a software ecosystem, called BusinessRedmine [27], for the full-cycle management of the integrated project management process. In this research the authors define the software ecosystem as follows:

Definition 1 Software ecosystem: a software ecosystem is formed from the system of relationships between different entities with the aim of sharing clients, market segments or increasing efficiency and effectiveness in the production processes they develop. It is supported by the definition of common architectures and platforms that facilitate the integration of solutions, the exchange of information, resources, artifacts and assets in general. These relationships must be supported by economic-financial relationships, work agreements or development models that promote the exchange of data and codes. Its main components are:

- Domain of application.
- Family of products that make up the portfolio of solutions.
- Ecosystem assets that comprise the set of all reusable elements of the organization, both code components and documentation.
- Architecture that enhances the integration of assets, specifying and development strategies.
- Set of the modes of production of the entities of the ecosystem.

- The set of relationships between the entities that comprise it (relational knowledge), the planned entry strategies and the licensing and marketing strategies of the solutions.

This Definition 1 constitutes the basis of the architectural proposal developed. Some elements that characterize the BusinessRedmine ecosystem are

- Allows help in decision-making in project management by combining traditional methods with different artificial intelligence techniques, facilitating help in decision-making.
- The processes that the BusinessRedmine ecosystem models are aligned with the elements that describe international standards such as: PMBOK [16], SCRUM, the ISO 21500 standard [28] and the ISO 19650 series [2].
- The BusinessRedmine ecosystem is based on web technologies and designed for installation in private or public clouds, guaranteeing a high level of security.
- The architecture that supports the ecosystem is based on free software and docker technologies that facilitate the updating and maintenance of solutions.

The ecosystem is made up of a set of assets organized in the following layers:

- Layer associated with the high availability of solutions.

    Asset 1. It is formed from the integration of HAProxy and KAlive tools that enhance high availability and platform performance. These facilities may be integrated into nginx or APACHE servers.

- Security layer:

    Asset 2. Includes the assembly of LDAP server with interface for management from the WEB or integration with existing LDAP servers in the organization.
    Asset 3. RuCaptcha that allows integration with captcha management that enhances secure access to information for each of the project participants.
    Asset 4. Logstash's security configured to capture the logs associated with ecosystem security.
    Asset 5. Elastic search security aimed at integrating with log stash and allowing the structuring of security logs with the platform for managing them.
    Asset 6. Grafana's security configured to manage alarms and display the platform's security logs. It allows the classification of logs associated with application errors, user access to the platform and user behavior within the platform.
    Asset 7. Firewall to control access to ecosystem components.

- Project management layer:

    Asset 8. Base platform for managing programs and projects (work objects). It allows the intelligent management of different stages of project development and their schedules. Facilitates integration with the first dimension of BIM.
    Asset 9. Subsystem for the feasibility analysis of programs, projects and project portfolios. It combines traditional methods with soft computing techniques

to better model uncertainty and ambiguity in feasibility analysis processes. It allows integration with the sixth dimension of BIM, associated with the analysis of sustainability of investments.

Asset 10. Subsystem for strategic management that allows the modeling of strategic objectives from the management of indicators of physical and financial execution of projects.

Asset 11. Subsystem for risk management at the organization level and at the project level with treatment of uncertainty. It facilitates the qualitative and quantitative analysis of risks using different artificial intelligence techniques.

Asset 12. Balanced scorecard subsystem to help business decision-making in project management. Includes integration with ERP systems. It takes the best practices proposed by Kaplan and Norton integrated with alert systems based on artificial intelligence techniques.

Asset 13. System for the management of control indicators and monitoring of support projects at different management levels. It includes facilities for augmented analytics and the reuse of different solutions reported in the bibliography. It includes a set of indicators aligned with the main project management standards and agile development methodologies. Furthermore, it incorporates the following graphs and indicators: Burn-up, Burn-Down, S Curve and CPI, SPI, EV, EAC indicators among others.

Asset 14. Subsystem for managing project participants and their competencies.

Asset 15. CRM subsystem for managing clients, suppliers, shareholders and contracts.

Asset 16. Subsystem for the construction of optimal or quasi-optimal project schedules. It integrates with the fourth dimension of BIM, associated with time.

Asset 17. System for the management of logistic stock to support decisions of productive chain. It enhances the integration with the fifth dimension BIM associated with the cost of investment projects.

Asset 18. Subsystem for billing and sales management with integration with point of sale that provides a friendly interface for the company's clients at all times of the negotiation process.

Asset 19. Subsystem for making financial decisions and integration with accounting management systems, enhancing the integration of the fundamental activity of the park with basic economic management.

Asset 20. Subsystem for pre-investment analysis and estimation of initial costs of projects integration with the first dimension of BIM.

- Data sharing environment layer:

Asset 21. Database manager subsystem that allows the assembly and administration of the data servers associated with the ecosystem.

Asset 22. Shared Data Environment Subsystem for integration with project file repositories. It includes facilities for the storage and development of advanced searches. Based on technologies that facilitate the integration of outputs from different BIM tools. Includes advanced document management and project files.

Asset 23. Subsystem for integration with distributed version control systems. That facilitate project management from the development process itself and the successful application of agile development techniques.

- Collaboration and continuous training layer:

  Asset 24. Subsystem for the assembly of conversational systems applicable to support management and with potential for its application in different sectors of society. This asset is expected to be developed from the reuse, adaptation and evolution of other solutions reported in the bibliography and from characterizing the principles and best practices of conversational systems.
  Asset 25. Subsystem for the support and virtual training of the clients of the system. It will allow the setting up of consulting and training services for client entities with facilities for remote work.

- Integration layer with BIM tools and advanced data analysis:

  Asset 26. Subsystem for integration with the BIM tools control and monitoring ecosystem. This facilitates the development of the second and third dimensions of BIM associated with the basic engineering and detailed engineering of work objects.
  Asset 27. Subsystem for the integration of the detailed schedules of the dashboard platform with BIM tools that relate the design models with the work execution schedules. And they allow the integration of BIM dimensions in the ecosystem.

The fundamental process that models the proposed ecosystem is the control and monitoring of the projects. This process has a strong integration with Artificial Intelligence techniques, as shown below. In this process, the ecosystem provides a guide based on the following steps:

Step 1. Calculation of indicators from data.
Step 2. Product quality review.
Step 3. Automatic evaluation of the project from the indicators.
Step 4. Detection of the fundamental difficulties of the project and areas of knowledge with difficulties.
Step 5. If there are no problems, finish the check and congratulate.
Step 6. In case of difficulties, go down the cascade of indicators and analyze their causes.
Step 7. Having identified the causes, proceed to decision-making and the final evaluation of the project.

In step 1, the ecosystem calculates the indicators proposed in Table 2.

In step 3 the BusinessRedmine platform introduces a fuzzy inference system FIS_{BR}, as a classifier for the automatic evaluation of projects that is described in details in [13]. Algorithm FIS_{BR} constitutes a hybrid neuro-fuzzy system of the Sugeno type. With this algorithm, we intend to take advantage of the potentialities

Table 2 Indicators for decision making

Analysis area	Indicators
Project integration management	• Comprehensive evaluation of the project • IAP (Compliance with court agreements)
Client management and commitments	• IE (Execution Index) • IRE (Execution Performance Index)
Time management	• SPI (Scheduling Performance Index)
Cost management	• CPI (Cost Performance Index) • PV, AC, EV (Earned Value)
Procurement management	• IRL (Logistics Performance Index)
Human resource management	• IRRH (Human Resources Performance Index)
Scope and quality management	• IREF (Efficacy Performance Index)
Information consistency	• ICD (Data Quality Index)

of neural networks and the interpretability of fuzzy inference systems. The proposed FIS_{BR} uses a node-oriented multilayer network architecture.

As part of the experimentation of this research, the results' analysis section presents the results of the comparison of the proposed algorithm with other classifiers reported in the bibliography.

The platform is being applied in several real scenarios. And the algorithms proposed for control and monitoring were validated by experimenting with their behavior in a database of completed projects. The design of the experiments and their results are listed below.

4 Results Analysis

As part of the validation of the impact of the proposed ecosystem, the following design of experiments was carried out:

- Experiment 1: Comparison of the BusinessRedmine ecosystem with other systems for managing projects and project-oriented organizations.
- Experiment 2: analysis of the implementation of the system in different scenarios.
- Experiment 3: comparison of the classifier used in the evaluation of projects, based on the fuzzy inference system, with other classifiers reported in the bibliography.

4.1 Experiment 1: Comparison of the Proposal with Other Tools

In this experiment, the proposed ecosystem is compared with other tools reported in the bibliography. It is compared to tools for project management or project-oriented organizations Table 3.

Table 3 Comparison of the proposal with other tools

	BusinessRedmine	Clarizen	MS project server	Jira	Basecamp	Redmine
Open source	Yes	No	No	No	No	Yes
Collaborative work	Yes	Yes	No	Yes	Yes	Yes
Task management	Yes	Yes	Yes	Yes	No	Yes
Resource management	Yes	Yes	Yes	No	No	No
Finance management	Yes	Yes	Yes	No	Yes	No
Stakeholders management	Yes	Yes	Yes	Yes	Yes	No
Contract management	Yes	No	No	No	No	No
Risk management	Yes	Yes	No	No	Yes	No
Feasibility analysis	Yes	No	No	No	No	No
Automatic notifications (email)	Yes	No	No	Yes	Yes	Yes
Project management indicators	Yes	Yes	Yes	Yes	Yes	No
Kanban method integration and SCRUM agile methods	Yes	No	No	Yes	Yes	Yes
Learning curve (easy, medium, complex)	Easy	Mean	Easy	Hard	Easy	Easy
Ability for customization and sovereign development	Yes	No	No	No	No	Yes

Table 3 identifies the versatility of the proposed platform and its capacity for process integration. Furthermore, the fact that it was developed using free software technologies allows stakeholders to have access to the four freedoms of free software. The proposed platform incorporates a greater number of functionalities than the rest of the platforms used in the comparison.

4.2 Experiment 2: Analysis of the System Implementation Process in Different Scenarios

The results of the system implementation process are presented in the following four scenarios:

- Implementation in the Network of Centers of the University of Informatics Sciences since 2010. In this scenario, it was installed under the commercial name GESPRO. The calculated availability of the overall assembled system is around 92%.
- Implementation in the CERVEM Company, for the management of business processes under a project approach since 2018.
- Implementation in Engineering and Project Management Network (RCGP) with an impact on the training of the Master in Project Management postgraduate program, since 2020.
- Implementation of the ETECSA company for the management of research and innovation projects in 2021.

As part of the analysis of the results, a survey was carried out with 32 clients of the system to analyze the variable "time spent analyzing projects and generating status reports." The distribution of the professionals surveyed was as follows: 14 project managers, 5 project planners, 7 project management office inspectors and 6 project quality reviewers. Measurements were made of the time spent analyzing projects and preparing status reports before and after the implementation of the solutions proposed by the BusinessRedmine ecosystem.

Respondents selected one of the following responses:

- less than 30 min.
- between 1 and 2 h to prepare a report.
- between 3 and 7 h to prepare a report.
- about 24 h to prepare a report.
- about 32 h to prepare a report.
- more than 40 h to prepare a report.

The results were satisfactory, achieving in some cases a reduction in time from 1 week to minutes. As part of the validation of the proposed model, the increase in productivity was calculated during the development of product customizations. Productivity is calculated from Eq. (1).

Table 4 Increase in productivity with the use of the ecosystem in customizations

Real sceneries	Product scope	Human resources	Time (days)	Efficiency
UCI	64	10	45	1.16
CERVEM	16	3	10	1.23
RCGP	32	3	10	2.46
Telecommunications company	32	2	7	3.56

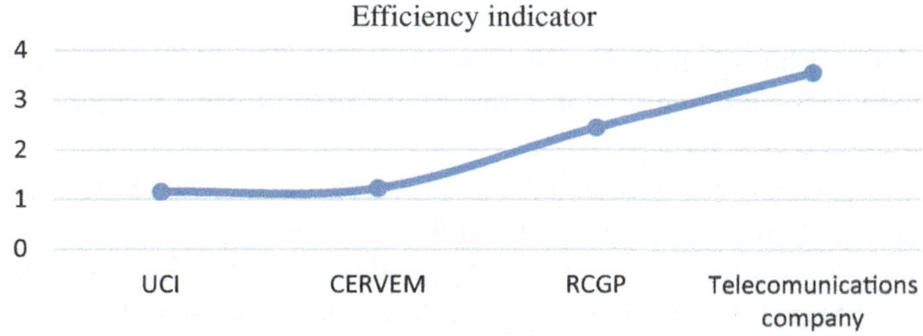

Fig. 1 Increase in productivity by reuse of BusinessRedmine software assets in different scenarios

$$productivity = \frac{product\ scope}{resources\ amount + time} \tag{1}$$

- Product scope is expressed in the number of use cases that were necessary to develop in the implementation process.
- Amount of Resources: refers to the amount of human resources that participated in the implementation project in a specific scenario.
- The time variable indicates the total time it took to set up the application in the specific scenario.

Table 4 represents a sample of the data associated with the development of the different customizations of the ecosystem products, ordered by the order of their occurrence over time.

As can be seen in Table 4 and in Fig. 1, there is an increase in productivity motivated by the use of software ecosystems and LPS that promote the reuse of all assets.

4.3 Experiment 3: Analysis of the Behavior of the Project Evaluation Subsystem

One of the key elements in the work of the platform is the introduction of artificial intelligence techniques to help decision-making by project managers, users of the platform. This element is specified in all the functionalities related to the visualization of indicators in each of the management processes that the ecosystem supports. One of these functionalities being the project evaluation process that is based on a fuzzy inference system that in this experimentation we will identify as FIS_{BR}. In this section, an experiment is carried out to compare FIS_{BR} with other classifiers that allow classification and can be applied in the automatic evaluation of projects.

For this experiment, a set of classifiers representing different AI techniques are chosen, including neural networks, the descending gradient or evolutionary algorithms, and compared with the proposal made based on fuzzy inference systems FIS_{BR}.

- FIR.DM algorithm suggested by Nomura et al. [29]. This algorithm uses a method based on the descending gradient to construct and optimize a Sugeno Degree Zero fuzzy inference system.
- FS.HGD algorithm recommended by Ishibuchi et al. [30]. It is based on the combination of the heuristic search and the descending gradient.
- GFS.LT.RS algorithm proposed in [31] for the generation of fuzzy rules in the evaluation of projects. This technique implements evolutionary lateral adjustment of the membership functions in order to obtain linguistic models with higher levels of precision. It uses a rule representation model based on 2-tuples linguistic representation that allows side-scrolling of labels (slight left/right shifts of the original membership function).
- GFS.THRIFT algorithm submitted in [32], This algorithm generates fuzzy rules and then optimizes them using a basic genetic algorithm (GA).
- HyFIS algorithm recommended by Kim and Kasabov [33] is a hybrid neuro-fuzzy model used to build and optimize fuzzy systems. It also combines the prediction capabilities of neural networks with fuzzy inference systems, generating Mamdani-type rules.
- Algorithm based on WM: recommended by Wang and Mendel [34] in our particular case, it generates a set of candidate fuzzy rules for the evaluation of projects that are refined in a second phase. The generated rules are of the Mamdani type. In the exploitation stage, it uses a fuzzy Mamdani-type inference system and is able to show the rules that influenced the final decision.

The three techniques used for the validation of the results are listed below:

- Pre-experiment technique without initial observation: analysis of the results of applying the model in a case study. The objective of this pre-experiment is to evaluate the potential of the platform with the end users. For this experiment, a group of end users of the application, experts in project control and monitoring, was selected. Then a survey was applied to them related to their experiences in the use of the system and the facilities for the control and monitoring of the projects. Finally, the word computing technique was applied to identify the experts' evaluation.
- Normality analysis technique: techniques were applied to analyze the normality of the data.

- Experimental technique with initial and final observation: Comparison of the different algorithms proposed in the control and monitoring module in an experimental database. The objective of this experiment is to find the most efficient algorithm for project evaluation. In our context, the most efficient algorithm is the one that makes the fewest errors in the evaluation of projects, taking as a reference the previous classification by a human expert.

4.3.1 Analysis of the Results of the Pre-experiment Technique Without Initial Observation

To evaluate the proposed model from experts, a questionnaire was developed in which 53 specialists dedicated to the control and monitoring of projects were involved. Specialists include project managers, project management office specialists, and development center managers. The objective of the questionnaire was to know the existing criteria about the quality of the proposed model and the indicators. This questionnaire was based on four criteria:

- Is the proposal for project evaluation applicable in different scenarios?
- Do you consider the proposal understandable?
- Is the proposed model usable, comfortable to be applied?
- Do the proposed indicators cover the knowledge areas of project management?

To validate the reliability of the prepared questionnaire, it was subjected to validation by 14 experts and using the SPSS tool, the Cronbach's Alpha coefficient test was executed on the data collected. This type of test is applicable to questionnaires with Likert-type responses, where each item or question is answered according to numbers that represent degrees of agreement or disagreement (Very High, Very High, High, Medium, Low, Very Low, None).

Cronbach's alpha coefficient was equal to 0.873. According to the literature, it is considered that alpha values greater than 0.7 are sufficient to guarantee the reliability of the scale. Given the positive result obtained in this test, it was considered valid to apply the aforementioned questionnaire.

Once the questionnaire is applied, computer techniques with words are applied to consolidate the criteria obtained in it. The following steps were followed:

1. A basic set of linguistic terms (LBTL) is defined for the evaluation of the criteria. LBTL = {not at all, very low, low, medium, high, very high, perfect}.
2. The specialists evaluate each criterion using some linguistic terms, see Table 5.

1. Expert preference is transformed into fuzzy sets based on the linguistic variable.
2. Following the 2-tuples model [35] of computation with words, the evaluations of the experts are added, consolidating them for each criterion to be evaluated. See Eq. (2)

$$x^e(x) = \left(\frac{1}{n}\sum_{i=1}^{n}{}^{-1}((s_i, \alpha_i))\right) = \left(\frac{1}{n}\sum_{i=1}^{n}\beta_i\right) \tag{2}$$

Table 5 Structure of the evaluation of the specialists for each criterion. C_i represents the criteria and e_j represents the specialists

Criteria	Experts		
	e_1	…	e_m
c_1	x_1^{11}	…	x_1^{1m}
⋮	⋮	⋮	…
c_p	x_1^{p1}	…	x_1^{pm}
c_1	x_2^{11}	…	x_2^{1m}
⋮	⋮	⋮	…
c_1	x_n^{11}	…	x_n^{1m}
⋮	⋮	⋮	…
c_p	x_n^{p1}	…	x_n^{pm}

Table 6 Evaluation results

Criteria	Summarized expert evaluation
Applicability in different scenarios	High
Comprehensibility of the proposal	Very high
Usability of the proposal	Mean
Coverage of the indicators of the areas of knowledge of project management	High
Final aggregation	**High**

3. Analyze the results see Table 6.

The criterion with the lowest evaluation was "Usability of the proposal" motivated by the use of non-traditional methods and the need to have platforms such as R that implement the proposed algorithms. The best evaluated criterion was the comprehensibility and applicability of the proposal. This is a positive element that shows the acceptance of the experts consulted for the application of the model in different scenarios. In general, the final evaluation given by the experts to the proposal was High.

4.3.2 Analysis of the Results of the Normality Analysis Technique

A knowledge base of finished projects from the software project development environment of the University of Computer Sciences was experimented with, with more than 10 production centers dedicated to this activity. Each record in the database corresponds to a cut of a project where control and monitoring indicators are collected. The cut of project is a periodic evaluation activity of a project, usually weekly. In this activity, the project's progress indicators are reviewed and decisions are made.

The knowledge base used presents the following distribution: total of 7843 projects, 40% evaluated as Good, 14% evaluated as fair and 46% evaluated as bad. Each project cut-off was evaluated by human experts in such a way that there is a standard for comparing the quality of the classification results [36].

The experts evaluated the projects considering the indicators described in Table 6. The assumption of normality of the samples was verified through the Shapiro–Wilk test. This showed that for each result of the metrics, at least one technique has probability values (*p*-value) lower than 0.05, so the idea of normal distribution is rejected with 95% confidence. Since not all samples follow a normal distribution, the use of non-parametric tests is chosen to verify significant differences.

4.3.3 Analysis of the Results of the Comparison Experiment Technique of the Different Algorithms Proposed

This experiment is applied on the same database of finished projects explained in the previous section and as a continuity of the analysis of the behavior of the algorithms suggested on it. For the comparison of the algorithms, the following set of metrics are defined as measurement criteria:

- Percent correct classifications: average achieved among the 20 partitions created for random cross-validation
- Number of false positives
- Number of false negatives
- MSE, following Eq. (3).

$$\mathrm{MSE} = \frac{1}{\mathrm{n}}\sum_{\mathrm{i}=1}^{\mathrm{n}}\left(\mathrm{Y}_{\mathrm{real}} - \mathrm{Y}_{\mathrm{pred}}\right)^2 \tag{3}$$

- SMAPE, following Eq. (4).

$$\mathrm{SMAPE} = 100 * \frac{1}{n}\sum_{i=1}^{n}\frac{\left|Y_{real} - Y_{pred}\right|}{\left(Y_{real} + Y_{pred}\right)/2} \tag{4}$$

where n is the number of validation cases, Y_{real} is the expected output value and Y_{real} is the output calculated by the system. The metrics are calculated on the cases selected for the tests, once the algorithm learned from the training cases.

For the comparison of the algorithms proposed in the model, the random cross-validation method is applied with k = 20 test iterations and n = 7, this last parameter representing the 7 proposed algorithms. For each iteration, the data set is randomly divided into a training set with 70% of the cases and a test set with the remaining 30%.

As they are related samples, the Wilcoxon and Friedman tests are used. In both cases, significant differences are considered when p-value < 0.05. In the case of the Wilcoxon test, the Monte Carlo method is applied for 99% of the confidence interval. During the application of the tests, in the cases where significant differences were found between the techniques, different groups of techniques were formed.

The groups formed meet the following property: the techniques of the same group do not have significant differences between them. Groups also have the following property: Group1 < Group2 < ... Group_n. The techniques in the smaller groups are those that report the best results, while those located in the higher groups have the worst results.

Regarding the metric "Percent correct classifications", the Friedman test was applied, comparing the samples obtained from applying the algorithms. In this context, the best result is considered to be the technique that statistically contributes the greatest value to the metric. As a result, significant differences are found between the techniques with p-value = "7.8547750364149e−18". The Wilcoxon test is applied, conveniently comparing two-by-two techniques, finally obtaining the following groups

Group 1: FIS_{BR}, FS.HGD, FIR.DM
Group 2: GFS.THRIFT
Group 3: GFS.LT.RS
Group 4: WM
Group 5: HyFIS

Regarding the metric "Number of false positives", the Friedman test was applied, comparing the samples obtained from applying the algorithms. In this context, the best result is considered to be the technique that statistically contributes the lowest value of this metric. As a result, there are significant differences between the techniques with p-value = "1.57230970124996e−17". The Wilcoxon test is applied, conveniently comparing two by two techniques, finally obtaining the following groups:

Group 1: FIS_{BR}
Group 2: WM, HYFIS, FIR.DM
Group 3: FS.HGD, GFS.LT.RS, GFS.THRIFT

In this case, the FIS_{BR} obtained the best results because it was the technique that obtained the least false positives. Getting false positives (rating Good when the project is actually rated Bad) in the area of project management is detrimental; Evaluating the Good of a project that is not excellent can cause damage in decision-making and cause unrealistic planning.

Regarding the metric "Number of false positives" in this context, the best result is considered to be the technique that statistically contributes the lowest value of this metric. The Friedman test was applied, comparing the samples obtained. As a result, significant differences are found between the techniques with p-value = "5.19919647410999e−14". The Wilcoxon test is applied, conveniently comparing two by two techniques, finally obtaining the following groups:

Grupo 1: WM
Grupo 2: FIS_{BR}, GFS.LT.RS
Grupo 3: FS.HGD, HYFIS, GFS.THRIFT
Grupo 4: FIR.DM

It can be affirmed that the WM technique obtains significantly better results in terms of the number of false negatives on the set of test cases.

Obtaining false negatives (rating Bad when the project is actually rated Good) in the project management area is not as damaging as false positives, but evaluating a project that is not actually rated as Bad may lead to taking bad decisions.

Regarding the "MSE" metric, in the context of this experiment, the technique that statistically contributes the least value to the metric is considered the best result. The Friedman test was applied, comparing the samples obtained. It was shown that there are significant differences between the techniques with p-value = "1.45922235866844e−16". The Wilcoxon test is applied, conveniently comparing two by two techniques, finally obtaining the following groups:

Group 1: FIS_{BR}, FIR.DM
Group 2: GFS.THRIFT, WM
Group 3: GFS.LT.RS
Group 4: FS.HGD
Group 5: HyFIS

It can be stated that the FISBR and FIR.DM techniques obtain significantly better results in terms of the mean square error over the set of test cases.

Regarding the "SMAPE" metric, in this context, the best result is considered to be the technique that statistically contributes the lowest value of this metric. The Friedman test was applied comparing the samples obtained. As a result, significant differences are found between the techniques with p-value = "1.45922235866844e−16". The Wilcoxon test is applied, conveniently comparing two by two techniques, finally obtaining the following groups:

Grupo 1: FIS_{BR}
Grupo 2: WM, FIR.DM
Grupo 3: GFS.LT.RS, GFS.THRIFT
Grupo 4: FS.HGD
Grupo 5: HYFIS

It can be stated that the FIS_{BR} algorithm obtains significantly better results in terms of the percentage error of the symmetric absolute mean over the set of test cases. FIS_{BR} technique generates the best system for project evaluation since it achieves the best results regarding the metrics: percentage of correct classifications (97%), number of false positives, MSE and SMAPE.

The fuzzy rules generated from the algorithms allow the projects to be evaluated efficiently, taking into account the high percentage of correct classification obtained in them. In addition, with the use of the proposed techniques, the treatment of phenomena such as imprecision, vagueness and uncertainty contained in the primary data is achieved, as well as the changing conditions in management styles and the level of maturity reached by the organization that provides the data.

5 Conclusions

From the analysis of the results previously presented, the following conclusions are reached:

- The development of a software ecosystem to aid decision-making for the integrated management of projects based on open source technologies and free software was identified as an opportunity. The BusinessRedmine ecosystem was developed with functionalities aligned with international standards in the domain of project management.
- The introduction of the results into social practice made it possible to significantly reduce time and costs in the control and monitoring of projects in the beneficiary organizations.
- Artificial intelligence techniques are incorporated into the proposed ecosystem to aid decision-making, enabling a better treatment of phenomena such as imprecision, vagueness and the uncertainty contained in the primary data. The proposed ecosystem makes intensive use of these techniques both for the classification of projects and for the visualization of the results.
- The use of algorithms such as FIS_{BR} allow the system to learn from evidence and can then use the knowledge to classify new instances. It facilitates the adaptation of systems to changing environments associated with the management styles and the level of maturity reached by the organizations that manage the projects.
- It was found that the FIS_{BR} technique had a significantly better performance than other techniques reported in the bibliography, showing 97% effectiveness in classification.

As future work, the authors propose the combination of systems to help the automatic classification of projects with recommendation systems that improve the usability of the models and platforms for project management.

References

1. Christophe, C., Daan, A., Ingrid, A.: BIM and ISO 19650 from a Project Management Perspective. European Federation of Engineering Consultancy Associations (2018) [Online]. Available: https://www.efcanet.org/sites/default/files/2020-01/390764_BIM%20booklet.pdf. Accessed: 12-May-2021
2. ISO: ISO 19650-1: Organization and Digitization of Information About Buildings and Civil Engineering Works, Including Building Information Modelling (BIM)—Information Management Using Building Information Modelling. Part 1: Concepts and Principles (2018). Available in: https://www.iso.org/standard/68078.html. Accessed: 12-April-2021

3. Mesároš, P., Mandičák, T., Behúnová, A.: Use of BIM technology and impact on productivity in construction project management. In: Wireless Networks (2020). ISSN 1572-8196. http://doi.org/10.1007/s11276-020-02302-6
4. Mahdi, G.S.: Algoritmos de estimación de distribuciones con tratamiento de restricciones para la construcción de cronogramas de proyectos, Tesis de Doctorado, Departamento de Investigaciones en Gestión de Proyectos, Universidad de las Ciencias Informáticas (2021)
5. Loyola, M.: Encuesta Nacional BIM 2019: Informe de Resultados. Universidad de Chile, Santiago (2019). ISBN: 978-956-401-111-0. Available: www.bim.uchilefau.cl. Accessed: 21-July-2021
6. Cudzik, J., Radziszewski, K.: Artificial intelligence aided architectural design, education and research in computer aided architectural design in Europe. In: Proceedings of the 36rd eCAADe Conference. Faculty of Civil Engineering, Architecture and Environmental Engineering, Lodz University of Technology (2018). ISBN 978-94-91207-15-0. Available: https://mostwiedzy.pl/pl/publication/artificial-intelligence-aided-architectural-design,146441-1. Accessed: 21-July-2021
7. González, C.F.: Metodología BIM (Building Information Modeling) aplicada a la prevención de riesgos laborales (PRL). J. BIM Constr. Manag. **1**(1), 20–30 (2019). ISSN 2659-6962
8. Hoar, C., Atkin, B., King, K.: Artificial Intelligence: What It Means for the Built Environment. Royal Institute of Chartered Surveyors (RICS), London (2017). Global/august 2017/DML/22396/RICS INSIGHTS. Disponible en: www.rics.org
9. Dave, B., Buda, A., Nurminen, A., Främling, K.: A framework for integrating BIM and IoT through open standards. Autom. Constr. **95**, 35–45 (2018). ISSN 0926-5805. http://doi.org/10.1016/j.autcon.2018.07.022
10. DunbaR, G.: Project Management Failures—Standish (Chaos) Reports (2016) [Online]. Available in: https://www.linkedin.com/pulse/project-management-failures-standish-chaos-report-2015-dunbar. Accessed: 21-April-2021
11. Lynch, J.: Project Resolution Benchmark for IBEX Financial Corp, p. 11. The Standish Group International, Inc (2018)
12. Johnson, J.: CHAOS Report: Decision Latency Theory: It is All About the Interval (2018). ISBN 978-0-692-04830-6
13. Pérez, I., García-Vacacela, R., Piñero, P.Y., Mahdi, G.S., Peña, M.: Experiencias en el uso de técnicas de softcomputing en la evaluación de proyectos de software. Investigación Oper. **41**(1), 108–119 (2020)
14. Mossalam, A., Arafa, M.: The role of project manager in benefits realization management as a project constraint/driver. HBRC J. 56–67 (2014)
15. Pacelli, L.: Grandes Errores en la Gestión de Proyectos. Resumen autorizado de: The Project Management Advisor: 18 Mayor Projetc Screw-ups, and Cut Them Off the Pass. Financial Times Prentice Hall (2004)
16. PMI: The Standard for Project Management and a Guide to the Project Management Body of Knowledge, 7th ed., p. 370. Project Management Institute (2021). ISBN: 978-1-62825-664-2
17. Northrop, L., Jones, M.G.: Introduction to software product lines adoption. In: SPLC (2013)
18. Krueger, C.W.: Mechanical product lifecycle management meets product line engineering. In: SPLC, pp. 316–320 (2015)
19. Bosch-Sijtsema, P.M., Bosch, J.: Plays nice with others? Multiple ecosystems, various roles and divergent engagement models. Technol. Anal. Strateg. Manag. **27**(8), 960–974 (2015)
20. Bosch-Sijtsema, P., Bosch, J.: Aligning innovation ecosystem strategies with internal, R&D. In: Proceedings of the 2014 IEEE ICMIT, 7th International Conference on Management, of Innovation and Technology, pp. 424–430, Singapore (2014). http://doi.org/10.1109/ICMIT.2014.6942464
21. Bosch, J.: ESAO: towards data and ecosystem driven R&D. In: Proceedings of the 18th International Software Product Line Conference, ACM, vol. 1, pp. 363–363 (2014)
22. Jansen, S., Handoyo, E., Frota, C.: Scientists' needs in modelling software ecosystems. In: Proceedings of the 2015 European Conference on Software Architecture Workshops, Dubrovnik/Cavtat, Croatia, pp. 1–44 (2015). http://doi.org/10.1145/2797433.2797479

23. Krueger, C., Clements, P.: Second generation systems and software product line engineering. In: SPLC 2015, pp. 388–389 (2015)
24. Maninkas, K., Marius, K.: Software ecosystems a systematic literature review. J. Syst. Softw. **86**(5), 1294–1306 (2013). https://doi.org/10.1016/j.jss.2012.12.026
25. CHAOS: Group 451, Open Source Is Not a Business Model, CHAOS Report [Online]. Available: https://www.google.com.cu/url?sa=t&rct=j&q=&esrc=s&source=web&cd=1&cad=rja&uact=8&ved=0CBsQFjAAahUKEwicyuimusfIAhWFJh4KHctEBSc&url=http%3A%2F%2Fdownload.microsoft.com%2Fdownload%2F0%2F4%2F2%2F04246FB1-0BF6-44F4-BC44-4CCB220E1711%2F451_-_08_OCT_-_CAOS_Report_9_Open_Source_is_Not_a_Business_Model.pdf&usg=AFQjCNHxtYWz4J9z-jVQDZteoXP180IxBQ&bvm=bv.105039540,d.dmo. Accessed: 4-Nov-2021
26. Eclipse Foundation: The Eclipse Foundation—open source community website. Ottawa, Ontario, Canada (2012)
27. Piñero, P.Y., Pérez, I., et al.: BusinessRedmine Sistema para la toma de decisiones empresariales y la gestión de proyectos. Número de Registro de Propiedad Intelectual: CENDA 4076-12-2019, Cuba
28. SEI: CMMI for Development v1.3, SEI CMMI Production, CMU/SEI-2010-TR-034. Technical report. Software Engineering Institute (2010)
29. Nomura, H., Hayashi, I., Wakami, N.: A learning method of fuzzy inference rules by descent method. In: IEEE International Conference on Fuzzy Systems, San Diego, CA (1992)
30. Ishibuchi, H., Nozaki, K., Nozaki, K., Tanaka, H., Tanaka, H., Hosaka, Y.: Empirical study on learning in fuzzy systems by rice taste analysis. Fuzzy Sets Syst. **64**(2), 129–144 (1994)
31. Alcalá, R., Alcalá, J., Herrera, F.: A proposal for the genetic lateral tuning of linguistic fuzzy systems and its interaction with rule selection. IEEE Trans. Fuzzy Syst. **15**(4), 616–635 (2007)
32. Thrift, P.: Fuzzy logic synthesis with genetic algorithms. In: Proceedings of the Fourth International Conference on Genetic Algorithms (ICGA91), San Diego, United States of America (1991)
33. Kim, J., Kasabov, N.: HyFIS: adaptive neuro-fuzzy inference systems and their application to nonlinear dynamical systems. Neural Netw. **12**(9), 1301–1319 (1999)
34. Wang, L.X., Mendel, J.: Generating fuzzy rule by learning from examples. IEEE Trans. Syst. Man Cybern. **22**(6), 1414–1427 (1992)
35. Herrera, F., Martınez, L.: A 2-tuple fuzzy linguistic representation model for computing with words. IEEE Trans. Fuzzy Syst. **8**(6), 746–752 (2000)
36. Torres, S., Piñero, P.Y., Capretz, L.F.: Creation and evaluation of software teams—a social approach. Int. J. Manuf. Technol. Manag. **28**(4) (2014). http://doi.org/10.1504/IJMTM.2014.066695

Part III
Knowledge and Human Resources Management Assisted by Artificial Intelligence

Chapter 12
Team Formation Integrating Various Factors: Model and Solution Approach

Ana Lilian Infante, Margarita André, and Alejandro Rosete

Abstract Nowadays team formation has become a very active research topic and it is recognized as a key process for the success of organizations. This article describes a model for multiple team formation that, unlike existing models, it integrates a diversity of factors, both individual and collective and therefore it can be used in various contexts. The model includes the following factors: competences, workload, cost of communication in the team, incompatibility between members, cost of working at a distance, interest in playing the role, interest in working in the team and balance of psychological characteristics in the team. The model responds to a combinatorial optimization problem, for that reason to solve the problem it is used a library of classes of meta-heuristic algorithms. To demonstrate the applicability of the model, the results of using the model in the formation of student teams are presented.

Keywords Individual and collective factors · Meta-heuristics · Multiple team formation

1 Introduction

It is recognized by numerous authors that team quality is a critical factor for project success [1–4]. PMBOK defines [3] a project team as a set of individuals with assigned roles and responsibilities who work together to achieve a common project objective.

Project resource management includes the processes of identifying, acquiring and managing the necessary resources for the successful completion of the project. The process of planning and estimating human resources includes identifying the roles to

---

A. L. Infante (✉) · M. André · A. Rosete
Universidad Tecnológica de La Habana "José Antonio Echeverría", Havana, Cuba
e-mail: ainfante@ceis.cujae.edu.cu

M. André
e-mail: mayi@ceis.cujae.edu.cu

A. Rosete
e-mail: rosete@ceis.cujae.edu.cu

P. Y. Piñero Pérez et al. (eds.), *Artificial Intelligence in Project Management and Making Decisions*, Studies in Computational Intelligence 1035,
https://doi.org/10.1007/978-3-030-97269-1_12

be filled, their responsibilities, the number of people needed and the competences or skills required for each role. When the human resources are planned and estimated, it is necessary to form the team [3].

When the team is formed, it's possible to improve its performance (e.g. developing the competences of team members and improving relationships among team members). However, the team formation is the key aspect.

Although PMBOK highlights the importance of the proper team formation and the multiple factors that must be considered in this process, it does not specify how to form a team.

The team formation problem is a complex process, due to it requires taking into account several factors, both individual and collective. Among the individual factors are [5–7]: the competences required to play a role in the team, workload, experience and preferences for playing a certain role or participating in a certain team. Among the collective factors are: the experience of collaborating together, the balance of psychological characteristics and the incompatibility among team members. This multitude of factors makes the formation process very difficult. The situation becomes more complex in larger organizations, with more people available, because it increases the number of combinations to be evaluated in order to assign the available people to the roles of the teams to be formed. To model the problem with the complexity described above, it's necessary to consider team formation as an optimization problem [8].

In the last decade, team formation has become a very active research topic, increasing the number of works in different environments such as: team formation in social networks [9–12], student team formation [13], software team formation [5, 14], sports team formation [2, 15–17] and others.

In the proposals mentioned above, the factors to be considered vary depending on the type of team to be formed and different factors can be considered in the same context.

Also, there are factors that are applicable in all contexts. However, none of the proposals include all the factors. A mono-objective solution approach is used even when considering multiple factors in the formation process [1, 9, 10, 14, 18–29]. In addition, few proposals include tools to support team formation and these are not configurable, which limits their applicability in different contexts [14, 20, 21, 27, 29–31].

As an antecedent of this work, in [5] it is developed a model for the Assignment of Human Resources to Software Project Teams that considers several factors, to form a software project team, such as: competences required to play the role, workload, incompatibilities between team members and cost of working remotely. It incorporates as a novel aspect, collective factors such as incompatibilities among members and elements associated with the psychological characteristics in the formation of the team as a whole. The importance of the psychological characteristics in the formation of the team has been discussed in several works [14, 32]. However, despite its advantages, it does not include certain factors as objective functions of the model, such as people's interest in playing a role or working in a given team, experience of collaborating together and the balance of psychological characteristics. This last aspect is treated as constraints in the model, but not as objective functions. These limitations imply that these factors cannot be taken into account in team formation.

As part of the work [5], the Teamsoft$^+$ tool [33] was developed to support the model. Originally, the model and the tool were conceived to form software teams, but it could be applied in other contexts due to the selection of factors to be considered, constraints and even the required competences and roles to be assigned to team members are configurable.

However, both the model presented in [5] and the tool that supports it were designed to form a single team; which limits their use in the formation of multiple project teams, such as in the formation of surgical teams, rescuers, students and professors. Using the tool developed to provide a solution to the above situations would involve forming teams one at a time (sequentially). As a consequence, as shown in [34], the first teams to be formed would be very good and the last ones would be bad, or it could even happen that the last teams are not formed because there are no people available according to the requirements established to perform the roles. Also, as the best available candidates are selected in each iteration, the teams formed would be very unbalanced in terms of the different factors considered.

Considering the limitations of the previous proposal and taking into account the research developed on the team formation problem, a new model was defined that allows the formation of multiple teams. The model takes into account three of the objective functions considered in the proposal of [5] (maximizing competences, balancing the workload and minimizing the cost of remote work). The communication cost is added as a new objective, which includes the incompatibilities among members and the experience of collaborating together in previous teams. The model incorporates the objective functions: maximizing the interest in playing the role, maximizing the interest in working in the team, maximizing the presence of Belbin roles and maximizing the diversity of psychological types (taking into account the MBTI psychological type of each person). In addition, it defines other objective functions with the purpose of balancing the teams based on the different factors.

The multiple team formation model responds to a multi-objective combinatorial optimization problem, for that reason meta-heuristics are used for its solution.

The objective of this work is to present a new model for the formation of multiple teams, which integrates a diversity of factors differing of the existing ones, so that it can be applied in different contexts. The results of the use of the model in the context of student team formation are presented, demonstrating its applicability.

The paper includes five sections. The first section describes the related work. The second section presents the model for the formation of multiple teams. The third section presents the solution of the model using a class library of meta-heuristic algorithms. The fourth section shows the experimental results that allow to demonstrate the applicability of the model in the context of student team's formation. Finally, the fifth section presents the conclusions of the work.

2 Related Works

In recent years there has been a notable increase in the number of papers dealing with team formation in different contexts. The following aspects were studied from the articles reviewed: factors taken into account in team formation, number of teams that form (one or multiple), solution approach (methods and algorithms), as well as if the solution is mono-objective or multi-objective in the case of works that model it as an optimization problem. In addition, it was analyzed how the authors perform the experiments (which data set is used and if it compares different algorithms) and whether the papers provide tools to perform the team formation process (indicated with an "*X*"), considering in each case whether the tool is configurable and applicable in different contexts.

There are papers that explain the application environment of the proposal and the factors considered but do not explain the solution approach (solution method or algorithm) or the use of the tool to support the proposed solution.

The individual factors considered in the analyzed proposals are:

1. Competences needed to play a role or to perform a task: It can be technical (specific to a job position) or generic (general for different job position).
2. Experience playing a role or performing a task: It considers the number of times he/she has played the role or performed the task and the evaluation received in this role or task.
3. Interest in playing the role or performing the task: Indicates the person's preference for playing a role or task.
4. Interest in working in a team: Indicates the person's interest in working in a team, taking into account the type of problem to be solved by the team.
5. Workload: Indicates the workload that the person already has as a result of being part to another team.
6. Cost of performing the function or task remotely: Considers the cost of the person performing the role according to their location with respect to the location of the rest of the team, as well as the cost of performing the role remotely.
7. Academic level of the person: Academic level achieved, for example: university, pre-university and others.

The collective factors are:

- Geographic proximity: Indicates how far away the team members are from each other.
- Experience of collaborating together: Indicates the experience that two people have been working together, taking into account the number of times they have worked together and the quality of the work done.
- Conflict among team members: Indicates the rate of compatibility that exists between two persons.
- Balance among psychological characteristics: Indicates that in the team there should be a diversity of psychological types and balance between team roles. The analyzed proposals consider MBTI psychological types and Belbin's team roles.

- Demographic diversity: Indicates diversity of gender, age, nationality and others.

The following is an analysis of the papers, grouped by application contexts.

2.1 Formation of Student Teams

Related with the formation of student teams fourteen papers were analyzed (Table 1), one of them is a systematic review.

These papers mainly consider the factors: competences, balance of psychological characteristics and conflict among members. Other factors considered, although in minor grade, are: experience in playing the role or performing the task, interest in playing the role or performing the task, interest in working in the team, academic level and demographic diversity.

The papers consider one, two or three factors for team formation. Only one paper [20] consider four factors. However, in all the papers that provide a solution, the problem is treated with a mono-objective approach, either because they consider only one of the factors during the team formation process or because they use a mono-objective solution method. The solution approach used in the papers are: meta-heuristics, heuristics, linear and integer programming, collision structure generation and constraint satisfaction algorithms.

The experimentation is carried out with generated data or real data from a specific educational institution. Only in two papers are compared the algorithms proposed by the author.

On the other hand, only three of the papers report the use of a tool. However, the tools are specified for the context of student teams.

In [41] the authors presented a systematic review of the literature about group formation for collaborative learning. The review includes 18 papers, from 2006 to 2016. The studies included in the review identify as factors to be considered in the formation of student groups: competences (used in six papers), learning styles (used in four papers), personality characteristics (used in four papers), incompatibilities among members, experience working together and Belbin roles.

The studies differ in which and how many factors they select. Five of the papers use a single factor, two papers use two factors and three studies use three factors. Regarding the techniques used in group formation, evolutionary algorithms (genetic algorithms) are the most used to solve the problem. In second place are clustering algorithms, frequently used to form homogeneous groups. Other techniques used although in minor grade are: multiagent, semantic ontologies and group technology [41].

Table 1 Papers about student team formation

Paper	Factors	Solution approach Algorithm	Experimentation		Tool
			Data set	Compare algorithms	
[18]	11	Evolutionary algorithms/mono-objective	10 generated data set	Two versions of evolutionary algorithm	–
[19]	10, 11	Constraint satisfaction algorithm/mono-objective	Real student data set	No	–
[20]	2, 7, 11, 12	GRASP, people sequential heuristic, genetic algorithm/mono-objective	Real student data set (Tuck Graduate Business School)	Three proposed algorithms	X
[35]	1	Not indicated	Does not include experiments		–
[21]	11	Collective intelligence, generation of collision structures and Bayesian learning/mono-objective	Real student data set (Universidad Politécnica de Valencia)	No	X
[22]	1	Genetic algorithm/mono-objective	Three generated data set	No	–
[36]	1	Not indicated	Does not include experiments		–
[37]	1, 10	Not indicated	Real student data set	No	–
[14]	1, 10, 11	Generation of collision structures-linear programming/mono-objective	Real student data set (Universidad Politécnica de Valencia)	No	X
[38]	1, 11	Not indicated	Does not include experiments		–
[39]	1, 4	Not indicated	Real data set (Capstone projects)	No	–
[40]	1	Integer programming/mono-objective	Does not include experiments		–
[23]	3, 10, 11	Integer linear programming/mono-objective	Real student data set	No	–

2.2 Formation of Experts Teams in Social Networks

Related with formation of expert teams in social networks ten papers were analyzed, as shown in Table 2. The factors considered are: competences, experience of collaborating together, conflict among members, experience playing the role or performing the task, interest in playing the role, workload and geographic proximity.

Table 2 Papers about team formation in social networks

Paper	Factors	Solution approach	Experimentation		Tool
			Data set	Compare algorithms	
[24]	1, 2	Heuristics/mono-objective	Marvel chronology project data set	Compare the proposed heuristics	–
[9]	1, 9	Heuristics/mono-objective	DBLP	Compare the three proposed heuristics	–
[25]	1, 2, 9	Genetic algorithm/mono-objective	Generated data set	Compare with exhaustive search	–
[10]	1, 9, 10	Heuristics (BRADO)/mono-objective	DBLP, IMDb	Compare with ICA, Genetic algorithm, PSO, RarestFirst, EnhancedStein	–
[13]	1	Not indicated	Does not include experiments		–
[42]	10	Not indicated	Does not include experiments		–
[26]	3	U2G/mono-objective	Real data of social networks CIA and EPINIONS	No	–
[43]	1, 8, 9	Genetic algorithm/mono-objective	GitHub data set	Compare with random algorithm, approximation rare algorithm and minimum cost contribution rare algorithm	–
[44]	1, 5, 9	Heuristics (RarestFirst, MinDisSol, Lbradius, EnStainer, MinLD, MinSD, MinAggrSol, MCC, LBSteiner)/multi-objective	Bibsonomy, DBLP, IMDB, StackOverFlow	Compare the heuristics	–
[27]	1, 9	Cultural algorithm/multi-objective	Real data set (team of patient care providers)	Compare with exact algorithm	X

The analyzed studies consider one, two and three factors. However, in most of the papers the problem is treated with a mono-objective approach. Only 25% of the developed papers in this context deal with the formation of several expert teams. The other 75% focus on the formation of a single team. The techniques and algorithms used in the solution are: heuristics and meta-heuristics.

For the experimentation, the authors use databases of known social networks and different algorithms are compared (either between the proposed algorithms or with the existing algorithms in other papers). One of the papers proposes a tool that supports this solution.

2.3 Formation of Sports Teams

Related with formation of sports teams four papers were analyzed. In all cases, the research focuses on the formation of a single team. The factors considered are: competences, experience playing the role and experience of collaborating together, as shown in Table 3. Two studies, take into account two factors and in the other two studies only one factor is taken into account. In most cases the problem is treated as a mono-objective. The algorithms used for the solution are: heuristic algorithms, integer programming and multi-criteria decision techniques.

For the experimentation the authors use real data and only in one of the works are two proposed methods compared. No tools are reported to support the solutions.

Table 3 Papers about sports team formation

Paper	Factors	Solution approach	Experimentation		Tool
			Data set	Compare algorithms	
[2]	1, 9	Heuristic/mono-objective	FIFA dataset of the year 2010–2011	Compare team harmony method with sum of the distance method	–
[15]	1, 2	Binary integer programming/mono-objective	Fifth season of Indian Premier League (IPL)	No	–
[16]	1	Heuristic/mono-objective	Turkish Volleyball League	No	–
[17]	1	Multicriteria decision making	Lithuanian Basketball League	No	–

2.4 Formation of Professional Teams

Related with formation of professional teams were analyzed eight papers, as shown in Table 4. In this context, the aim is the formation project teams or expert teams. The factors considered are: competences, experience in performing the role or task, workload, experience of collaborating together, psychological characteristics, conflict

Table 4 Papers about professional team formation

Paper	Factors	Solution approach	Experimentation		Tool
			Data set	Compare algorithms	
[28]	1, 2, 3, 5, 10	Decision method/mono-objective	Real data set of Department of Industrial and Systems Engineering, University of Tennessee	Compare with manual team formation	–
[46]	1, 5, 11	Not indicated	Does not include experiments		–
[47]	1	Meta-heuristics/mono-objective	Does not include experiments		–
[48]	1, 2, 7, 11	Not indicated	Does not include experiments		–
[6]	1, 5, 9	Heuristics/mono-objective	Generated data set	Compare with constraint programming approach, local search heuristics and variable neighbourhood metaheuristics	–
[49]	1, 9	Heuristics/mono-objective	DBLP	Compare different heuristics (FindSkillAgg, EnSteiner, MinAggrSol, FindSkillDaim, MinDiaSol, RarestFirst)	–
[50]	1, 2, 9	Heuristics/mono-objective	Dataset of Microsoft Research Department, November 2015	Compare with IterL&F, RarestFirst, RF-Level and Kmeans	–
[30]	1, 10	Collaborative filtering recommendation algorithm	Does not include experiments		X

among members and interest in performing the role or task and academic level. The papers typically included one to five factors. Despite considering several factors, all the papers use a mono-objective technique to solve the problem. The algorithms used are: heuristics, meta-heuristics, integer linear programming and collaborative filtering recommendation algorithm. The experiments are generally carried out with real data and different algorithms are compared. Only one proposal is supported by a recommendation tool.

2.5 *Formation of Software Teams*

A software team is a kind of professional team. Related with formation of software teams six papers were analyzed; two of them are a systematic review of the topic. The factors considered are: competences, experience in playing the role or performing the task, workload, psychological characteristics, academic level, conflict among members and cost of working remotely, see Table 5.

In formation of software team's papers, all the proposals consider two or more factors. However, only one of the proposals [31] uses the multi-objective solution approach. The solution approach used are: meta-heuristics, heuristics and ontology. Most of the papers form a single team, only the proposal of [45] proposes to form multiple teams. In experimentation the authors generally use generated data and compare various algorithms.

Table 5 Papers about formation of software teams

Paper	Factors	Solution approach	Experimentation		Tool
			Data set	Compare algorithms	
[31]	1, 2, 5, 6, 10, 11	Meta-heuristics/multi-objective	Generated data set	Compare six multi-objective meta-heuristic algorithms	X
[45]	1, 2	Not indicated	Does not include experiments		–
[5]	1, 2, 5, 6, 10, 11	Meta-heuristics/mono-objective	Generated data set	Compare different variants of path meta-heuristic algorithms	–
[51]	1, 7	Ontology	Real dataset (3 finance, 3 e-learning, 2 e-health, 2 social networks)	No	–

One of the papers proposes a tool that supports this solution [31]. The tool is flexible and configurable, allowing the definition of competences, roles, people and projects.

In [52] is presented a systematic mapping of software team formation. In this paper 40 articles were analyzed. The factors considered in this context are: technical competences (80% of the papers), the cost of the person (calculated using the salary per hour of each team member) (35% of the papers), psychological characteristics (using the Belbin and Myers-Briggs tests) (27.5% of the papers), generic competences (25% of the papers) and social attributes (17.5% of the papers), which indicate the relations among members, for example: experience working together, compatibility among members and others.

In [52] more than 30 different techniques were identified to solve the team formation problem. These can be grouped into five categories. The most popular is search and optimization (62.5% of the papers), which includes genetic algorithms, dynamic programming and others. In particular, genetic algorithms represent the most used technique, present in 20% of the studies. The other categories are: mathematics and statistics (30% of the papers; including regression models, distance measures and statistical methods), probabilistic reasoning (27.5% of the papers; including fuzzy logic and Bayesian networks), decision making (12.5% of the paper; including gray decision theory, analytic hierarchy process and ontology techniques) and data mining (5% of the paper; including clustering and association rules) [52].

Otherwise, the authors in [53] has identified the vital team composition characteristics together with their attributes when forming an effective team. The elements were collected from various past studies regarding team formation for software development teams (from 2007 to 2019), including any related models. Using contents analysis, the study has identified six composition characteristics namely knowledge, skill, ability, attitude, experience and personality.

2.6 *Formation of Medical Teams*

Related with formation of medical teams were analyzed two papers (Table 6). In both of them is considered the formation of multiple teams. The factors considered are: competences, conflict among members, experience in the performance of the role and workload. In both cases, three factors are considered during the team formation process.

In [54] is used bio-objective integer programming and e-constraint method. In [29] are used meta-heuristics (genetic algorithm) and a mono-objective solution approach. In this paper, a free software tool is proposed that allows the formation of multiple surgical teams and is applicable in the context of medical team formation.

The papers carry out experiments with real data set and in one of the cases they compare different algorithms.

From the analysis of these works, several factors were identified that have been taken into account in the formation of groups/teams. As shown in Table 7, there

Table 6 Papers about medical team formation

Paper	Factors	Solution approach	Experimentation		Tool
			Data set	Compare algorithms	
[54]	1, 5, 10	Bio-objective integer programming/e-constraint method	Real data set of Belgian Hospital	Compare EWBB and Augmecon2 algorithm	–
[29]	1, 2, 10	Genetic algorithm/mono-objective	Real data set of 6065 orthopedic surgeries	No	X

Table 7 Number of studies that consider each factor by team formation context

Context (Type of team)/factors	1	2	3	4	5	6	7	8	9	10	11	12
Students	8	1	1	1	0	0	1	0	0	4	8	1
Social networks	8	2	1	0	1	0	0	1	6	2	0	0
Sports	4	1	0	0	0	0	0	0	1	0	0	0
Professionals	8	3	1	0	3	0	1	0	3	2	2	0
Software	4	3	0	0	2	2	1	0	0	2	2	0
Medical	2	1	0	0	1	0	0	0	0	2	0	0
Total	34	11	3	1	7	2	3	1	10	12	12	1

are no factors that are considered in only one context; even there are factors such as: competences, experience in playing a role or performing a task, workload and interest in playing a role or performing a task, experience of collaborating together, conflict among team members and the balance between psychological characteristics that are considered in most of the context analyzed.

Although not all factors have been considered in all contexts, there are factors that are applicable to any context; for example, the balance of psychological characteristics, considering the Belbin and Myers-Briggs tests, is applicable to any type of team [14, 32]. The fact of forming teams whose members have different psychological types and that there are people in the team with a preference for team roles (according to Belbin) favors having people with different strengths, which places the team in a better position to face complex projects or challenging objectives. On the other hand, a variety of factors are applied in the same context, depending on the environment of use. However, there is no proposed solution that includes the diversity of factors.

Approximate algorithms such as heuristics and meta-heuristics are mostly used as solution methods to solve optimization problems, as shown in Table 8.

Although most of the analyzed papers consider a diversity of factors in team formation, the papers use a mono-objective solution approach (28 papers). Only in

Table 8 Number of papers by solution approaches used in team formation contexts

Solution approach/context	Students	Social networks	Sports	Professionals	Software	Medical	Total
Heuristics	1	8	2	3	0	0	14
Meta-heuristics	3	2	0	1	2	1	9
Linear and integer programming	2	0	1	1	0	1	5
Multicriteria decision making	0	0	1	0	0	0	1
Generation of collision algorithm	1	0	0	0	0	0	1
Constraint satisfaction algorithm	1	0	0	0	0	0	1
Collaborative filtering recommendation algorithm	0	0	0	1	0	0	1
Ontology	0	0	0	0	1	0	1

two papers the authors use a multi-objective solution approach. However, the team formation problem is a multi-objective combinatorial optimization problem, so the multi-objective solution approach reflects the reality of the problem.

Due to the diversity of aspects to be considered in team formation, it is particularly useful to have a tool to support the solution, since it is difficult for a human to take into account the diversity of factors without the help of a decision support system. However, an analysis of the papers shows that only six of them have a tool to support them. The tools are tied to the context and not configurable, which limits their application in other environments.

3 Multiple Team Formation Model

The model presented can be applied in different contexts and the objectives and constraints to be considered can be selected according to the specific scenario of use. It includes a variety of factors such as: competences, experience, interest in playing the role, interest in working in the team, workload, cost of distance work, experience of collaborating together, conflict among members and balance of psychological characteristics (considering Belbin's team roles and psychological types such as MBTI).

The model can be applied to the formation of any number of teams. For the formation of multiple teams, it includes objectives that allow to create balanced teams taking into account the selected factors.

The model of multiple team formation is presented below.

t	Number of teams to be formed.
r_k	Number of roles to be assigned in the team k, $k = 1\ldots, t$.
n	Number of people available.
c_{ijk}	Competence of the person i to play the role j in the team k; $i = 1\ldots n$, $j = 1\ldots r_k$, $k = 1\ldots t$.
mlc_{cjk}	Minimum required level of the competence c to play the role j in the team $k, j = 1\ldots r_k, k = 1\ldots t$.
lc_{ic}	Level of the person i in the competence c.
s_{hi}	Incompatibility relations between the person h and the person i; $h, i = 1\ldots n$.
P_{ih}	Set of teams in which the person h and the person i have worked together; $h, i = 1\ldots n$.
e_{hi}	Number of times where the person h and the person i have worked together; $h, i = 1\ldots n$.
g_i	Workload of the person i according to the teams in which he/she currently work; $i = 1\ldots n$.
b_{jk}	Workload to play the role j in the team k; $j = 1\ldots r_k$, $k = 1\ldots t$.
ME	Average workload including the workload of the teams under execution and the workload generated to play the role in the team(s) under analysis, Eq. (1).

$$ME = \frac{\sum_{i=1}^{n}(g_i) + \sum_{k=1}^{t}\sum_{j=1}^{r_k} b_{jk}}{n} \tag{1}$$

l_{ijk}	Cost of the person i to play the role j in the team k; $i = 1\ldots n, j = 1\ldots r_k, k = 1\ldots t$.
NE_{jk}	Number of people required to play role j in team k; $j = 1\ldots r_k, k = 1\ldots t$.
$MaxR$	Maximum number of roles that a person can play, considering all the teams to which he/she has been assigned.
R	Number of considered incompatible sets of roles.
IR_r	Set of incompatible roles, $r = 1\ldots R$
$MaxWL$	Maximum workload for each person.
NP_k	Number of people that must have the team k.
$D = (d_{ij})$	a matrix of values 0 or 1, taking the value 1 if the person i has the Belbin's role j as preferred, $1 \leq i \leq n$; $1 \leq j \leq 9$.

$B = (b_{ij})$ — a matrix of values 0 or 1, taking the value 1 if the person i has the psychological type j (according to MBTI test), $1 \leq i \leq n$; $1 \leq j \leq 16$.

$InterestRole = (interestR_{ij})$ — a matrix of values 0, 0.5 or 1, taking the value 1 if the person i is interested in playing role j, 0 if the person i avoids playing the role j and 0.5 if the person i is indifferent to playing the role j, $1 \leq i \leq n$; $1 \leq j \leq r$.

$InterestTeam = (interestT_{ik})$ — a matrix of values 0 or 1, taking the value 1 if the person i is interested in belong to the team k and 0 otherwise, $1 \leq i \leq n$; $1 \leq k \leq t$.

$eval_k$ — Evaluation of the team k.

$x_{ijk} = 1$ — if the person i is assigned to the role j in the team k and 0 otherwise, where $i = 1 \ldots n$; $j = 1 \ldots r_k$, $k = 1 \ldots t$.

$u_{ik} = 1$ — if the person i is assigned to at least one role in the team k and 0 otherwise; where $i = 1 \ldots n$, $k = 1 \ldots t$.

$f_{qk} = 1$ — if there is at least one person with a preference for the Belbin role q in the team k and 0, otherwise; where $q = 1..9$, $k = 1 \ldots t$.

$m_{zk} = 1$ — if there is at least one person have the MBTI subtype z in the team k and 0, otherwise; where $z = 1 \ldots 16$, $k = 1 \ldots t$.

The functions to optimize are:

$$\max \sum_{k=1}^{t} \sum_{j=1}^{r_k} \sum_{i=1}^{n} c_{ijk} x_{ijk} \tag{2}$$

$$\min \sum_{k=1}^{t} \sum_{h=1}^{n-1} \sum_{i>h}^{n} \left(s_{hi} + \frac{1}{1 + e_{ih} + \sum_{y \in P_{ih}} eval_y} \right) u_{hk} u_{ik} \tag{3}$$

$$\min \sum_{k=1}^{t} \left(\sum_{i=1}^{n} \left[\left(g_i + \sum_{j=1}^{r_k} b_{jk} x_{ijk} \right) - ME \right]^2 \right) \tag{4}$$

$$\min \sum_{k=1}^{t} \sum_{j=1}^{r_k} \sum_{i=1}^{n} l_{ijk} x_{ijk} \tag{5}$$

$$\max \sum_{k=1}^{t} \sum_{j=1}^{r_k} \sum_{i=1}^{n} interestR_{ij} x_{ijk} \tag{6}$$

$$\max \sum_{k=1}^{t} \sum_{i=1}^{n} interestT_{ik} u_{ik} \tag{7}$$

$$\max \sum_{k=1}^{t} \sum_{q=1}^{9} f_{qk}, \ (1 - f_{qk}) \sum_{i=1}^{n} d_{iq} u_{ik} = 0, \quad (1 - f_{qk}) + \sum_{i=1}^{n} d_{iq} u_{ik} > 0 \tag{8}$$

$$\max \sum_{k=1}^{t} \sum_{z=1}^{16} m_{zk}, \ (1 - m_{zk}) \sum_{i=1}^{n} b_{iz} u_{ik} = 0, \quad (1 - m_{zk}) + \sum_{i=1}^{n} b_{iz} u_{ik} > 0 \tag{9}$$

Equation (2) maximizes the competences of the team members, which takes into account all the roles required and the possible people to play them. Equation (3) minimizes the cost of communication in the team, considering the incompatibility relationships among team members, the experience of collaborating together (number of times they have worked together) and the evaluation received each time they have collaborated together. Equation (4) balances the workload, minimizing the possibility that there are people who are unburdened or overburdened. Equation (5) minimizes the cost of remote work, taking into account the cost of developing the team's roles remotely. Equation (6) maximizes the interest of people to play the assigned role(s) in the team(s). Equation (7) maximizes people's interest in being members of the team or teams to be formed. Equation (8) maximizes the number of different Belbin roles present in the team. Equation (9) maximizes the number of different psychological types (according to the Myers-Briggs test) present in the team.

Another set of functions are applicable in the scenario in which multiple teams are formed, with the objective of balancing the teams considering different factors. The goal is to form balanced teams, avoiding the existence of teams with high indexes in the factors and others with low indexes.

$$\min \sum_{k=1}^{t} \left(\left| \frac{\left(\sum_{l=1}^{t} Comp_l\right)}{t} - Comp_k \right| \right), \quad \mathrm{Comp_k} = \sum_{j=1}^{r_k} \sum_{i=1}^{n} c_{ijk} x_{ijk} \tag{10}$$

$$\min \sum_{k=1}^{t} \left(\left| \frac{\left(\sum_{l=1}^{t} CommunicationCost_l\right)}{t} - CommunicationCost_k \right| \right),$$

$$\mathrm{CommunicationCost_k} = \sum_{h=1}^{n-1} \sum_{i>h}^{n} \left(s_{hi} + \frac{1}{1 + e_{ih} + \sum_{y \in P_{ih}} eval_y} \right) u_{hk} u_{ik} \tag{11}$$

$$\min \sum_{k=1}^{t} \left(\left| \frac{\left(\sum_{l=1}^{t} Workload_l\right)}{t} - Workload_k \right| \right),$$

$$Workload_k = \sum_{i=1}^{n} \left[\left(g_i + \sum_{j=1}^{r_k} b_{jk} x_{ijk} \right) - ME \right]^2 \tag{12}$$

$$\min \sum_{k=1}^{t} \left(\left| \frac{\left(\sum_{l=1}^{t} DistanceC_l\right)}{t} - DistanceC_k \right| \right),$$

$$DistanceC_k = \sum_{i=1}^{n}\sum_{j=1}^{t_k} l_{ijk}x_{ijk} \tag{13}$$

$$\min \sum_{k=1}^{t}\left(\left|\frac{\left(\sum_{l=1}^{t} RoleInterest_l\right)}{t} - RoleInterest_k\right|\right),$$

$$RoleInterest_k = \sum_{j=1}^{r_k}\sum_{i=1}^{n} interestR_{ij}x_{ijk} \tag{14}$$

$$\min \sum_{k=1}^{t}\left(\left|\frac{\left(\sum_{l=1}^{t} TeamInterest_l\right)}{t} - TeamInterest_k\right|\right),$$

$$TeamInterest_k = \sum_{i=1}^{n} interestT_{ik}u_{ik} \tag{15}$$

$$\min \sum_{k=1}^{t}\left(\left|\frac{\left(\sum_{l=1}^{t} RolesBelbin_l\right)}{t} - RolesBelbin_k\right|\right),$$

$$RolesBelbin_k = \sum_{q=1}^{9}(f_{qk}) \tag{16}$$

$$\min \sum_{k=1}^{t}\left(\left|\frac{\left(\sum_{l=1}^{t} MBTIType_l\right)}{t} - MBTIType_k\right|\right),$$

$$MBTIType_k = \sum_{z=1}^{16}(f_{zk}) \tag{17}$$

Equation (10) balances the competences index between the different teams. Equation (11) balances the cost of communication index between the different teams. Equation (12) balances the workload in the different teams. Equation (13) balances the cost of working at distance between the teams. Equation (14) balances the rate of interest in the assigned role(s) between the different teams. Equation (15) balances the interest rate in the team among the different teams. Equation (16) balances the Belbin role index among the different teams. Equation (17) balances the MBTI psychological type index among the different teams.

The model includes a set of individual constraints:

$$\sum_{i=1}^{n} x_{ijk} = NE_{jk}, \quad NE_{jk} \in \mathrm{N}, \forall k = 1 \ldots t \tag{18}$$

$$\sum_{j \in IRr} x_{ijk} \leq 1, \quad r = 1 \ldots R, \forall k = 1 \ldots t \tag{19}$$

$$\sum_{k=1}^{t}\sum_{j=1}^{r_k} x_{ijk} \leq MaxR, \quad MaxR \in \mathrm{N}, \forall k = 1 \ldots t \tag{20}$$

$$u_{ik}\left(mlc_{cjk} - lc_{ic}\right) \leq 0, \quad \forall c \in C_{jk}, \forall i = 1 \ldots n, \forall j = 1 \ldots r_k, \forall k = 1 \ldots t \tag{21}$$

$$g_i + \sum_{k=1}^{t}\sum_{j=1}^{r_k} b_{jk}x_{ijk} \leq MaxWL, \quad i = 1 \ldots n, \forall k = 1 \ldots t \tag{22}$$

$$(1 - u_{ik})\sum_{j=1}^{r_k} x_{ijk} = 0, (1 - u_{ik}) + \sum_{j=1}^{r_k} x_{ijk} > 0, \quad i = 1 \ldots n, \forall k = 1 \ldots t \tag{23}$$

$$\sum_{h=1}^{n-1}\sum_{i>h}^{n} s_{hi} * u_{hk} * u_{ik} = 0 \tag{24}$$

$$\sum_{i=1}^{n} d_{iq}u_{ik} > 0, \quad q = 1 \ldots 9 \tag{25}$$

$$\sum_{k=1}^{t} g_{i1} * u_{ik} = 0, \quad i = 1 \ldots n \tag{26}$$

The constraint (18) ensures that the roles in the teams are filled according to the number of people required to play them. The constraint (19) ensures that a person in a team cannot play at the same time roles that are considered incompatible with each other. The constraint (20) restricts the maximum number of roles that can be played by any person. The constraint (21) requires that the person playing a role must have the minimum competence requirements for that role in the team. The constraint (22) ensures that the total workload assigned to a person must not be greater than a maximum value. The constraint (23) guarantees that the variable u_i takes value 1 or 0. The constraint (24) does not allow certain incompatibilities selected by the decision maker, it guarantees that if the value of s_{hi} is 1 for any two persons, *h* and *i*, at least one of the two must stay out of the team. The constraint (25) requires the presence of all Belbin roles. The constraint (26) limits the inclusion in the team of a person if is already assigned as a leader in other teams.

A set of collective constraints are included in the model:

$$\sum_{q=1}^{3}\left(\sum_{i=1}^{n} d_{iq}u_{ik}\right) > 0, \sum_{q=4}^{6}\left(\sum_{i=1}^{n} d_{iq}u_{ik}\right) > 0,$$

$$\sum_{q=7}^{9}\left(\sum_{i=1}^{n} d_{iq}u_{ik}\right) > 0, \quad \forall k = 1 \ldots t \tag{27}$$

$$\sum_{q=1}^{3}\left(\sum_{i=1}^{n} d_{iq}u_{ik}\right) > \sum_{q=4}^{6}\left(\sum_{i=1}^{n} d_{iq}u_{ik}\right), \quad \forall k = 1\ldots t \tag{28}$$

$$\sum_{q=4}^{6}\left(\sum_{i=1}^{n} d_{iq}u_{ik}\right) > \sum_{q=7}^{9}\left(\sum_{i=1}^{n} d_{iq}u_{ik}\right), \quad \forall k = 1\ldots t \tag{29}$$

$$x_{i1k} \le d_{i1} + d_{i7}, \quad \forall k = 1\ldots t \tag{30}$$

$$\sum_{i=1}^{n} d_{i4}u_{ik} \ge 1, \quad \forall k = 1\ldots t \tag{31}$$

$$x_{i1k} \le \frac{\left(\left(\sum_{z=1}^{8} b_{iz}\right) + b_{i1} + b_{i3} + b_{i5} + b_{i17} + b_{i9} + b_{i11} + b_{i13} + b_{i15}\right)}{2}, \quad \forall k = 1\ldots t \tag{32}$$

The set of constraints (27) ensures that the three categories of roles proposed by Belbin (action roles, mental roles and social roles) are represented in the team. The constraint (28) ensures that in the team the preference for action roles must be higher that the preference for mental roles. The constraint (29) ensures that in work teams, the preference for mental roles must be higher that the preference for social roles. The constraint (30) guarantees that the person who play the Team Leader role must have a preference for Belbin's roles: Shaper or Coordinator, assuming that in a matrix D the preference to play the Shaper and Coordinator roles are recorded in columns 1 and 7, respectively. The constraint (31) guarantees that at least one person in the team must have as preferred the mental role Brain, assuming that in column 4 of matrix *D* the preference or not for the role Brain is recorded. The constraint (32) guarantees that the person who develops the Team Leader role must be extroverted and planned, that is, the E_ _J subtype, considering the psychological types (according to the Myers-Briggs test), assuming that in a B matrix, the subtypes that include the *E* dimension (Extroverted) are recorded in columns 1 to 8 and those that include the *J* dimension (Judgment) in columns 1, 3, 5, 7, 9, 11, 13 and 15.

Another set of constraints are associated with the formation of multiple teams:

$$\sum_{k=1}^{t}\sum_{i=1}^{n} u_{ik} = n \tag{33}$$

$$\sum_{i=1}^{n} u_{ik} = NT_k, \quad NT_k \in \mathrm{N}, \forall k = 1\ldots t \tag{34}$$

$$\sum_{k=1}^{t} u_{ik} = 1, \quad \forall i = 1\ldots n \tag{35}$$

The constraint (33) ensures that all available persons are assigned to some team. This constraint is used in student team formation context, where the instructor must group all students in teams. The constraint (34) restricts the number of people of the team. The constraint (35) restricts that each person to be assigned to only one team.

Although the model is extensive due to the number of factors and constraints, is possible select the functions and constraints would be used in a given context. This feature allows its use in different contexts and scenarios.

4 Solution Approach to the Multiple Team Formation Model

Multiple team formation is a multi-objective combinatorial optimization problem. There are different ways to represent the preferences of decision makers in multi-objective optimization problems [55]. The weighted factors method is based on the idea of converting the multi-objective problem into a scalar problem by constructing a single objective function which is the sum of the initial objective functions, weighted by a weight. The lexicographic method only requires establishing the priority of the objectives, it is not necessary to define the weight for each objective. Although it is easier for the decision-maker to prioritize objectives than to assign them weights, both methods require that these be defined early in the process, in many cases without any basis for doing so. In many situations, the decision-maker is clear about the objectives to be considered, but not about the priority or weights to be assigned to each objective. With the pure multi-objective method, it is not necessary to define neither the priority nor the weights. This method reflects the multi-objective reality of the problem because it treats the objectives separately [55].

Due to the feasibility of using meta-heuristics in the solution of optimization problems, such algorithms are used in the solution of the proposed model. The solution design includes the use of the BICIAM class library [56], which provides a variety of trajectory and population-based meta-heuristic algorithms, both mono-objective and multi-objective. The use of the library allows the reuse of already implemented algorithms, employing good problem implementation practices, by separating the logic of the meta-heuristics from the logic of the problem, using design patterns.

In BICIAM and optimization problem follows the approach proposed in [57], where a codification or representation of a solution of the problem (Codification class), operators (Operator class) and objective functions (ObjectiveFunction class) must be defined, as shown in Fig. 1. It also includes: The Problem class, in charge for manipulating the problem and the State class that represents a solution.

As shown in Fig. 1, to configure the team formation problem, it is necessary to define a set of classes (presented in orange). The TeamFormationProblem class handles the problem. The TeamFormationCodification class represents the codification of the problem and manipulates the constraints. The TeamFormationState class

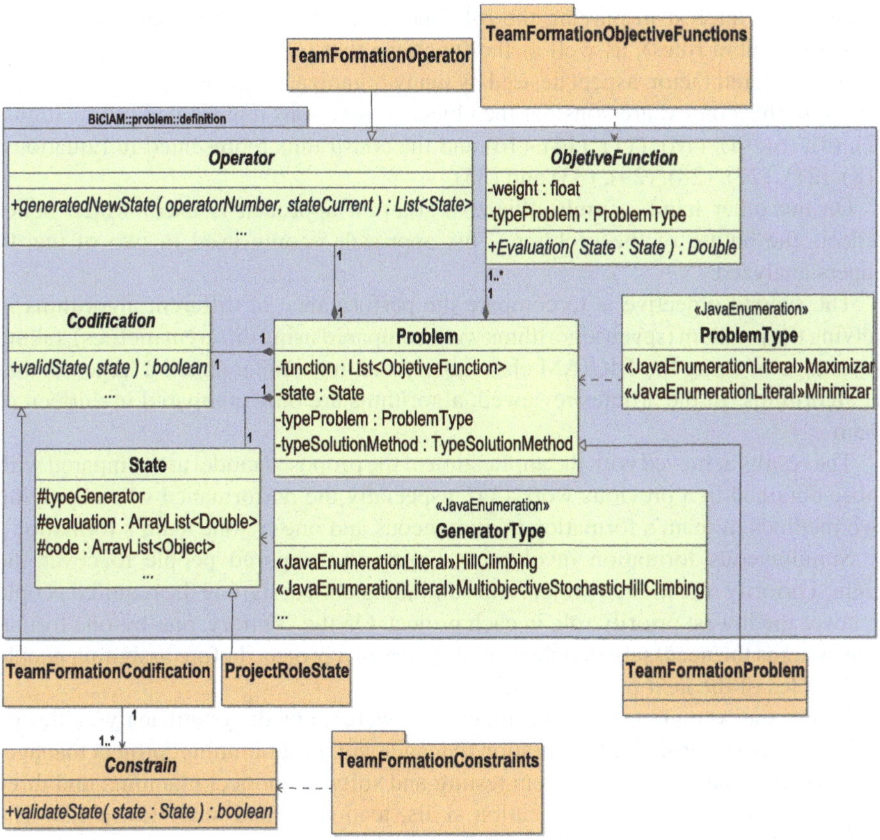

Fig. 1 Class diagram representing the multiple team formation problem in BICIAM

represents a solution. The TeamFormationOperator class implements the operator. There are a class per each objective function (which inherit from the ObjectiveFunction class) and there are a class per each constraints of the problem (which inherit from the Constraint class).

5 Experiments

The experiment has several objectives. In the first place, apply the model to form teams in a real environment, such as the student's team formation to develop projects. The paper used a real dataset of students for the Software Engineering III course at the *Universidad Tecnológica de La Habana*, presented in [34]. For the formation of the teams, different factors are taken into account (competences, cost of communication in the team, interest in playing the role and psychological characteristics considering the Belbin roles), as well as the functions that balance the teams in terms of each considered factor, aspect desired by many organizations when forming multiple teams. In this context are consider the objective functions represented in Equations: (2), (3), (6), (8), (10), (11), (14), (16) and the constraints represented in Equations: (18), (21), (27), (28), (29), (33) and (35).

On the other hand, a multi-objective solution approach is used, which better reflects the nature of the problem. This approach is only used in two of the 44 papers analyzed.

The second objective is to compare the performance of different algorithms in solving the problem (seven algorithms were compared using different metrics), taking advantage of using the BICIAM class library that has implemented a large number of algorithms. In the articles reviewed, algorithms are only compared in thirteen of them.

The results achieved with the application of the proposed model are compared with those obtained in a previous work [34], especially the performance when applying two methods of team's formation (simultaneous and one-by-one team's formation).

Simultaneous formation involves assigning the required people to cover the highest priority role in each project and continuing until assigning the required people to cover the lowest priority role in each project. On the contrary, one-by-one formation, assigns the necessary people to all the roles of a project, before assigning people to the roles of the next project.

In the context eight technical competences were defined: system analysis, design and architecture, project configuration management, programming, project management, quality management, system testing and software project planning; and three generic competences: communication skills, teamwork and analytical skills. The technical competences were obtained from the student grades in different subjects. The generic competences were self-assessed using instruments created for this purpose.

In order to ensure that all students are placed in a role in a team, a total of eleven projects were proposed to be covered: nine of them with eight roles to cover (project manager, planner, quality specialist, configuration manager, analyst, architect, programmer and tester); one with seven roles (in this case the role of project manager is unified with the planner); one with six roles (the role of project manager is unified with the planner and the roles of quality specialist and tester are unified).

The multi-objective algorithms used in the experimentation are provided by the BICIAM class library and are: Multi-objective Stochastic Hill Climbing (MSHC), Multi-objective Stochastic Hill Climbing with Reset (MSHCR), Multi-objective Stochastic Hill Climbing by Greater Distance (MSHCD), Multi-objective Tabu Search (MTS), Multi-case Multi-objective Simulated Annealing (MCMOSA), Ulungu Multi-objective Simulated Annealing (UMOSA) and Non-dominated Sorting Genetic Algo-rithm II (NSGAII).

The parameters of algorithm are:

- Executions: 30 (applies to all)
- Iterations: 30,000 (for all algorithms, except NSGAII)
- UMOSA/MCMOSA: iterations with the same temperatura: 100, alpha: 0.9, initial temperatura: 20, final temperatura: 0.
- MSHCD/MSHCR: size of the neighborhood: 2.
- MTS: size of tabu list: 20.
- NSGAII: iterations: 6000, population size: 5, tournament selection, uniform crossover, one-point mutation, mutation probability: 0.9 and crossover probability: 0.5.

To evaluate the performance of multiobjective algorithms, three metrics are used:

- Error rate: indicates the percentage of solutions that are not members of the true Pareto front. The ideal value is 0.
- Generational distance: Indicates how far the elements are from the current Pareto front with respect to the true Pareto front. The ideal value is 0.
- Dispersion: measures the variance of the distance of each member of the Pareto optimal set found so far with respect to its closest neighbor, the ideal value is 0.

Non-parametric statistical techniques were used to compare the results obtained in the metrics by the different algorithms. For the analysis of the results, a hypothesis test is formulated in order to determine whether or not there are differences between the behaviors of the algorithms in general, where the null hypothesis states that there are no differences between the algorithms. All hypothesis tests are performed with a significance level of $\alpha = 0.05$. The RMST-Tool is used to calculate the metrics and the non-parametric tests.

In the problem, the true Pareto front is unknown, so an approximation to it was sought. This approximate known Pareto front is the non-dominated solutions of all the solutions obtained by all the algorithms.

The results of the error rate, generational distance and dispersion metrics are shown in Table 9. The best results of the metrics by team formation method and algorithm are in bold.

The first significant result is that all eleven project teams are fully created, unlike the results obtained in [34] where it was not possible to obtain complete solutions in most cases.

Taking into account the generational distance metric, the best algorithm is MTS. Regarding the error rate, the best algorithms are MSHC and MCMOSA. Regarding the dispersion metric, the best algorithms are NSGAII and MSHCD. It should be noted how the multi-objective variants of trajectory algorithms obtain better results in most cases than the population variant NSGAII, as in [34].

Table 10 shows an analysis by team formation method of the error rate, generational distance and dispersion metrics using the Friedman test.

As can be seen in Table 10, forming the teams simultaneously obtains better results in generational distance and dispersion metrics and forms the team one by one obtain better results in error rate metric.

Table 9 Error rate, generational distance and dispersion by algorithm and method for forming the teams

Algorithms/metrics	Generational distance		Error rate		Dispersion	
	Simultaneous	One by one	Simultaneous	One by one	Simultaneous	One by one
MSHC	0.00115	0.00103	**0.61818**	0.72727	0.04434	0.04566
MSHCR	0.00111	0.00150	0.70370	0.69091	0.04191	0.05420
MSHCD	0.00112	0.00123	0.71429	0.79630	0.04793	**0.03949**
MTS	**0.00097**	**0.00100**	0.70370	0.73585	0.04255	0.03979
MCMOSA	0.00120	0.00135	0.65957	**0.55102**	0.05255	0.05450
UMOSA	0.00132	0.00102	0.75000	0.62963	0.05018	0.05490
NSGAII	0.00167	0.00170	0.70000	0.66667	**0.03511**	0.05505

Table 10 Friedman test by team formation method

Team formation method	Generational distance	Error rate	Dispersion
Simultaneous	1.29	1.57	1.29
One by one	1.71	1.43	1.71
p-value	0.257	0.705	0.257

Table 11 shows two of the 217 solutions of the true Pareto front. As can be seen in the proposal made, it is possible to obtain balanced solutions considering the various factors. The balance index represents the deviation of each index with respect to the average, so it is ideal value is 0. The normalized balance index shows the balance ratio value in a range from 0 to 1.

6 Conclusions and Future Works

Team formation is a very active research area today as the importance of the team in the success of organizations is recognized.

As a main result of this research, a formal model was defined for the formation of multiple teams. As a novel element, the model integrates a large number of factors (competences, workload, communication cost, cost of distance work, interest in playing the role, interest in the team and psychological characteristics), both individual and collective and incorporates objectives that allow balancing the teams according to the different factors.

The use of multi-objective algorithms in solving the problem supports the work of decision-makers who face the team formation process, allowing to show a set of equally good possible solutions.

Table 11 Evaluation of each factor by team and balance index

Team/factors	Solution 1				Solution 2			
	Competence	Communication cost	Interest by role	Belbin roles	Competence	Communication cost	Interest by role	Belbin roles
Team 1	0.8125	0.5	0.625	0.8889	0.8348	0.5	0.75	0.7778
Team 2	0.8437	0.5	0.8125	0.8889	0.8304	0.5	0.9375	1.0
Team 3	0.8125	0.5	1.0	0.8889	0.8304	0.5	0.625	0.8889
Team 4	0.8661	0.5	0.8125	0.8889	0.8214	0.5178	0.6875	0.8889
Team 5	0.9018	0.5	0.75	0.8889	0.8794	0.5	0.75	1.0
Team 6	0.7946	0.5	0.625	0.8889	0.8304	0.5	0.75	0.8889
Team 7	0.8884	0.5	0.8125	0.8889	0.8616	0.5	0.75	1.0
Team 8	0.7991	0.5	0.8125	1.0	0.8080	0.5	0.8125	1.0
Team 9	0.8839	0.5	0.8125	0.8889	0.8616	0.5	0.8125	0.8889
Team 10	0.8189	0.5238	0.6428	1.0	0.8386	0.5	0.7857	0.8889
Team 11	0.8594	0.5	0.9167	1.0	0.8564	0.5333	0.6667	1.0
Balance index	**0.3620**	**0.0433**	**0.9849**	**0.4848**	**0.1885**	**0.0836**	**0.6401**	**0.7070**
Normalized balance index	**0.0021**	**0.0003**	**0.0226**	**0.0099**	**0.0011**	**0.0005**	**0.0147**	**0.0144**

The use of BICIAM library allows reusing a large number of metaheuristic algorithms and the performance of all the algorithms in the library can be compared. The multi-objective methods and algorithms incorporated into BICIAM facilitated the development of experiments.

The experimentation proves that it is possible to form complete multiple teams considering different factors and achieving a balance in the teams formed. The proposal allows balancing the teams taking into account the different factors considered in the team formation or only some of them, which offers many facilities to the decision-makers.

In experimentation, the multi-objective variants of the trajectory algorithms perform better in most of the metrics and methods used when compared to the NSGAII population variant.

Different trends are identified in this area of research:

- Development of configurable tools to assist decision makers in the multiple team formation process by allowing them to take into account different factors.
- Development of experiments to evaluate the performance of different methods and meta-heuristic algorithms in the solution of other team formation problem using a multi-objective approach.

References

1. Gutiérrez, J.H., et al.: The multiple team formation problem using sociometry. Comp. Oper. Res. **75**, 150–162 (2016)
2. Afshar, J., et al.: Harmonic mean based soccer team formation problem. In: 7th International Conference on Emerging Databases: Technologies, Applications and Theory, vol. 461, pp. 240–246. Springer, Berlin (2018)
3. PMI: The Standard for Project Management and a Guide to the Project Management Body of Knowledge, 7th ed., p. 370. Project Management Institute (2021). ISBN: 978-1-62825-664-2
4. Pavlov, D., Yahontova, I.: Formation of effective leading project teams: a multi-objective approach. In: Proceedings of the 6th International Conference on Social, Economic and Academic Leadership, pp. 84–90. Atlantis Press (2020)
5. André, M., Baldoquín, M.G., Acuña, S.T.: Formal model for assigning human resources to teams in software projects. Inf. Softw. Technol. **53**, 259–275 (2011)
6. Gutiérrez, J., et al.: The multiple team formation problem using sociometry. Comput. Oper. Res. **75**, 150–162 (2016)
7. Boehm, B.: Chapter 4. Human resource allocation and scheduling for software project management. In: Ruhe, G., Wohlin, C. (eds.) Software Project Management in a Changing World. Springer, Heidelberg (2014)
8. Zakarian, A., Kusiak, A.: Forming teams: an analytical approach. IIE Trans. **31**, 85–97 (1999)
9. Ashenagar, B., et al.: Team formation in social networks based on local distance metric. In: 12th International Conference on Fuzzy Systems and Knowledge Discovery, FSKD 2015, pp. 946–952. Institute of Electrical and Electronics Engineers Inc. (2016)
10. Basiri, J., Taghiyareh, F., Ghorbani, A.: Collaborative team formation using brain drain optimization: a practical and effective solution. World Wide Web **20**, 1385–1407 (2017)
11. Han, Y., et al.: Exploiting geographical location for team formation in social coding sites. In: 21st Pacific–Asia Conference on Knowledge Discovery and Data Mining, PAKDD 2017, vol. 10234, pp. 499–510. LNAI Springer (2017)

12. Zhou, Y., et al.: On participation constrained team formation. J. Comput. Sci. Technol. **32**, 139–154 (2017)
13. Bredereck, R., et al.: Parameterized complexity of team formation in social networks. Theor. Comput. Sci. (2016)
14. Alberola, J.M., et al.: A general framework for testing different student team formation strategies. In: 6th International Conference on Methodologies and Intelligent Systems for Technology Enhanced Learning, mis4TEL, vol. 478, pp. 23–31. Springer (2016)
15. Bhattacharjee, D., Saikia, H.: An objective approach of balanced cricket team selection using binary integer programming method. Opsearch **53**, 225–247 (2016)
16. Budak, G., et al.: New mathematical models for team formation of sports clubs before the match. Cent. Eur. J. Oper. Res. 1–17 (2017)
17. Dadelo, S., et al.: Multi-criteria assessment and ranking system of sport team formation based on objective-measured values of criteria set. Expert Syst. Appl. **41**, 6106–6113 (2014)
18. Yannibelli, V., Amandi. A.: A hybrid evolutionary algorithm based on adaptive mutation and crossover for collaborative learning team formation in higher education. In: 18th International Conference on Intelligent Data Engineering and Automated Learning, IDEAL 2017, vol. 10585, pp. 345–354. LNCS Springer (2017)
19. Balmaceda, J.M., Schiaffino, S., Díaz-Pace, J.A.: Using constraint satisfaction to aid group formation in CSCL. Inteligencia Artif. **17**, 35–45 (2014)
20. Bergey, P., King, M.: Team machine: a decision support system for team formation. Decis. Sci. J. Innov. Educ. **12**, 109–130 (2014)
21. Del Val, E., et al.: A team formation tool for educational environments. In: 12th International Conference on Practical Applications of Agents and Multi-agent Systems, PAAMS 2014, vol. 293, pp. 173–181. Springer (2014)
22. Pinninghoff, M.A., et al.: Collaborative group formation using genetic algorithms. In: International Work-Conference on the Interplay Between Natural and Artificial Computation, IWINAC 2015, vol. 9108, pp. 330–338. Springer (2015)
23. DuPont, B., Hoyle, C.: Automation and optimization of engineering design team selection considering personality types and course—specific constraints. In: 2015 122nd ASEE Annual Conference and Exposition, ASEE Annual Conference and Exposition: Making Value for Society, vol. 122. American Society for Engineering Education (2015)
24. Appel, A.P., et al.: Building socially connected skilled teams to accomplish complex tasks. In: 8th Workshop on Social Network Mining and Analysis, SNAKDD 2014. Association for Computing Machinery, Inc. (2014)
25. Awal, G.K., Bharadwaj, K.K.: Team formation in social networks based on collective intelligence—an evolutionary approach. Appl. Intell. **41**, 627–648 (2014)
26. De Meo, P., et al.: Forming time-stable homogeneous groups into online social networks. Inf. Sci. **414**, 1339–1351 (2017)
27. Selvarajah, K.: Investigation of team formation in dynamic social networks. Doctor of Philosophy, University of Windsor, Ontario, Canada (2020)
28. Aviles-Gonzalez, J., Smith, N.R., Sawhney, R.: Decision making method to select team members applying personnel behavior based lean model. Ind. Eng. Manag. Syst. **15**, 215–223 (2016)
29. Ebadi, A., et al.: DisTeam: a decision support tool for surgical team selection. Artif. Intell. Med. **76**, 16–26 (2017)
30. Patil, S.: Team formation using recommendation systems. Master of Science in Data Science, New Jersey Institute of Technology, Newark, N.Y. (2020)
31. Infante, A.L., et al.: Formation of project teams applying software multiobjective metaheuristics algorithms path. Intel. Artif. **17**, 1–16 (2020)
32. Capretz, L.F., Varona, D., Raza, A.: Influence of personality types in software tasks choices. Comput. Hum. Behav. **52** (2015)
33. André, M., Baldoquín, M.G.: Un sistema de soporte a la decisión para la asignación de recursos humanos a equipos de proyectos de software. Revista Investigación Operacional **31**, 61–69 (2010)

34. Infante, A.L., et al.: Métodos para la formación de múltiples equipos de estudiantes aplicando un enfoque multiobjetivo. Rev. Ing. Ind. **41** (2020)
35. Costaguta, R., De Los Angeles, M.: An assistant agent for group formation in CSCL based on student learning styles. In: 7th Euro American Conference on Telematics and Information Systems, EATIS 2014. Association for Computing Machinery, Valparaiso (2014)
36. Čarapina, M., Botički, I.: Exploring technology supported collaborative and cooperative group formation mechanisms. In: 12th International Conference on Cognition and Exploratory Learning in the Digital Age, pp. 377–378. CELDA, IADIS (2015)
37. Chen, Y.R., et al.: Team formation for collaborative learning with social network consideration based on edX's online discussion board. In: 8th International Conference on Ubi-Media Computing, UMEDIA, pp. 146–151. Institute of Electrical and Electronics Engineers Inc. (2015)
38. Adetunji, A.: Adaptive support for group formation in computer supported collaborative learning. In: 25th ACM International Conference on User Modeling, Adaptation and Personalization, UMAP, pp. 109–110. Association for Computing Machinery, Inc. (2017)
39. Agrawal, V., Jariwala, A.S.: Web-based tools for supporting student-driven capstone design team formation. In: 124th ASEE Annual Conference and Exposition. American Society for Engineering Education (2017)
40. Amarasinghe, I., Hernandez-Leo, D., Jonsson, A.: Intelligent group formation in computer supported collaborative learning scripts. In: 17th IEEE International Conference on Advanced Learning Technologies, ICALT, pp. 201–203. Institute of Electrical and Electronics Engineers Inc. (2017)
41. Maqtary, N., Mohsen, A., Bechkoum, K.: Group formation techniques in computer-supported collaborative learning: a systematic literature review. Tech. Knowl. Learn 1–22 (2017)
42. Buccafurri, F., et al.: Driving global team formation in social networks to obtain diversity. In: Lecture Notes in Computer Science (including subseries Lecture Notes in Artificial Intelligence and Lecture Notes in Bioinformatics), vol. 8541, pp. 410–419. Springer, Berlin (2014)
43. Han, Y., et al.: Exploiting geographical location for team formation in social coding sites. In: 21st Pacific–Asia Conference on Knowledge Discovery and Data Mining, PAKDD 2017, vol. 10234, pp. 499–510. LNAI, Springer, Berlin (2017)
44. Wang, X., Zhao, Z., Ng, W.: A comparative study of team formation in social networks. In: 20th International Conference on Database Systems for Advanced Applications, DASFAA 2015, Vol. 9049. Shahabi, C., et al, pp. 389–404. Springer Verlag. (2015)
45. De Aguiar, G., Kemczinski, A., Gasparini, I.: The automated formation of corporate groups for software projects: a systematic mapping. In: 12th Brazilian Symposium on Information Systems, SBSI 2016, pp. 605–612. Universidade Federal de Santa Catarina, Florianopolis—UFSC/Departamento de Informatica e Estatistica (2016)
46. Bach-Dabrowska, I.: Fuzzy Methods and Models for a Team—Building Process. Computation for Humanity: Information Technology to Advance Society, ed. J. Zander and P.J. Mosterman. pp. 381–408. (2014)
47. Bhowmik, A., et al.: Submodularity in team formation problem. In: 14th SIAM International Conference on Data Mining, SDM 2014, vol. 2, pp. 893–901. Society for Industrial and Applied Mathematics Publications (2014)
48. Chaikovska, I., et al.: Economic–mathematical tools for building up a project team in the system of company's knowledge management. East. Eur. J. Enterp. Technol. **3**, 29–37 (2017)
49. Yang, J., et al.: Forming a research team of experts in expert-skill co-occurrence network of research news. In: 2016 IEEE/ACM International Conference on Advances in Social Networks Analysis and Mining, ASONAM 2016, pp. 1143–1150. Institute of Electrical and Electronics Engineers Inc. (2016)
50. Zhang, J., Yu, P.S., Lv., Y.: Enterprise employee training via project team formation. In: 10th ACM International Conference on Web Search and Data Mining, WSDM 2017, pp. 3–12. Association for Computing Machinery, Inc. (2017)
51. Paredes, M.A., et al.: An ontology-based approach with which to assign human resources to software projects. Sci. Comput. Program. **156**, 90–103 (2018)

52. Costa, A., et al.: Team formation in software engineering: a systematic mapping study. IEEE Access **8** (2020)
53. Aryanee, D., Razali, R., Mansor, Z.: Team formation for agile software development: a review. Int. J. Adv. Sci. Eng. Inf. Technol. **10**, 2088–5334 (2020)
54. Di Martinelly, C., Meskens, N.: A bi-objective integrated approach to building surgical teams and nurse schedule rosters to maximize surgical team affinities and minimize nurses' idle time. Int. J. Prod. Econ. **191**, 323–334 (2017)
55. Coello, C., Lamont, G.B., Veldhuizen, D.A.: Evolutionary Algorithms for Solving Multi-Objective Problems. Springer, New York (2007)
56. Fajardo, J.: Soft computing en problemas de optimización dinámicos. Tesis doctoral, Universidad de Granada, España (2015)
57. Talbi, E.G.: Metaheuristics: From Design to Implementation. Wiley, Hoboken (2009)

Chapter 13
A TOPSIS-Based Method for Personnel Selection in Software Projects

Ana del Carmen Espinosa Robert, Yamilis Fernández-Pérez, and Yeleny Zulueta-Veliz

Abstract Personnel selection is considered as a multicriteria decision-making problem due to its complex, time-demanding and multifaceted structure which usually includes both qualitative and quantitative criteria. Personnel selection is also a critical strategic problem in business context and complexity and importance of the problem demand for analytical demands well-founded decisions instead of intuitive and non-systemic decisions. The aim of this paper is to support adequately the decision-making process in personnel selection for developing software projects by means of a decision analysis method based on the Technique for Order Preference by Similarity to an Ideal Solution (TOPSIS).

Keywords Decision-making · Personnel selection · Project management · TOPSIS

1 Introduction

The main objective of human resource management is to contribute to the growth and development of the company and although it is known that there may be many changes within projects, their success does not depend on the number of changes, but on the team's ability to deal with them.

Human resource is a key success factor for any organization to raise their competitive strength and productivity in the market. But the input quality of these human resources is directly related to the recruitment and selection processes. Ultimately, hiring the right people at the right time, that is, from the beginning of a project,

A. del Carmen Espinosa Robert
XETID Information Technology Enterprise, 269th Street, between 217 and 203, Torrens, Boyeros, Havana, Cuba

Y. Fernández-Pérez (✉) · Y. Zulueta-Veliz
Universidad de las Ciencias Informáticas, San Antonio de los Baños highway, Km 2½, Torrens, Boyeros, Havana, Cuba
e-mail: yamilisf@uci.cu

Y. Zulueta-Veliz
e-mail: yeleny@uci.cu

P. Y. Piñero Pérez et al. (eds.), *Artificial Intelligence in Project Management and Making Decisions*, Studies in Computational Intelligence 1035,
https://doi.org/10.1007/978-3-030-97269-1_13

could save a company a lot of headaches, could save a company financial loss, dissatisfaction with work, diminished prestige, among other difficulties. An effective personnel selection method should support smart decisions for choosing an appropriate individual for a given job in an organization. Personnel selection is a methodical process to hire individuals. Usually the term is applied to all aspects of the process including recruitment, selection, hiring, onboarding and acculturation, the most common meaning focuses on the selection of workers. This is the meaning adopted in this paper.

Planning, recruiting and selecting people, hiring, training, retention and compensation are vital functions of human resources [1]. Special attention is paid to the recruitment and selection processes due to the relationship between them. The first one includes actions to motivate and attract the attention of the greatest number of candidates who meet the requirements for a position; and the second one aims to better filter, choose and decide which of the candidates fit the needs of the company.

Today, important issues related to the subject are still being analyzed since management varies in terms of the selection and evaluation of personnel, despite the fact that they are administered in a general way by several fundamental processes as recruitment, selection, evaluation and hiring [2]. The personnel recruitment and selection processes have evolved in accordance with the emergence of new and diverse forms of technological development, not only due to the use of the Internet or the social media websites, but also due to the strong impact of the way both recruiters and job applicants work.

The recruitment and selection processes are characterized by meeting a need in a certain area, which requires choosing, classifying and attracting the best human talent with the characteristics and knowledge required for their good performance in the assigned position. Therefore, to understand the importance it is necessary to study and analyze the different research approaches that some authors have made on the importance of both processes [3]. Its execution depends a lot on the characteristics of the organization, the level of hierarchy, the position and the cost that each stage represents.

The selection techniques phase supports the determination of whether the knowledge meets the requirements of the role, making it easier for the project leader to identify which would be the best alternative to select. Similarly, one of the mechanisms that helps to identify the productivity and efficient performance of candidates are personal competencies, because "they comprise those knowledge, skills and attitudes needed to perform like one" [4].

Competence is understood as "the relationship between human beings and work tasks, that is, the concern is not about knowledge and skills per se, but about what knowledge and skills are required to perform a specific job or task efficiently" [5]. According to People CMM, competency management is "a collection of workforce practices that is used to enhance the ability of the workforce to perform its assigned tasks and responsibilities and to achieve specific growth objectives".

In general, job competencies are considered to be nothing more than the experience, capabilities, personal attributes, knowledge and/or skills that the candidate needs to perform well in the role and can be classified as soft and hard competencies (or skills). "Soft skills refer to the competencies that aim to identify the individual's ability to interact with other people and successfully, refers to interpersonal relationships and the ability to work in a team. Instead, the hard skills are framed in the technical skills or required to perform the tasks or functions within a project. They are those that are acquired through training, qualification and, sometimes, in the same exercise of the functions of the position" [4].

It can be inferred that competencies refer to specific elements that characterize people for their job performance. In the study carried out, competencies can be considered as criteria to be used during the selection process and they are the ones that will allow the evaluation of suitable candidates for project management. Within project management, the selection of people is one of the most controversial issues today. It starts with a group of candidates and a set of criteria that depends on their performance and the participation of experts to evaluate them. That is, the results of the evaluation of the candidate will depend on the selection criteria of the experts involved, so it is necessary to select algorithms or methods that solve the problem in the best possible way.

Software development, as a technology industry, impacts the progress of society. However, the number of software projects that are not completed successfully is still revealing. According to the studies of the CHAOS Report, belonging to the Standish Group on the success rates in projects where information technologies and best project management practices are used, it is found that 19% of projects fail, 50% are modify and only 31% succeed. Software development companies depend on rapidly evolving technologies and paradigms, which has led to a high dependence on a workforce with specialized knowledge and skills. As a consequence, critical factors for the success of project management have been identified, including those related to human resources: specifically, leadership, emotional maturity of the development team, communication, experience in project management, among others.

Growing globalization and technological advances in the digital world are increasingly forcing business managers to seek more and more qualified human resources. In the Information Technology company, there is a computer application that contains a selection function in the HR module, where an empirical mathematical model is used, with a lack of methods that take advantage of the stored information. Currently, for the solution of personnel selection problems, the use of Multi-Criteria Decision-Making methods is a trend. MCDM methods are deemed to be the most suitable decision support tools to handle the complex, multicriteria structure of the personnel selection problem.

The multi-criteria methods allow establishing a balance between the proposed objectives and the simplification of the decision-making problems according to those objectives and in a transparent way [6]. They are executed on the basis of the selection of criteria and the ranking of the results that is very easy to understand by the group that performs the analysis.

The aim of this paper is to support adequately the decision-making process in personnel selection for developing software projects by means of a decision analysis method based on the Technique for Order Preference by Similarity to an Ideal Solution (TOPSIS). The rest of the paper is organized as follows: Sect. 2 presents a background on the main MCDM methods and some relevant studies on the personnel selection problem. Section 3, presents the proposed TOPSIS method to support the personnel selection and its empirical application for solving a problem in the IT context. Finally, conclusions and research challenges are discussed.

2 Background on MCDM Process and Methods

Decision-making becomes in complex situation when the number of criteria and alternatives for a problem is high. In the context of the present: the selection of several candidates to join a software company, where the competencies of the individuals and the number of applications for them are considered, is something difficult to handle. Multi-criteria methods are very advantageous tools in these cases. According to the structure of the decision problems, the data is easy to obtain, provides an extraordinary solution and, in most cases, a ranking of the possible alternatives. As a decision-making problem, the following characteristics were identified:

- It has one or more goals to achieve.
- A decision-maker, either an individual or a group of individuals with responsibility for making the decision.
- It contains a set of alternatives, actions or possible decisions in which the decision-maker can select and must contain at least two elements.
- It has a set of states of nature called Environment that is made up of the situations (or states) that the decision-maker can encounter when choosing the alternatives, but that he cannot control.
- It has a set of results or consequences associated with each action and each state.
- It has the criterion that classifies the consequences.
- The nature of the information available.
- The process duration.
- There is a social framework in which the decision-making process is developed and it establishes the rules to guide the decision-maker in the selection [7].

Taking into account the characteristics of the elements of the problem, the number of candidates identified during the selection processes can be defined as alternatives. The evaluation is carried out by a Human Resource professional team, led by the head of the Department as decision-maker. During the analysis process, the criteria were identified from the classification of hard and soft skills defined in the previous section.

Based on the number of attributes analyzed and the experts involved, the problem is classified as multi-attribute and multi-expert. According to the decision environment, it is classified as an environment of certainty because the nature of the information is known. According to the consideration of changes over time, this problem is classified as static because preferences are issued in a single decision moment, on a single set of alternatives, according to a single set of criteria of fixed importance. According to the expression domains used, this problem is classified as homogeneous, since

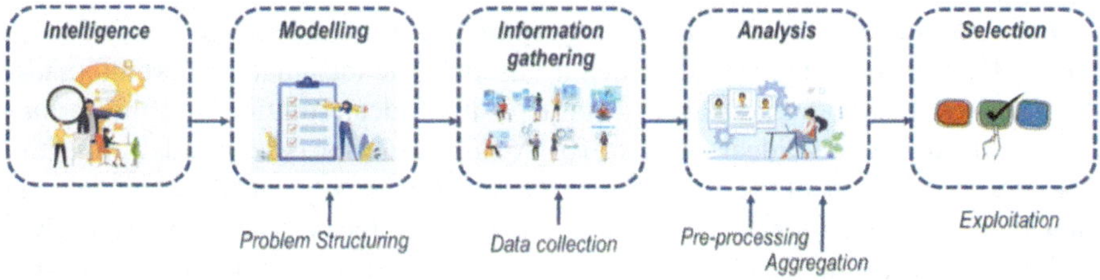

Fig. 1 Phases of the decision-making process

only numerical will be applied to express an order of preference among the different alternatives, depending on the usefulness of each one.

The decision-making process follows a general problem-solving scheme that is divided into five stages summarized in Fig. 1.

In the Intelligence phase, the problem, alternatives and objectives to be achieved through the resolution process are identified. In the Modelling stage, the structure of the problem is defined, establishing preferences and uncertainty, among other elements. In the Information gathering stage, the information, knowledge and preferences provided by the decision-makers are obtained according to the previously defined model. It is also known as Data collection stage. In the stage of Analysis, the information collected is managed according to the objectives and limitations and results are obtained to be considered in the selection phase. Finally, in the Selection stage, an exploitation process is carried out in which decision-makers can choose alternative solutions for decision-making, according to the results obtained in the analysis (or aggregation) stage [8].

Among the Multi-Criteria Decision-Making Methods (MCDM) most used in personnel selection solutions are: Analytical Hierarchical Process (AHP), the Weighted Aggregate Sum Product Evaluation WASPAS (Weighted Aggregate Sum Product Evaluation) and the Technique for Order of Preference by Similarity to the Ideal Solution (TOPSIS). All these methods offer mechanisms that establish an order between the alternatives, providing the criteria for the final decision making.

WASPAS is a combination of two multi-criteria decision methods: the weighted sum method (WSM) and the weighted product method (WPM). This method determines the relative importance of each criterion, then evaluates and prioritizes the alternatives [9]. In calculations, attributes are independent and all qualitative attributes are made into quantitative attributes, although not all decision-making problems require a quantitative output. In the case of AHP, it allows obtaining the most appropriate solution through a hierarchical structure: it reduces a hard decision to a series of pairwise comparisons and synthesizes those results, where both objective and subjective are used in the decision-making process [9]. However, if there is a large number of criteria or alternatives, the evaluation becomes cumbersome since the number of comparisons experiences a quadratic behavior with both the criteria and the alternatives.

The technique for order preference by similarity to ideal solution (TOPSIS) is a classic decision-making method, which is based on the idea that the best alternative should have the shortest distance from a positive ideal solution and the farthest distance from a negative ideal solution [10]. An ideal solution is defined as a collection of scores or values in all the attributes selected in the decision and it can happen that such a solution is unattainable. The vector composed of the best values of the attribute with respect to all the possible alternatives is called the "positive ideal solution"; reciprocally the "negative ideal solution" will be the one whose vector contains the worst values in all the attributes [11].

In the literature two main versions of this method can be found: relative TOPSIS vs absolute TOPSIS. The relative TOPSIS refers to the traditional application of the method where the positive ideal solution and the negative ideal solution are calculated using the values from the original decision matrix. However, in the absolute mode, the positive ideal solution and the negative ideal solution are calculated based on the original matrix with two new alternatives, obtained as the extreme values of the range of the different criteria [12].

For the personnel selection problem, the use of the distance method is considered adequate and, therefore, absolute TOPSIS is the one used in this paper. In addition, TOPSIS is used for the following main reasons: it directly adds the values of the competencies, the applied logic is rational and understandable, the computer processes are simple, the concept allows the search for the best alternatives for each criterion presented in a mathematical form simple and the importance of weights is incorporated into the comparison processes.

Several works deal with the application of TOPSIS method for solving personnel selection problems. For example: there are the following articles of An Integrated Neutrosophic-TOPSIS Approach and Its Application to Personnel Selection: A New Trend in Brain Processing [13], A systematic approach for an application of personnel selection in assembly line balancing problem [14], A Two-Phase Model for Personnel Selection Based on Multi-Type Fuzzy Information [15], Internal auditor selection using a TOPSIS/non-linear programming model [16] and An Empirical Comparison of the Usefulness Between TOPSIS Method and Fuzzy TOPSIS Method for Personnel Selection [17].

3 The Proposed TOPSIS-Based Method for Personnel Selection in Software Projects

The conceptual flow of the proposed method is depicted in Fig. 2.

The measurement of the distances between the best alternatives (d_i^+), the worst solution (d_i^-) and the coefficient of relative proximity R_i is calculate by Eqs. (1), (2) and (3).

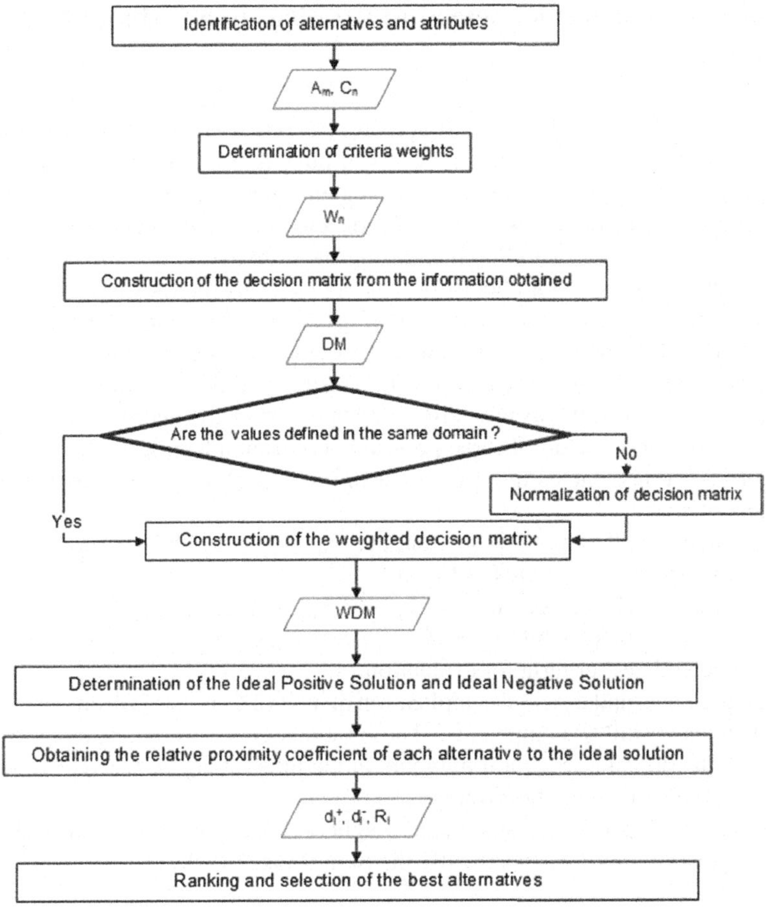

Fig. 2 TOPSIS method steps

$$d_i^+ = \sqrt{\sum \left(x_{ij} - x_j^+\right)^2} \quad (1)$$

$$d_i^- = \sqrt{\sum \left(x_{ij} - x_j^-\right)^2} \quad (2)$$

$$R_i = \frac{d_i^-}{d_i^+ + d_i^-} \quad (3)$$

R_i is the value of the relative proximity coefficient: the closer its value is to 1, the higher the priority of the i-th alternative. By definition, the order of preference applied in the method is descending. The following epigraph illustrate an example made in the Company of Information Technologies for Defence.

4 Solving a Personnel Selection Problem in a Cuban IT Project

The Company of Information Technologies for Defense, is a Cuban company founded in 2013. Its thematic areas cover processes of human resources management, planning, supply management, security management and communications. As part of it, the Centre of Technologies for systems development and Integration deals with the develop of systems that allow increasing the effectiveness of the promotion of products and services of our clients through applications for displaying information in large format, as well as the interaction of users with the content represented. It is also in charge of automating management processes through microservices.

The purpose of this empirical application of the method, is to illustrate its applicability. The experiment was basically setup upon a real-life decision situation. The objective is to form a team of three programmers from a bag of twelve individuals based on the competencies tree presented in Fig. 3 and determined by the project managers.

As explained in the previous section, the first step is to identify the elements of the problem which are depicted in Fig. 4.

$A = \{A_1, A_2, A_3, A_4, A_5, A_6, A_7, A_8, A_9, A_{10}, A_{11}, A_{12}\}$ represents the set of candidates' alternatives while $C = \{C_1, C_2, C_3, C_4, C_5, C_6, C_7, C_8, C_9\}$ represents the set of candidates' criteria for evaluation.

The candidates underwent the application of a set of tests and instruments in order to evaluate each skill or competence. These ratings $\{x_{i,j} \in N : 1 \leq x_{i,j} \leq 5\}$ were defined in the domain of natural numbers in the interval [1, 5], where 1 represents the worst rating and 5, the best rating.

Based on the expert's assessment, the weight of the criteria is assigned by applying the Classification Method, specifically the Sum Range (Table 1).

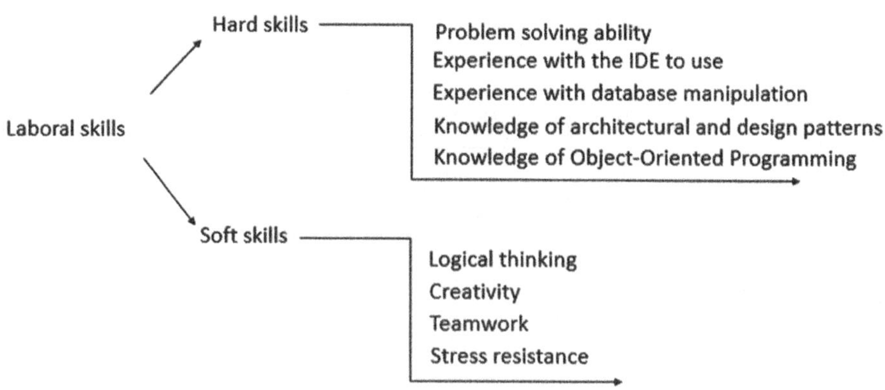

Fig. 3 Determination of criteria from the profile of a programmer

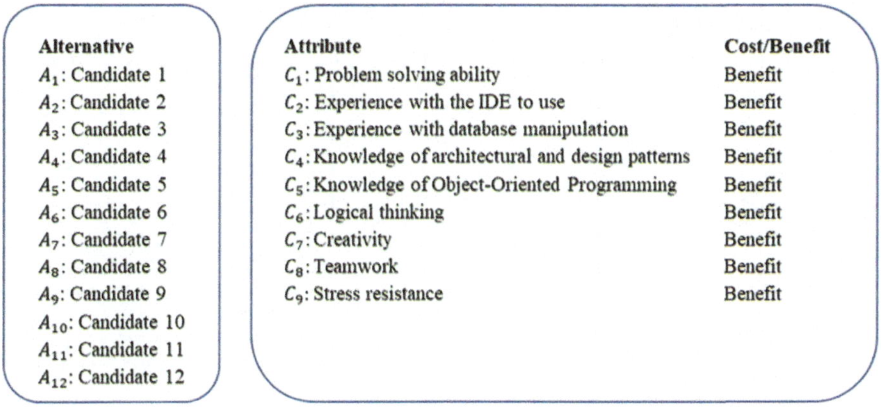

Fig. 4 Alternatives and attributes in the MCDM problem

Once the elements of the problem have been identified, the decision matrix (Table 2) is generated from the preferences provided by the Admission Committee, which is integrated with area leaders and others experts selected.

The absolute TOPSIS variant is applied to obtain the following matrix (Table 3) in which the weights corresponding to each criterion are identified, as well as

Table 1 Ranking application method—ranking sum

Attributes		Straight rank	$(n - r_i + 1)$	Normalized	Weights
C_1	Problem solving ability	1	9	0.39	w_1
C_2	Experience with the IDE to use	4	6	0.26	w_2
C_3	Experience with database manipulation	8	2	0.09	w_3
C_4	Knowledge of architectural and design patterns	5	5	0.22	w_4
C_5	Knowledge of object-oriented programming	9	1	0.04	w_5
C_6	Logical thinking	2	8	0.35	w_6
C_7	Creativity	3	7	0.30	w_7
C_8	Teamwork	6	4	0.17	w_8
C_9	Stress resistance	7	3	0.13	w_9
Sum			**23**	**1**	

Table 2 Preparation of the decision matrix

	C_1	C_2	C_3	C_4	C_5	C_6	C_7	C_8	C_9
A_1	4	3	5	3	3	5	3	3	3
A_2	2	5	4	4	4	4	4	4	4
A_3	3	1	3	5	5	5	5	5	5
A_4	2	2	3	2	5	4	3	3	4
A_5	4	3	3	3	5	4	4	4	3
A_6	5	5	5	5	5	5	5	3	3
A_7	4	4	4	4	4	4	4	4	4
A_8	3	4	5	3	3	3	3	3	3
A_9	4	3	5	1	1	1	1	1	1
A_{10}	4	3	5	4	4	5	5	5	5
A_{11}	5	5	2	4	3	3	3	3	3
A_{12}	4	3	5	3	3	3	4	4	4

Table 3 Preparation of the normalized matrix

	C_1	C_2	C_3	C_4	C_5	C_6	C_7	C_8	C_9
A_1	4	3	5	3	3	5	3	3	3
A_2	2	5	4	4	4	4	4	4	4
A_3	3	1	3	5	5	5	5	5	5
A_4	2	2	3	2	5	4	3	3	4
A_5	4	3	3	3	5	4	4	4	3
A_6	5	5	5	5	5	5	5	3	3
A_7	4	4	4	4	4	4	4	4	4
A_8	3	4	5	3	3	3	3	3	3
A_9	4	3	5	1	1	1	1	1	1
A_{10}	4	3	5	4	4	5	5	5	5
A_{11}	5	5	2	4	3	3	3	3	3
A_{12}	4	3	5	3	3	3	4	4	4
A^{max}	5	5	5	5	5	5	5	5	5
A^{min}	1	1	1	1	1	1	1	1	1
Weights	0.39	0.26	0.09	0.22	0.04	0.35	0.3	0.17	0.13

the determination of the ideal Positive Solution and the Ideal Negative Solution, respectively.

The corresponding calculations are applied to obtain the weighted normalized matrix. From the established weights, the weighted decision matrix is constructed by multiplying each of the values by the corresponding alternatives, see Table 4. Following, the measurement of the distances between the best alternatives (d_i^+)

Table 4 Final results

	C_1	C_2	C_3	C_4	C_5	C_6	C_7	C_8	C_9	d_i^+	d_i^-	R	Ranking
A_1	1.56	0.78	0.45	0.66	0.12	1.75	0.9	0.51	0.39	1.079	2.114	0.662	6
A_2	0.78	1.3	0.36	0.88	0.16	1.4	1.2	0.68	0.52	1.298	2.020	0.608	9
A_3	1.17	0.26	0.27	1.1	0.2	1.75	1.5	0.85	0.65	1.312	2.3608	0.642	7
A_4	0.78	0.52	0.27	0.44	0.2	1.4	0.9	0.51	0.52	1.749	1.433	0.450	11
A_5	1.56	0.78	0.27	0.66	0.2	1.4	1.2	0.68	0.39	0.978	2.032	0.675	4
A_6	1.95	1.3	0.45	1.1	0.2	1.75	1.5	0.51	0.39	0.428	2.833	0.868	1
A_7	1.56	1.04	0.36	0.88	0.16	1.4	1.2	0.68	0.52	0.732	2.196	0.75	3
A_8	1.17	1.04	0.45	0.66	0.12	1.05	0.9	0.51	0.39	1.381	1.606	0.537	10
A_9	1.56	0.78	0.45	0.22	0.04	0.35	0.3	0.17	0.13	2.314	1.33	0.364	12
A_{10}	1.56	0.78	0.45	0.88	0.16	1.75	1.5	0.85	0.65	0.687	2.520	0.785	2
A_{11}	1.95	1.3	0.18	0.88	0.12	1.05	0.9	0.51	0.39	1.077	2.235	0.674	5
A_{12}	1.56	0.78	0.45	0.66	0.12	1.05	1.2	0.68	0.52	1.117	1.918	0.631	8
A^{max}	1.95	1.3	0.45	1.1	0.2	1.75	1.5	0.85	0.65				
A^{min}	0.39	0.26	0.09	0.22	0.04	0.35	0.3	0.17	0.13				

and the worst solution (d_i^-) was determined respectively, the coefficient of relative proximity R and the *Ranking*, see Eqs. (1), (2) and (3).

After analysing the factors that they use to the success of the projects, it is considered that companies give the greatest importance to soft skills. According to the results obtained, the order of selection for the present case study is candidates 6, 10 and 7.

In order to confirm the stability and reliability of the proposed model, the gained results are compared with the results obtained using the following MCDA methods: the weighted aggregated sum product assessment (WASPAS) method, relative TOPSIS. The comparison of the gained ranking orders of the alternatives is shown in Fig. 5 and Table 5.

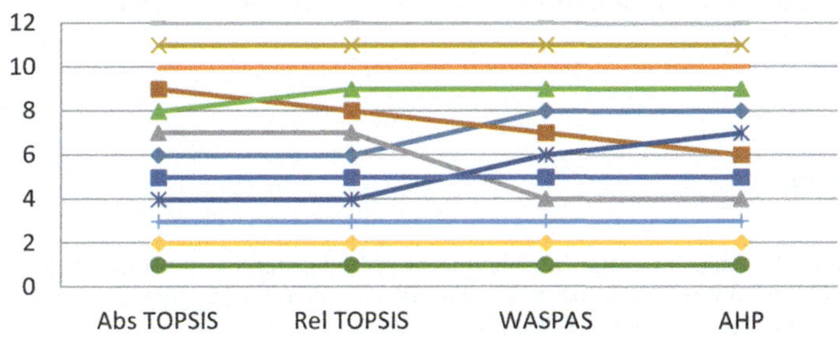

Fig. 5 Comparison of gained ranking

Table 5 Comparison of gained ranking

	Abs TOPSIS	Rel TOPSIS	WASPAS	AHP
A1	6	6	8	8
A2	9	8	7	6
A3	7	7	4	4
A4	11	11	11	11
A5	4	4	6	7
A6	1	1	1	1
A7	3	3	3	3
A8	10	10	10	10
A9	12	12	12	12
A10	2	2	2	2
A11	5	5	5	5
A12	8	9	9	9

5 Conclusions

Decision-making methods are becoming more and more important in business because they enable recruitment to become an organized, transparent and fair hiring process that helps human resources staff to put the right people in the right offices. The aim of this paper is to support adequately the decision-making process in personnel selection for developing software projects by means of a decision analysis method based on TOPSIS.

This TOPSIS-based method for personnel selection in software projects was created inspired by real practical experiences in Cuban software development companies. It is capable of dealing with elements such as the soft and hard skills of a given role. Results show that it can be applied in various situations, both to assess the competencies of the staff and to select the most suitable person for a given role. The method is operational, reliable and easy to interpret. These characteristics make it applicable in decision situations that are usually presented in business and industrial contexts.

References

1. Mishra, R., Sisodia, A.G., Raj, K., Sharma, K.: Multi-criteria IT personnel selection on intuitionistic fuzzy information measures and ARAS methodology. Iran. J. Fuzzy Syst. **17**(4), 55–68 (2020). https://doi.org/10.22111/ijfs.2020.5406
2. Nikolaou, I.: What is the role of technology in recruitment and selection? Span. J. Psychol. **24** (2021). http://doi.org/10.1017/SJP.2021.6

3. Yagama, C., Helena, C.: Caso de estudio tienda deportiva Dashstore: diseño del proceso de selección y atracción del talento humano para el área de ventas. Repositorio Institucional UJTL (2021)
4. Anón, S.: Interpersonal abilities soft skill people project management (2021) [Online]. Available https://www.pmi.org/learning/library/interpersonal-abilities-soft-skill-people-project-6095. Accessed: Aug 2021
5. Ramírez- Díaz, J.L., et al.: An approach by competencies and its current relevance: considerations from occupational guidance in educational context. Rev. Electrónica Educare **24**(2), 475–489 (2020). https://doi.org/10.15359/ree.24-2.23
6. López-Cadavid, D.A., Vanegas-López, J.G., Baena-Rojas, J.: Aplicación de un método multicriterio en la enseñanza de la investigación de mercados internacionales. Información Tecnológica **31**(1), 113–122 (2020). https://doi.org/10.4067/S0718-07642020000100113
7. Chernoff, H., Moses, L.E.: Elementary Decision Theory. Courier Corporation (2012)
8. Antanas, Z., Victorio, K.: Making Hard Decisions with Decision Tools (2012)
9. Slebi-Acevedo, C.J., Pascual-Muñoz, P., Lastra-González, P., Castro-Fresno, D.: A Multi-criteria Decision-Making Analysis for the Selection of Fibres Aimed at Reinforcing Asphalt Concrete Mixtures. Int. J. Pavement Eng. **22**(6), 76–79 (2021). http://doi.org/10.1080/10298436.2019.1645848
10. Ching-Lai, H., Yoon, K.: Multiple Attribute Decision Making, vol. 186. Springer, Berlin (1981)
11. Karabašević, D., Stanujkić, D., Kazimieras, E., Stanimirović, P., Popović, G., Predić, B., Ulutaş, A.: A novel extension of the TOPSIS method adapted for the use of single-valued neutrosophic sets and hamming distance for e-commerce development strategies selection. Symmetry **12**(8), 12–63 (2020)
12. Ceballos, B., Lamata, M., Pelta, D., Sánchez, J.: El método TOPSIS relativo vs absoluto. Recta **14**, 181–192 (2013)
13. Nabeeh, A., Smarandache, F., Abdel-Basset, M., El-Ghareeb, H.A., Aboelfetouh, A.: An integrated neutrosophic–TOPSIS approach and its application to personnel selection: a new trend in brain processing and analysis. IEEE Access **7**(297), 34–44 (2019). https://doi.org/10.1109/ACCESS.2019.2899841
14. Burak, E., Kurt, M.: A systematic approach for an application of personnel selection in assembly line balancing problem. Int. Trans. Oper. Res. **25**(3) (2018). http://doi.org/10.1111/itor.12439
15. Chen-Tung, C., Wei-Zhan, H.: A two-phase model for personnel selection based on multi-type fuzzy information. Mathematics **8**(10), 1703 (2020). https://doi.org/10.3390/math8101703
16. Konstantinos, P., Drogalas, G., Zografidou, E.: Internal auditor selection using a TOPSIS/non-linear programming model. Ann. Oper. Res. **296**(1), 513–539 (2021). https://doi.org/10.1007/s10479-019-03307-x
17. Kwok, A.P., Yan, M., Xuan, Y., Zhi, H., Bin, R., Yang, C., Heng, N.: An empirical comparison of the usefulness between TOPSIS method and fuzzy TOPSIS method for personnel selection. In: Advances in Decision Science and Management, Advances in Intelligent Systems and Computing, pp. 589–595. Springer, Singapore (2022)

Chapter 14
Combining Artificial Intelligence and Project Management Techniques in Ecosystem for Training and Innovation

Pascual Verdecia Vicet, Pedro Y. Piñero Pérez, Iliana Pérez Pupo, Roberto García Vacacela, and Nicolas Villavicencio Bermúdez

Abstract Training and innovation in project management has a high impact in different scenarios of society. In this context, one way to achieve sustainable development is the creation of ecosystems for training and innovation. This article presents an ecosystem proposal that combines the teaching of project management techniques with artificial intelligence techniques. It is an ecosystem created to contribute to the introduction of emerging technologies in the management of investment projects. To achieve sustainability, the ecosystem itself develops computer solutions such as BusinessRedmine that facilitate practical teaching. These same IT solutions are marketed in different environments, generating funds that are reinvested in the development of the ecosystem. For the validation of the research, a set of indicators based on different international experiences are used. In addition, multi-criteria analysis methods are applied for the evaluation of the training ecosystem by the beneficiaries. The applicability of the ecosystem is demonstrated.

Keywords Artificial intelligent · Project management · Training programs

P. Verdecia Vicet (✉) · P. Y. Piñero Pérez · I. Pérez Pupo
Grupo de Investigaciones en Gestión de Proyectos, Universidad de las Ciencias Informáticas, CP 54830 La Habana, Cuba
e-mail: pverdecia@uci.cu

P. Y. Piñero Pérez
e-mail: ppp@uci.cu

R. García Vacacela · N. Villavicencio Bermúdez
Facultad de especialidades empresariales, Universidad Católica Santiago de Guayaquil, Guayaquil, Ecuador
e-mail: roberto.garcia@cu.ucsg.edu.ec

N. Villavicencio Bermúdez
e-mail: nicolas.villavicencio@cu.ucsg.edu.ec

P. Y. Piñero Pérez et al. (eds.), *Artificial Intelligence in Project Management and Making Decisions*, Studies in Computational Intelligence 1035,
https://doi.org/10.1007/978-3-030-97269-1_14

1 Introduction

Artificial intelligence has impacted many fields of human development, including finance, telecommunications, the automotive industry, medicine and logistics. But there are knowledge areas, such as investment management, where many knowledge areas converge and where numerous problems of unstructured decision-making coexist. On the other hand, the success or failure in the management of investment projects has a high economic and social impact.

In the case of construction projects in [1, 2] it is stated that:

- Around 34% of the resources invested in a project are wasted
- Approximately 95% of the information that generated later is underused, and is not used as lessons learned in new projects.
- In these same sources, it is stated that inefficient planning is one of the main factors that affects the inappropriate use of resources. In this context, remains a necessity the development of new algorithms and techniques that allow: the optimal or quasi-optimal construction of project schedules, considering the limitations of resources, and the variability in the ways of executing the tasks.

Similarly, it happens in the development of software projects. In this area of knowledge, it is identified by several sources that, in medium-sized projects, approximately 26% need to be renegotiated and 31% are canceled [3–5].

Both scenarios are characterized by having a high dependence on human resources and their skills. For the solution of these problems, the training of human resources is required both in project management and in Artificial Intelligence techniques. Both project management training and artificial intelligence training is a necessity, but also a complex task, considering the particularities of both areas. In this context, we formalize these two areas of knowledge as shown below:

- Project management as an interdisciplinary area where elements of psychology, management techniques, economic management, logistics, mathematical and computer science converge with applications in numerous sceneries.
- Artificial Intelligence as the set of Software and Hardware techniques that simulate different human reasoning methods and the collective intelligence of natural ecosystems, to solve complex problems where the algorithm for its solution has non-polynomial complexity, or there are no algorithms for its solution.

On the other hand, in project management training it is necessary to combine theory and practice. For this, it is necessary to have an information system that supports the training processes from a practical approach.

In this context, the bibliography reports around 125 solutions [6] associated with project management, but only 20% are based on free software. From the analysis of the tools, the following elements are identified:

- 60% allow the planning of tasks, but only 44% of the tools allow the management of project portfolios.

- 56% allow the management of human and non-human resources, and 32% incorporate control and monitoring reports.
- Only 60% include facilities for document management and the management of project files.
- Among the open-source or free solutions, only 8 include enough control and monitoring reports that they represent. In this way, a significant insufficiency is identified in the decision-making support processes in project management.

Based on the problem raised, the development of an investigation was proposed with the aim of building an ecosystem for online training that combines the teaching of project management techniques and artificial intelligence. This particular work describes the experiences in the implementation of the ecosystem and its impact on helping decision-making in investment projects.

The work is organized by sections as follows. The second section presents a proposal for an ecosystem of training and innovation in Project Management. The third section presents the analysis of the results and application of the program. Finally, the last section presents the conclusions of the work.

2 Proposal for an Ecosystem of Training and Innovation in Project Management

As part of the strategy to solve the problem raised, an ecosystem was developed for postgraduate training in Project Management and emerging technologies to help decision-making in the management of investment projects. The proposal developed is focused on the satisfaction of those interested in the program. It is conceived as an innovation system as shown in Fig. 1.

As shown in the figure, the ecosystem created for innovation training is made up of the following components:

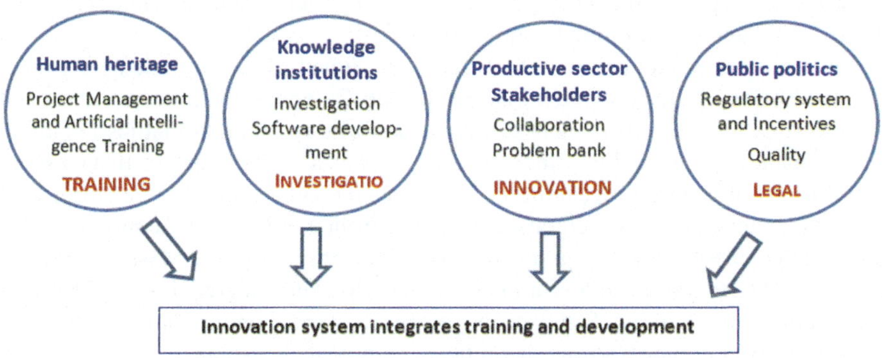

Fig. 1 Components of the innovation system that integrated training and development

- Human heritage centered on a system of courses designed for online and face-to-face training. The training is carried out assisted by the following IT solutions: Moodle, institutional repository with master's and doctorate theses, telegram and twitter social networks, Picta media management platform, BusinessRedmine project management platform [7].
- Knowledge institutions that focus the work of the research system and the development of software ecosystems such as the BusinessRedmine platform. The BusinessRedmine system constitutes a platform for project management that incorporates different artificial intelligence techniques, facilitating decision-making in the different areas of project management. The use of a project management tool such as BusinessRedmine constitutes a fundamental element for the training of practical skills from training. This component enables interaction with a tool for project management and simulation based on real or fictitious projects of the processes of initiation, planning, execution, control, monitoring and closure of projects, covering the processes described in the good practices.
- Productive sector component where the needs of the interested parties identified in a bank of real problems are managed. The center of collaboration between the productive sector, research and training is centered on the Collaborative Network of Engineering and Project Management [8]. This component encourages exchanges with professionals from different regions about best practices in project management. In addition, as part of this component, the BusinessRedmie platform is introduced to solve real problems identified in the problem bank.
- System of public policies that encourage innovation that are also part of the regulatory system and quality control. The work from this component focuses on aligning the entire training ecosystem with the main standards and guides for project management, particularly in the PMBOK version 7 guide [9], the ISO 21500 [10] and 19650 standards [11]. In addition, a set of checklists were developed for the assembly of postgraduate courses and a system of indicators for the evaluation of courses and other activities by interested parties and experts that uses computer techniques with words to collect the preferences of these.

In its conception, the proposed training ecosystem includes undergraduate, master's and doctoral training as shown below:

The ecosystem includes a set of activities aimed at undergraduate training, designed to be taught in careers associated with Computer Engineering and Civil Engineering, although its content can be generalized to other specialties related to project management. In this context, it is presented as a set of courses that complement the basic training of engineers and cover the fundamental competencies of project management. Undergraduate students enrolled in this set of courses receive training that enables them to graduate from engineering, join the Master of Project Management program, and a year later acquire a Master's degree. From undergraduate training, the training of students begins by applying the four components of the innovation system with a strong theoretical and practical training.

The fundamental nucleus of the proposed Training Ecosystem is the Master in Project Management. In this dimension, the joint system of courses that combine traditional project management techniques with artificial intelligence techniques with high added value is essential.

The ecosystem also includes a doctoral training dimension as continuity in the master's training in project management. The master's program is designed so that graduates of the program can continue with the development of their research and within a period of 2.5–3 years defend a doctorate associated with decision-making in project management. In this dimension they include training in good research practices, research topics and combined training in project management and artificial intelligence.

Some characteristics of the Master's program in Project Management are described below:

- The program introduces a set of compulsory and elective courses. In all courses, traditional project management techniques are combined with artificial intelligence techniques in project management.
- To obtain the master's degree, students must acquire a total of 62 credits with the following distribution.
  - Obligatory Courses:12 credits that include training in Integrated Project Management, Scope and Time Management, Cost and Procurement Management and Good Research Practices.
  - Optative courses: each course provides 3 credits and students must acquire at least 15 credits, that is, take at least 5 courses from the block. The following courses are included: Quality Management, Organizational Transformation, Human Resources and leadership, Negotiation, Risk Management, Sustainability and feasibility of projects, Trends and BIM Standards, Artificial intelligence applied to projects and Investment projects with BIM.
  - Free credits: four credits that can be acquired through specialized courses or training that are part of other training programs.
  - Publication in scientific journals and events: 10 credits that must be acquired for the publication of research results. Works indexed in WoS or Scopus (level 1) provide 4 credits. Works indexed in Scielo, or in other recognized sources (level 2) provide 3 credits. The works in reports of refereed events contribute 2 credits, while the scientific series only 1 credit.
  - Research seminar: participation in research seminars provide 12 credits. Where each participation in the seminar can contribute up to 3 credits to the student.
  - Presentation and defense of thesis: 9 credits.

The courses that are part of the master's degree in project management combine traditional techniques for decision-making in project management with artificial intelligence techniques, as shown below:

- Integrated Project Management: in addition to traditional techniques, fuzzy logic knowledge is incorporated for the analysis of indicators and knowledge-based systems (CBR, FIS) for project evaluation [7].

- Scope and Time Management: incorporates knowledge of metaheuristics and other artificial intelligence techniques for the solution of the "Next Release Problem" and the construction of optimal or quasi-optimal schedules in the solution of planning problems RCPSP, RCMPSP, MMRCPSP, MMRCMPSP [12–14].
- Cost and Procurement Management: in this course, knowledge-based systems are introduced as one of the strategies for estimating resources.
- Good practices for research: in this course, qualitative and quantitative methods of experiment design are combined. In addition, word computing techniques are introduced as tools to apply qualitative methods.
- Quality Management: knowledge-based systems and prediction models are introduced.
- Organizational transformation: in this course, elements of business architectures are introduced under the approach of improving processes and systems simultaneously. The introduction of emerging technologies is proposed to aid decision-making in project-oriented organizations.
- Human Resources and Leadership: this course introduces elements of text mining, opinion mining, graph theory, and neural networks for the evaluation of competencies and the analysis of human resources performance.
- Negotiation Course: new negotiation methods are introduced, and it is proposed to accompany these with techniques for analyzing the learning styles of the participants.
- Risk Management Course: in this course, traditional methods for risk management are complemented with fuzzy logic techniques, neutrosophic theory and impact prediction models.
- Course on Sustainability and Feasibility of Projects: in this course, traditional methods are complemented with elements of fuzzy logic, neutrosophic theory, and knowledge-based systems.
- Artificial intelligence applied to projects: this course presents trends in artificial intelligence techniques and their applicability in decision-making in project management.
- BIM Trends and Standards: this course analyzes the ISO 19650 series and its impact on the virtual construction of works. In addition, this course presents trends in artificial intelligence and augmented reality in its application in BIM.
- Investment projects with BIM: in this course, they analyze the particularities of investment projects. Non-venous techniques for decision-making and prediction are introduced.

An important element in the Knowledge Institutions component is the research system, made up of the following elements:

- Bank of real problems that need to be solved through research and innovation actions.
- The repository of databases for the development of research in project management [15]. This repository is nourished by the data obtained in the own master's and doctoral investigations that the training program encourages. It also draws on the information systems developed to solve the bank's problems.

- Event system established since 2008 with annual events. Often biannual, these events are organized as an international event.
- Project management research seminar with monthly frequency that focuses on the research training of masters and doctors.

Another fundamental element is the conception of the BusinessRedmine project management information system as a software ecosystem.

A Software ecosystem is defined as a set of relationships between different entities with the aim of sharing clients, market segments or increasing efficiency and effectiveness in the production processes they develop. It is supported by the definition of common architectures and platforms that facilitate the integration of solutions, the exchange of information, resources, artifacts and assets in general. These relationships must be supported by economic-financial relationships, work agreements or development models that promote the exchange of data and codes. The software ecosystem consists on the following components: application domain, Family of products, set of reusable software actives, architecture, production modes of the ecosystem entities and the set of relationships between the entities (relational knowledge).

A continuous development system has been achieved where the BusinessRedmine platform is used for practical training. At the same time, the BusinessRedmine system is developed from the own master's and doctoral research. The BusinessRedmine platform is commercialized as project management tool, the funds from its commercialization are used for to develop the own training ecosystem. In Sect. 3 it is presented the results of the continuous develop.

As a strategy for controlling the quality of the ecosystem proposed in the public policy system component, a system of indicators is introduced that allow the effectiveness and efficiency of each of the components to be measured. In addition, a method based on word computing is designed for the analysis of the preferences of the evaluators of the training process.

For the construction of this system of indicators, different international approaches and trends were analyzed in postgraduate evaluation. Evaluation indicators of the Ibero-American Graduate University Association, AUIP, and experiences in the United States, Canada, Brazil, Mexico, Argentina were analyzed. France and other European countries [16–18]. A summary of the analyzed models is presented below:

- The evaluation model proposed by CALED [19] is structured in 4 areas, 16 subareas and 80 indicators. This model evaluates programs with a focus on the quality and didactics of the learning resources they propose.
- The generic ADDIE [20] model used to design the contents proposes the following phases: analysis, design, development, implementation, evaluation.
- The model used by the UAPA [21] university is organized into the following areas: technology, training, instructional design, services and support. In the technology area, 37 indicators associated with: technological availability, performance, capacity, security, privacy, accessibility, usability, navigability and maintenance are measured. The training area evaluates the availability and execution of the pedagogical and technical training plans that teachers and students must

have. The area of instructional design, with 32 indicators, has a weight of approximately 50% in the evaluation this area focuses on the structure of the contents of the courses, the design and the methodology used for their development. The services and support area evaluate the availability of information and participant care services to normally carry out its activities.

- Another model analyzed is the one proposed by the Distance Education Accreditation Commission in the United States known as DAEC [22]. This organization dictates policies, procedures, standards and guidance for the implementation of the best distance education practices in the United States. Its proposal includes elements such as: institutional mission, strategic planning, its effectiveness, curriculum and teaching materials, program outputs, student satisfaction, academic leadership of the team, bibliography, documentation and admission, financial management and conflict management.
- The model proposed by Gilly Salmon focuses on the design of virtual activities [23]. It establishes five stages in the management of virtual environments: access and motivation, socialization, information sharing, knowledge construction and conclusion.
- The Kirkpatrick Model [24] focuses on impact analysis and establishes four stages aimed at measuring participant satisfaction, acquired knowledge, applied knowledge, and impact.
- The model proposed by the JAN National Accreditation [25] is based on a system of indicators of effectiveness and efficiency of the programs. Organize the evaluation process in the following six variables. Variable 1 is aimed at measuring relevance. Variable 2 measures the tradition of the institution and its level of collaboration. Variable 3 includes indicators to evaluate the quality of the faculty, considering teachers and tutors. Variable 4 focuses on didactic, material and administrative assurance. Variable 5 measures the level of student satisfaction. While variable 6 evaluates the curriculum and focuses its analysis on eight aspects.

From this analysis of the models, two fundamental approaches are identified in the evaluation processes. The first approach focused on the efficiency of the programs, where the fundamental indicators are located in the number of graduates and the quality of the faculty. The second approach focused on the quality of the process, where the quality of learning resources and training services is measured [23, 26]. However, despite the differences in the identified approaches, both include indicators and instruments to measure stakeholder satisfaction, relevance, visibility in research results, and the impact of programs on society [27, 28].

In this research, a system of indicators is proposed that considers the main international trends and extends the certification system proposed by the National Accreditation Board for postgraduate studies in Cuba JAN [25]. A set of questions aimed at measuring the new proposed indicators are listed below for each variable.

Variable 1 relevance is complemented by the following elements:

- Indicate the level of application today of the knowledge acquired by you in the program

- You have experienced the opening of new job opportunities from the master's degree
- Indicate the level of application today of the results of your research in social practice
- Indicate qualitatively how many people benefit today from the results of your research
- Did your resume improve as a result of the program?
- To what degree was your research related to projects in prioritized lines of research in the country or the university?
- To what degree did your research have a social impact?
- To what degree did your research have an economic impact?

Variable 2, measures the tradition of the institution and its level of collaboration is extended with the indicators:

- Indicate the level of participation of the program in national collaboration networks
- Indicate the level of participation of the program in international collaboration networks
- Indicate the level of exploitation of social networks by the program
- To what extent does the master's program stimulate and enhance your participation in events
- To what extent participation in the research group and its seminars facilitates their training and development of their research.

Variable 3, teachers and tutors is extended from the following elements:

- Indicate the extent to which teachers focus the guidance on learning resources on the platforms
- Indicate the practical level of the discussions on the platform and of the content
- How do you classify the frequency of communication of the tutors and the master's coordination team throughout the process?
- How do you rate the communication frequency of the course teachers with you?

Variable 4 measures the didactic, material and administrative assurance of the program, it is extended as shown:

- At what level does the institution offer access to the learning resources designed for the courses?
- To what degree is the course platform available for submitting assignments?
- To what extent is the platform available in access to learning resources and bibliography?
- To what degree is the platform available for interaction between team members, participation in discussion forums?
- Evaluate the availability of the platform for the management of GESPRO projects oriented to the assembly of the course work.
- To what extent are the repositories for research accessible?

- To what extent are the discussion forums of the thematic lines accessible to cover your information needs?
- To what extent does the training program exploit social networks for training?
- To what extent are videos and other supplementary visual materials available for your training?
- To what extent do you consider that the technologies that you have in your environment (home, work, city) are affecting or affecting your work with training platforms?

Variable 5 is aimed at measuring student satisfaction from the following extensions:

- Indicate the level of need for your access to the course platform to clarify doubts about the program
- Indicate the level of need for your access to the social networks associated with the program for clarification of doubts
- To what extent do you use video conferencing to acquire information in the program?
- Evaluate the application of peer evaluation techniques in the program?
- To what degree are you satisfied with the result of the evaluation processes?
- Evaluate the degree of diversity in the forms of presentation of the contents
- To what degree do you consider that the way of presenting the content is comfortable for you
- Assess the level of organization of teamwork on the training platforms.

Finally, variable 6 measures the curriculum by extending the original JAN indicators as shown:

- Assess the level of initial presentation and orientation in the graduate program
- To what extent has the program organization allowed you to become familiar with the platform before formally starting the program?
- To what degree do the methodological orientations of the activities allow you to orient yourself in the path that you must follow for the acquisition of knowledge
- To what extent are the activities offered to the participant organized logically, with indications of the space of the technological platform where each activity will take place, a calendar with the relevant dates for the delivery of tasks and exams?
- To what degree are professional information from the facilitator, email, and counseling hours incorporated into the activities?
- To what degree are the learning resources adequate to support you in assimilating the course content?
- Evaluate the quality of the presentation of the content of the lessons, regarding the volume of information they transmit and the relevance of it
- Evaluate the diversity of the forms of evaluation used
- Evaluate the quality of the term glossaries available in the program
- To what degree is the presentation of the content personalized for you?
- At what level do the methodological guidelines and guides help you to write the thesis?

- To what extent does the knowledge transmitted in the program help you to write scientific articles?
- To what extent did the self-evaluation activities contribute to your preparation for the final evaluation of the course?
- Evaluate the level of effort you had to make to understand the content proposed by the course
- Evaluate how laborious it was for you to carry out the evaluation activities
- Indicate how much approximate time it took to develop the evaluation activities
- To what extent are or were the available repositories useful for your research?
- Indicate how often you appreciate that the learning resources are reviewed to ensure that they respond to the needs of the participants and facilitators
- To what extent does the program promote your participation in research seminars and events
- Analysis of results and application of the program.

3 Analysis of Results and Application of the Program

The proposed training ecosystem is applied in the training of masters and doctors at the "Universidad de las Ciencias Informáticas". It is used as part of the Master's degree program in Project Management and in the development of information systems for decision-making in project management.

3.1 Analysis of Results in the Application in the Master's Program in Project Management

The training program was delivered to a group of stakeholders that involved 151 people organized into four groups in the role of students and 16 people in the role of teachers and facilitators. The first version of the quality system was applied by applying a set of surveys to 25% of those involved, obtaining the following results, see Tables 1, 2, 3 and 4.

In the analysis of these variables, it is necessary to identify that 53 people dropped out of the program for various reasons, the most important reason being difficulties

Table 1 Criteria analysis regarding the relevance variable

Criteria	Valoration
Indicate the level of application today of the knowledge acquired by you in the program	High, 0.02
To what degree was your research related to projects in prioritized lines of research in the country or the university?	Medium, 0.4
To what degree did your research have a social or economic impact?	High, −0.03

Table 2 Criteria analysis regarding the institutional tradition variable

Criteria	Valoration
Indicate the level of participation of the program in national collaboration networks	Medium, −0.02
Indicate the level of participation of the program in international collaboration networks	High, 0.01
Indicate the level of exploitation of social networks by the program	Medium, 0.04
To what extent does the master's program stimulate and enhance your participation in events	Very high, −0.02
To what extent participation in the research group and its seminars facilitate their training and development of their research	High, 0.04

Table 3 Analysis of criteria regarding the variables teachers and curriculum by activities taught

Criteria	Attention to teachers	Quality of content presentation	Approved %	Promotion quality (%)
Integrated project management	High, 0.13	High, 0.21	43	24
Risk management	High, 0.24	High, 0.24	82	73
Artificial intelligence applied to project management	High, −0.48	High, −0.06	64	40
Negotiation	Very high, −0.4	High, 0.38	92.86	85

Table 4 Analysis of criteria regarding the variables teachers and curriculum by activities taught

Criteria	Valoration
Completeness and information access to the course platform	High, −0.43
Moodle course platform availability	Medium, −0.43
Platform availability during the Moodle evaluation peak periods	Medium, −0.48
Complexity of work on the Moodle course platform	Low, 0.36
Availability of BusinessRedmine project management platform	High, −0.37
Satisfaction of the repository system help videos	Medium, 0.29
Accessibility to research repositories	High, −0.3
To what extent does the training program exploit social networks for training?	Medium, 0.25

with access to technology. Difficulties in access to video conferences by participants were also identified during the experiment. As part of the ecosystem, the defense of 105 master's theses and 9 doctoral theses have been achieved.

From the analysis of the experiment, the following elements are also recommended:

- It is necessary to guide the students through various ways which is the learning route, designing adjustable routes for each interested party, in such a way that they meet the needs of the participants with different learning styles. Proper guidance will improve the planning of the study time of the interested parties and will set the pace of work.
- Create mechanisms to identify the main causes of desertion and attack them from the early stages. It has been identified that among the main causes is the lack of communication and feedback from the facilitator towards the participant, it is important to generate interaction between all the actors in the learning process, enhancing the participant's confidence.
- The didactic learning units represented by the lessons, the evaluative activities, and each of the learning resources available on the platform, must contain enough content to incorporate new knowledge to the student, but not too extensive to overwhelm him. In addition, diversity must be guaranteed in the didactic learning units to cover the different learning styles and achieve a visually pleasing presentation for the participants.
- Indicators must be included that allow evaluating the cost in time dedicated to the development of activities, allowing them to continuously improve their design.
- It is important that there is an evaluation rubric for each of the activities that are programmed, and they must also be public for the interested parties, in this way the interested party will be able to concentrate on the elements of the highest priority or relevance for the evaluation.

3.2 Analysis of Results in the Development of the BusinessRedmine Ecosystem and Its Application in Different Environments

The main results of the training ecosystem have been precisely the synergy be-tween development and commercialization of services. The ecosystem provides services from the use of the BusinessRedmine platform to different organizations. The income obtained from this commercialization is in turn reinvested in the development of the training ecosystem itself. On the other hand, the research system, thesis and doctorate results contribute to the development of the BusinessRedmine platform, supporting its continuous development.

The platform was installed in the Network of Software Development Centers of the University of Computer Sciences from 2010 to the present. This network is made up of more than 10 software development centers specialized in different application areas. The calculated availability of the overall assembled system is around 92%. The introduction of ecosystem products to this organization significantly helped it achieve its CMMI level 2 international certification in October 2015.

In this environment, a survey was conducted with 32 clients of the system to analyze the variable "time spent analyzing projects and generating status reports." The distribution of the professionals surveyed was as follows: 14 project managers,

5 project planners, 7 project management office inspectors and 6 project quality reviewers. Measurements were made of the time spent analyzing projects and preparing status reports before and after the implementation of the BusinessRedmine platform. Respondents selected one of the following responses:

- Less than 30 min.
- Between 2 and 7 h to prepare a report.
- About 24 h to make a report.
- 40 h or more to prepare a report.

The results were satisfactory, achieving in some cases a reduction in time from 1 week to minutes.

Another element that we measured was the efficiency in the assembly of the platform in different real scenarios over time, see Eq. (1). In this sense, the commercial name of GESPRO was used, and the platform was installed in: Software development centers of "Universidad de las Ciencias Informáticas" (UCI), in the software development company "XETID", in the "Parque Científico Tecnológico de la Habana" (3CE Technological Scientific Park of Havana) and in the "Red Colaborativa de Ingeniería y Gestión de Proyectos" (RCGP). Table 5 shows the data associated with the development and customizations of the platform in the four scenarios proposed.

$$efficiency = \frac{product\ size}{amount\ of\ resources + time} \tag{1}$$

As can be seen in Table 5 and in Fig. 2, there is an increase in productivity motivated by the use of software ecosystems that promote the reuse of all assets.

Table 5 Efficiency in the assembly of BusinessRedmine over time

Product	Product size	Number of people	Time duration	Efficiency indicator
1st UCI	64	10	45	1.16
2nd XETID	16	3	10	1.23
3rd RCGP	32	3	10	2.46
4th 3CE	32	2	7	3.56

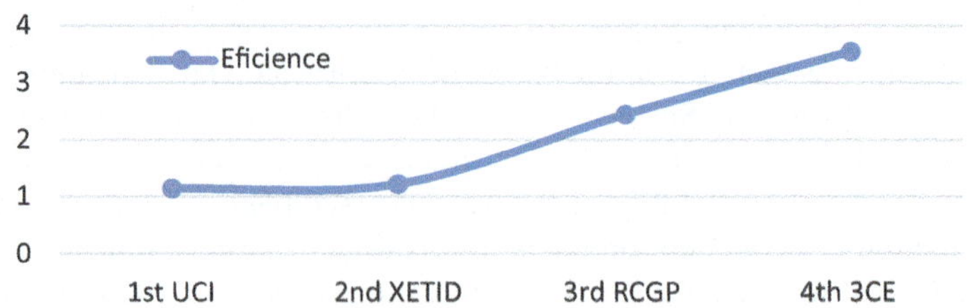

Fig. 2 Analysis of the efficiency in the development and deployment of the platform in different scenarios

4 Conclusions

- The design and operation of online training programs requires a high effort in time and cost for the institutions involved. In particular, it requires the assembly of a technological infrastructure that supports the entire training platform required to satisfy the needs of those involved.
- It is necessary to guarantee the quality in the design and presentation of the contents and the diversity in the way of presenting them in order to satisfy the needs of the participants with different learning styles.
- It is important to establish a quality system based on the principles of total quality with the involvement of all members of the organizations and based on indicators and forms of evaluation that allow the continuous improvement of organizations to be promoted.
- Furthermore, it is important to create mechanisms to identify the main causes of desertion and attack them from an early stage, enhancing the acceptance of the program by the interested parties.
- A work system was created that guarantees the sustainability of the training ecosystem. The Ecosystem's products are marketed as services and tools (BusinessRedmine) for decision-making in project management. The income generated by the commercialization is reinvested in the development of the system itself.
- The development and assembly of the BusinessRedmine platform allowed the agile development of a family of products and its implementation in different scenarios. The introduction of the results in social practice made it possible to significantly reduce time and costs in the control and monitoring of projects in the beneficiary organizations. In addition, it allowed support in international quality certification processes in one of these entities.
- The BusinessRedmine platform conceived as a software ecosystem allowed to take advantage of free software and open source solutions. Development costs were lowered, and high-quality solutions were achieved by being based on previously proven assets.

References

1. Dave, B., Buda, A., Nurminen, A., Främling, K.: A framework for integrating BIM and IoT through open standards. Autom. Constr. **95**, 35–45 (2018). https://doi.org/10.1016/j.autcon.2018.07.022
2. Herczeg, M., et al.: Resource efficiency in the building sector. ECORYS Ned. BV Cph. Resour. Inst. (2014)
3. Dunbar, G.: Project management failures—Standish (Chaos) reports (1994–2015) (2016). Available: https://www.linkedin.com/pulse/project-management-failures-standish-chaos-report-2015-dunbar. Accessed: 21-July-2021
4. Lynch, J.: Project Resolution Benchmark for IBEX Financial Corp, p. 11. Standish Group Int. Inc (2018)
5. Johnson, J.: CHAOS report: decision latency theory: it is all about the interval. Lulu.com (2018)

6. Gamarra, G.D.: Plan Agrupado de Formación Continua Comunidades de Práctica, buenas prácticas para la dinamización de entornos virtuales disponibles (2010) [Online]. Available: http://puntodeencuentro.utn.edu.ar/wp-content/uploads/2010/12/dinamizacion.pdf. Accessed: 21-May-2021
7. Pérez, I., García-Vacacela, R., Piñero, P.Y., Mahdi, G.S., Peña, M.: Experiencias en el uso de técnicas de softcomputing en la evaluación de proyectos de software. Rev. Investig. Oper. **41**(1), 106–117 (2020)
8. Piñero, P.Y.: Estatutos Red Colaborativa de Gestión de Proyectos. Departamento de Investigaciones en Gestión de Proyectos, Universidad de las Ciencias Informáticas Maestría en Gestión de Proyectos (2021) [Online]. Available: http://gespro.uci.cu. Accessed: 21-May-2021
9. PMI: The Standard for Project Management and a Guide to the Project Management Body of Knowledge, 7th ed., p. 370. Project Management Institute (2021). ISBN: 978-1-62825-664-2
10. Stellingwerf, R., Zandhuis, A.: ISO 21500 Guidance on Project Management: A Pocket Guide (Best Practice). Van Haren (2013)
11. ISO: ISO 19650 Building Information Modelling (2019)
12. Mahdi, G.S.: Estimation of distribution algorithm for solving the multi-mode resource constrained project scheduling problem. EAI Endorsed Trans. Energy Web **7**(27) (2020). http://doi.org/10.4108/eai.13-7-2018.164111
13. Mahdi, G.S., Piñero, R.Y., García-Vacacela, R., Quintana, J.M., Al-subhi, H., Pérez, I.: Nuevo algoritmo de estimación de distribuciones con diferentes métodos de aprendizaje aplicado a planificación de proyectos. Rev. Investig. Oper. **41**(7), 945–960 (2020)
14. Mahdi, G.S., Al-subhi, H., García-Vacacela, R. Pérez, I., Madera, J., Piñero, P.Y.: Algoritmo de estimación de distribución con tratamiento de restricciones en el modelo probabilístico en problemas de scheduling. Rev. Investig. Oper. **40**(5), 674–686 (2019). ISSN 0257-4306
15. Piñero, P.Y., Pérez, I., Rivero, C., Rojas, C., Sosa, R., Torres, S.: Repositorio de datos para investigaciones en gestión de proyectos. Rev. Cuba. Cienc. Informáticas **13**(1), 176–191 (2019)
16. Sarif, S., Ramly, S., Yusof, R., Fadzillah, N.A., Sulaiman, N.Y.: Investigation of success and failure factors in IT project management. In: Proceedings of the 7th International Conference on Kansei Engineering and Emotion Research, Singapore, pp. 671–682 (2018). http://doi.org/10.1007/978-981-10-8612-0_70
17. Aguiar, D., Fenner, G., Sampaio, A.: A risk management methodology proposal for information technology projects. IEEE Lat. Am. Trans. **12**(4), 643–656 (2014). https://doi.org/10.1109/TLA.2014.6868866
18. Kurek, E., Johnson, J., Mulder, H.: Measuring the Value of Enterprise Architecture on IT Projects with CHAOS Research. MSCI, Orlando, USA (2017)
19. CALED: Instituto Latinoamericano y del Caribe de Calidad en Educación Superior y a Distancia (2021) [Online]. Available: http://www.caled-ead.org/. Accessed: 24-Nov-2021
20. Universidad de Valencia: Modelo ADDIE modelo básico de Diseño Instruccional (2013)
21. UAPA: Universidad Abierta para Adultos. Departamento de Orientación Profesional y Académica (2021) [Online]. Available: www.uapa.edu.do. Accessed: 24-Nov-2021
22. DEAC: Accreditation Handbook. Policies, Procedures, Standards and Guides of the Distance Education Accrediting Commission, 30th ed. (2021)
23. Salmon, G.: E-actividades: el factor clave para una formación en línea activa. Editorial UOC (2004)
24. Alsalamah, A., Callinan, C.: The Kirkpatrick model for training evaluation: bibliometric analysis after 60 years (1959–2020). Ind. Commer. Train. (2021). https://doi.org/10.1108/ICT-12-2020-0115
25. JAN: Sistema de Evaluación y Acreditación de Maestrías. Junta de Acreditación Nacional, Ministerio de Educación Superior (2014)
26. Marciniak, R., Gairín, J.: Dimensiones de evaluación de la calidadde de educación virtual: revisión de modelos referentes. Rev. Iberoam. Educ. Distancia RIED **21**(1), 216–231 (2018). https://doi.org/10.5944/ried.21.1.16182

27. Mark, W., Steven, V.: Directrices de la UNESCO para las políticas de aprendizaje móvil. UNESCO (2013)
28. Villar, L.B., Herrero, L.L., López, G.Á., Gil, E.P.: Caracterización de las mejores prácticas educativas: UNESCO y el paradigma del aprendizaje móvil. Digit. Educ. Rev. **39**, 336–355 (2021)

Chapter 15
Evaluation of an Accreditation Variable for University Institutions Using 2 Tuple Linguistic Representation Model

Julio Alfredo Telot González, Aylín Estrada Velazco, Maylí Estopiñán Lantigua, and Camilo Armando Suárez Álvarez

Abstract The evaluation of higher education institutions and their corresponding accreditation constitute tools of proven effectiveness for the insurance of the quality in this level. There are different ways to develop these accreditation processes, with different visions and parameters and there are quantitative and qualitative approaches for their evaluation. In Cuba, there is a System of Evaluation and Accreditation for High Education, which establishes the variables, components, standards and indicators that define the main activities to be evaluated in a university. The present paper is directed to propose the use of the pattern of the linguistic representation model based on 2-tuples to achieve a more objective evaluation according to the different parameters that are evaluated and the experts' characteristics. It is analyzed the variable of Human resources as a case of study to apply this model to their different indicators and to propose a valuation of this variable, through the use of the mentioned tools.

Keywords Decision making · Linguistic representation model 2-tuples · Universities institutional evaluation

1 Introduction

Evaluation and accreditation process for programs and institutions in higher education is a necessary condition for quality insurance in these institutions. It allows a greater objectivity and guarantees to feedback to the society about their results [1–4].

Accreditation is the process that recognizes the quality of an institution or educational program and it is based on a previous evaluation.

In order to conduce and advice these processes, there are organizations, such as Accreditation Board for Engineering and Technology (ABET), which covers

J. A. T. González (✉) · M. E. Lantigua · C. A. S. Álvarez
Universidad de Matanzas, Matanzas, Cuba

A. E. Velazco
Universidad de las Ciencias Informáticas, CP 54830 La Habana, Cuba

P. Y. Piñero Pérez et al. (eds.), *Artificial Intelligence in Project Management and Making Decisions*, Studies in Computational Intelligence 1035,
https://doi.org/10.1007/978-3-030-97269-1_15

different fields of engineering education such as: ABET-CS in Computer Sciences [5, 6]. There are other interesting experiences such as the ones given by the Council for the Evaluation, Accreditation and Quality Assurance of Higher Education in Ecuador [7], the National Accreditation Commission in Chile [8], Ibero-American University Association for Postgraduate Studies, known as AUIP (Spanish acronym) [9, 10] and the National Accreditation Board (JAN, Spanish acronym) of Cuba. They are the main sources for the present paper.

When the results of institutional evaluations made in Cuban universities between 2015 and 2016 are analyzed, there is a lack of homogeneity in the characterization of strengths and weaknesses, which can impact on the judgment of the experts during the evaluation process, which starts in self-evaluation [11].

The experts who participate in the accreditation process in Cuba are specialists from different Cuban and other countries' institutions and they are selected for their experience and academic and scientific level. These experts use an evaluation guide, where the variables, indicators and the established measurement criteria are described. Each variable contains some indicators, which may be quantitative or qualitative and any quantitative criterion may contain a qualitative analysis about its significance and relevance. This guide, however, does not establish any specific scale to evaluate, this is the reason why each expert carries out his/her own evaluation keeping in mind their experience, opinion and their assessments are subject to subjectivity. At the present, in order to achieve a greater objectivity in the issuance of the criteria, work is being done on the establishment of a breakdown by components and standards, from which the indicators are derived. From the fact that it is practically impossible that every expert had the same vision about strength or a weakness, the lack of a mechanism to treat the preferences of multiple experts on the criteria is a negative element. It brings a great loose of time and information to achieve a convergence among the different valuations.

On the other hand, in general, the institutional evaluation is not considered as a Decision-Making (MD) problem, this provokes that there is not a right use of the multiple alternatives offered by the information handling in uncertain environments, where the experts can model their preferences in different expression domains, since it is a problem where the available information doesn't always reflect the reality, or it is imprecise, or there are situations whose nature recommends the use of qualitative valuations.

From the analysis of the difficulties that were previously indicated, it is proposed as the investigation problem the following question: How to evaluate the quality of Institutions of Higher Education with a greater level of objectivity that guarantees the multigranular information handling emitted by different experts?

The objective of the present paper is the use of representation model 2-linguistic tuple in order to handle the multigranular information offered by different experts to evaluate a variable in the quality evaluation system for superior education institutions.

2 Materials and Methods

2.1 Characteristics of the Quality Evaluation Process of Higher Education Institutions in Cuba

The process of evaluation and certification of quality in Cuba is opened to the professional and productive world, although it is designed and conducted by representatives from the academic world, in correspondence with the specific characteristics of Higher Education Institutions (HEI) in the Republic of Cuba [12].

This work is developed in Cuba according to what it is established by the Evaluation and Accreditation System for Higher Education Institutions (SEA-IES, Spanish acronym) and this is a fundamental component of Evaluation and Accreditation System for Higher Education (SEAES, Spanish acronym) [13]. This system has, as the main objective, to promote an enhancement of the management in the Cuban HEIs. The main specifications and legal regulations about this question are contained in the SEA-IES and they constitute the most important references for the development of the whole evaluation process of a HEI in Cuba [14, 15].

According to the documents established by JAN, there are six quality variables for the system: (1) Institutional context and integral relevance, (2) Human resources, (3) Undergraduate formation (4) Investigation and postgraduate formation (5) Infrastructure and (6) Economic and social impact.

These variables are defined in order to determine the degree in which an institution satisfies the corresponding quality standards. In principle, they are attributes that can take different values or be expressed in categories. Each variable is subdivided into components, which are fundamental parts to understand the meaning of the variables and these components are broken down into standards and the standards contain the different indicators [16].

For the Cuban higher education, the indicators contained in the evaluation guide allow to measure the degree of adjustment to the quality criteria within each of the standards. These indicators may be quantitative or qualitative, and, in principle, they express an aspiration or a challenge.

The procedure for the development of the evaluation processes of the HEIs quality in Cuba may be resumed in the following steps:

1. Improvement plan (short, medium and long term).
2. Self-evaluation (systematic and staged).
3. External evaluation (face-to-face and/or remote).
4. Analysis at the Technical Committee for Institutional Evaluation (opinion proposal).
5. Analysis at the National Accreditation Board (opinion).

At the fourth step, the documents, elaborated by the commission that participated in the external evaluation (step 3) are analyzed by the Technical Committee for Institutional Evaluation (CTE-IES in Spanish), which is an entity composed by specialists from different centers (universities, research centers and other places) in Cuba or other countries and from the report prepared after the external evaluation, it issues an opinion proposal to be assessed by JAN. This opinion contains the main strengths and weaknesses that were identified in the evaluated institution and it is proposed, if it proceeds, a higher category of accreditation, in correspondence with the quality level of the management of the institution according to the scale of Excellence, Certified or Qualified.

Although each step in this process is appropriately legislated and documented, there are elements that allow a heterogeneity of criteria, such as the fact that the evaluation guide does not define the scales to carry out the evaluations as rules. This causes that each expert makes his/her assessment from his/her knowledge level on the elements to be evaluated. Besides more than one evaluator participates in each step and there is not a mechanism to treat the preferences of the multiple experts about the criteria. This leads to loss of time and information to achieve a convergence among their evaluations.

After that process, JAN analyzes the opinion proposal and the rest of documents, in order to approve, in plenary session, the category that will be assigned to the institution. This aspect is not covered in the present paper, although it might be interesting for further researches.

2.2 *The Evaluation of the Quality of HEIs as a Decision-Making Problem*

The institutional evaluation process can be considered as a decision-making (DM) problem, in which one or some evaluators must analyze a set of criteria to classify the institutions according to the different accreditation categories (Qualified, Certified, or Excellence). This consideration allows to take advantage of the multiple alternatives for the management of information in environments of uncertainty during this evaluation process, where the experts can model their preferences in different expression domains, or using scales of different granularity, since this is a problem in which the available information does not always reflect the reality, it is imprecise, or there are aspects whose nature recommends to use qualitative assessments [17].

In the quality assessment processes, it is feasible to express the estimates of the experts by means of linguistic values rather than exact numerical values. This way of approaching a decision-making problem is based on fuzzy set theory and is known as fuzzy linguistic approach (FLA). It is used when the variables that appear in the problem have a linguistic character.

At Fuzzy Decision Making, the FLA is in charge of modeling the experts' preferences from the use of linguistic variables, which are characterized by their syntactic value or label and a semantic value or meaning. The label is a word that belongs to a set of linguistic terms and its meaning is given by a fuzzy subset in a universe of discourse. Since words are less precise than numbers, the concept of linguistic variable seems to be adequate to characterize objects that are too complex, or not completely defined to be evaluated through precise numerical values.

In the literature related to FLA there are different models for performing computational processes with words. To treat the outlined problem, the use of the linguistic model based on 2-tuple is proposed [18–21], because it is easy to understand, it maintains a fuzzy representation of the linguistic information where its results have a syntax and semantics, as defined by FLA, which is feasible for the treatment of uncertainty, associated with the evaluation of the quality of the HEI.

3 Results and Discussion

3.1 Description and Classification of the Problem

In the present paper the variable 2 of SEA-IES (Human Resources) has been selected, in order to evaluate, with the linguistic model based on 2-tuple, the accreditation level that would reach the HEI, if this would be the only variable to be evaluated. Later on, by means of aggregation methods, the other variables could be added to obtain the proposal of accreditation level for the studied institution.

To the effects of the experiment, three experts are selected, they will be identified as A, B and C. The criteria to carry out the evaluation would be the indicators of variable 2, which will be described later. This problem is considered as multi-attribute, multi-expert, heterogeneous, multi-granular and static.

Next, the evaluation process of the variable 2 for the accreditation of the HEI is described, through the stages or phases of solution to a basic decision-making problem with the use of the representation model based on 2-tuple.

Stage 1. Definition of the framework.

- Set of alternatives: in this case only one alternative is considered, one HEI, since it is all about to obtain a qualification that classifies the institution in an accreditation category $s_a = \{s_0, s_1, s_2, s_3\}$, where ($s_0$) No category, ($s_1$) Qualified, ($s_2$) Certified, ($s_3$) Excellence.
- Set of Experts: the evaluation commission is composed by three experts, $E = \{e_i \mid i \in (1, \ldots, p)\}$, such as: $E = \{A, B, C\}$. The group of experts is reduced in this case in order to explain how to approach the problem. In subsequent applications, where all the variables established for the institutional evaluation are used, the number of experts must be between 20 and 30.
- Set of Attributes: The number of attributes in a multi-attribute decision problem, like this one, is finite and it is defined as $C = \{c_j \mid j \in (1, \ldots, 16)\}$, where, for each indicator, there will be an evaluation y_{ij}, which indicates the preference of the expert e_i, with respect to the attribute c_j.

Table 1 Components of the variable 2 (Human resources)

No.	Component description
CO1	Human resources planning
CO2	Human resources selection and integration
CO3	Work organization
CO4	Preparation and development of the human resources
CO5	Control and evaluation of the results of strategic objectives

JAN establishes five components for the Human Resources variable, which are shown in Table 1, but for the purposes of this paper, the 16 indicators that correspond to this variable that appear in Table 2 are considered.

Table 2 shows the description of the criteria that correspond to the Indicators, as well as the corresponding component and the weight assigned to each one based on the definition that each component has the same relevance, therefore, same weight (0.2), that is the reason why the weight vector is $W = \{0.07, 0.06, 0.06, 0.025, 0.025, 0.025, 0.025, 0.025, 0.025, 0.025, 0.025, 0.1, 0.1, 0.1, 0.1, 0.2\}$. This table also shows the expression domain for each Indicator, since it is assumed that multiple linguistic scales are used $S_3 = \{s_0, s_1, s_2\}$, in order to evaluate the criteria two, five, seven, eight, twelve and thirteen and $S_5 = \{s_0, s_1, s_2, s_3, s_4\}$, to evaluate criteria one, three, four, six, nine, eleven, fourteen, fifteen and sixteen, as shown in Fig. 1.

These scales were selected by expert criteria based on the fact that the criteria framed in the S_3 scale are considered to have few levels of distinction of uncertainty, whereas in the case of greater granularity (S_5) there is a greater possibility of uncertainty.

Stage 2. Compilation of the experts' preferences

Each evaluator ei will express his/her considerations, using the utility vector X $= x_{1k}, \ldots, x_{jk}$, where x_{jk} represents the preference about the institution studied in variable 2, according to the ck criterion. The information, emmitted by the experts, is shown in Table 3, where are used the criteria VG (Very Good), G (Good), R (Regular), B (Bad), VB (Very bad), F (Fulfills), I (Indifferent) and BR (breaches).

Subsequently, a transformation of the criteria issued by the experts to a linguistic 2-tuple is carried out, in such a way that, when an assessment is issued, $s_i \in S$, a transformation is made with $a = 0$, hence the tuples (s_i, a) are created as shown in Table 4. It should be noted that the criteria VG (Very good), G (Good), R (Regular), B (Bad), VB (Very bad) correspond to the domain s_5 and the criteria F (Fullfills), I (Indifferent), BR (Breaches) correspond to s_3.

Stage 3. Analysis (aggregation)

During this stage the appropriate method to be used in solving the problem must be selected, in this case the linguistic model based on 2-tuple, with its extension of the Extended Hierarchical Linguistic approach [21]. After having the evaluations of the experts and the transformation into 2-tuple, the collective preferences for each

Table 2 Indicators for variable 2, with the corresponding component, assigned weight and expression domain

Indicator	Description	Weight	Component	Expression domain
c1	Vision of human resources in the strategic planning of the institution	0.07	CO1	S_5
c2	Projection of workers in relation to student enrollment	0.07	CO1	S_3
c3	Design of the functions of the positions and structures	0.06	CO1	S_5
c4	Work system with the executive staff	0.025	CO2	S_5
c5	Coverage of the executive staff	0.025	CO2	S_3
c6	Verification of the validity of the actions for the sustainability of the faculty and the increasing obtaining of higher teaching categories, scientific and academic degrees	0.025	CO2	S_5
c7	Annual stability of cadres in office	0.025	CO2	S_3
c8	Number of cadres designed per exception	0.025	CO2	S_3
c9	Leadership of management cadres	0.025	CO2	S_5
c10	Situation, preparation and projection of the reserve including specificities by type of HEI	0.025	CO2	S_5
c11	Cadres who when promoted, come from the reserve of the position	0.025	CO2	S_3
c12	Verification of the direct participation of teachers with the main teaching and scientific categories in the formation process	0.1	CO3	S_3
c13	Specification, in the individual plan of each worker, of the activities that are his responsibility according to their teaching categories	0.1	CO3	S_3
c14	Integral improvement system for the entire university community; definition of actions according to characteristics, work performed, interests and specific needs of workers	0.1	CO4	S_5
c15	Training system for non-teaching workers for their specific performance and professional development	0.1	CO4	S_5

(continued)

Table 2 (continued)

Indicator	Description	Weight	Component	Expression domain
c16	Correspondence, with the objectives of the HEI, of the individual and collective action plans to face the indications of the evaluation process	0.2	CO5	S5

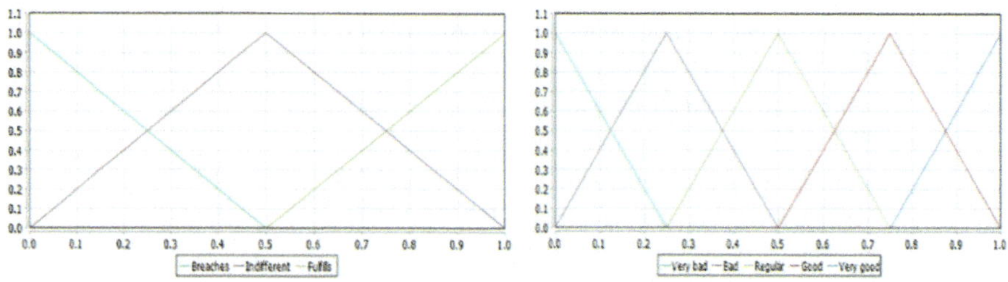

Fig. 1 Expression domains used in the evaluation

Table 3 Experts' preferences

Criteria	Experts		
	A	B	C
C1	Very good	Good	Regular
C2	Fulfills	Fulfills	Breaches
C3	Very good	Good	Regular
C4	Very good	Good	Bad
C5	Fulfills	Fulfills	Indifferent
C6	Very good	Good	Bad
C7	Fulfills	Fulfills	Breaches
C8	Fulfills	Fulfills	Breaches
C9	Very good	Good	Bad
C10	Very good	Good	Bad
C11	Very good	Good	Bad
C12	Fulfills	Fulfills	Indifferent
C13	Fulfills	Fulfills	Breaches
C14	Very good	Good	Bad
C15	Very good	Good	Bad
C16	Very good	Good	Bad

Table 4 Preferences transformed into 2-tuple linguistic

Criteria	Experts		
	A	B	C
C1	(Very good, 0)	(Good, 0)	(Regular, 0)
C2	(Fulfills, 0)	(Fulfills, 0)	(Breaches, 0)
C3	(Very good, 0)	(Good, 0)	(Regular, 0)
C4	(Very good, 0)	(Good, 0)	(Bad, 0)
C5	(Fulfills, 0)	(Fulfills, 0)	(Indifferent, 0)
C6	(Very good, 0)	(Good, 0)	(Bad, 0)
C7	(Fulfills, 0)	(Fulfills, 0)	(Breaches, 0)
C8	(Fulfills, 0)	(Fulfills, 0)	(Breaches, 0)
C9	(Very good, 0)	(Good, 0)	(Bad, 0)
C10	(Very good, 0)	(Good, 0)	(Bad, 0)
C11	(Very good, 0)	(Good, 0)	(Bad, 0)
C12	(Fulfills, 0)	(Fulfills, 0)	(Indifferent, 0)
C13	(Fulfills, 0)	(Fulfills, 0)	(Breaches, 0)
C14	(Very good, 0)	(Good, 0)	(Bad, 0)
C15	(Very good, 0)	(Good, 0)	(Bad, 0)
C16	(Very good, 0)	(Good, 0)	(Bad, 0)

criterion are calculated by means of the weighted arithmetic average aggregation operator, considering the weight of each criterion.

Stage 4. Selection (exploitation)

After the preferences aggregation, according to the criteria, the value that reaches the Variable 2 (Human Resources) is obtained, according to the evaluations of the experts, this value permits to identify the best or the worst alternative. In this case, it is obtained the value (Good, −0.27), it means that, in the Domain s_5, would be on the left of the Good evaluation, what implies that, if it is assumed that the Good evaluation corresponds to Certified Institution, it corresponds to a proposal that, if it is only considered the Human Resources Variable, the possibility that this HEI obtains that accreditation category, or the recommendation of the Qualified Institution category must be taken into account.

3.2 *Solution of the Problem by Means of FLINSTONES*

The example described above was developed with the help of the FLINSTONES software (Fuzzy Linguistic Decision Tools Enhancement Suite) [22].

Figure 2 shows the structure of the framework with the domain assignments to the different criteria for all domains, where it is observed the assignment of the domains s_3 and s_5 to the different criteria, as was described previously.

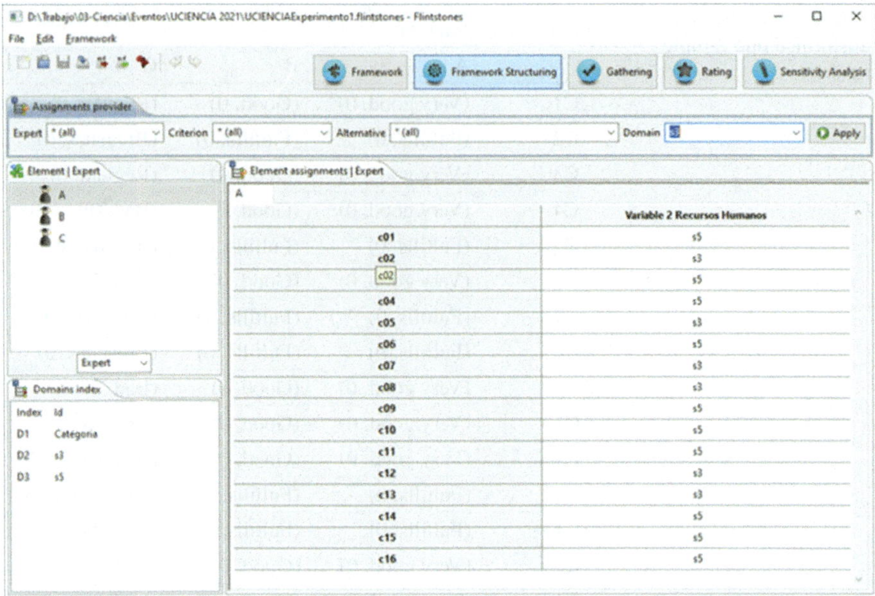

Fig. 2 The structuring of the framework for the studied example

Figure 3 shows the collecting information process from the experts in each of the criteria (in this case it is expert A).

The method used was the extension of the Extended Hierarchical Linguistic approach, which is selected in the corresponding FLINTSTONES stage.

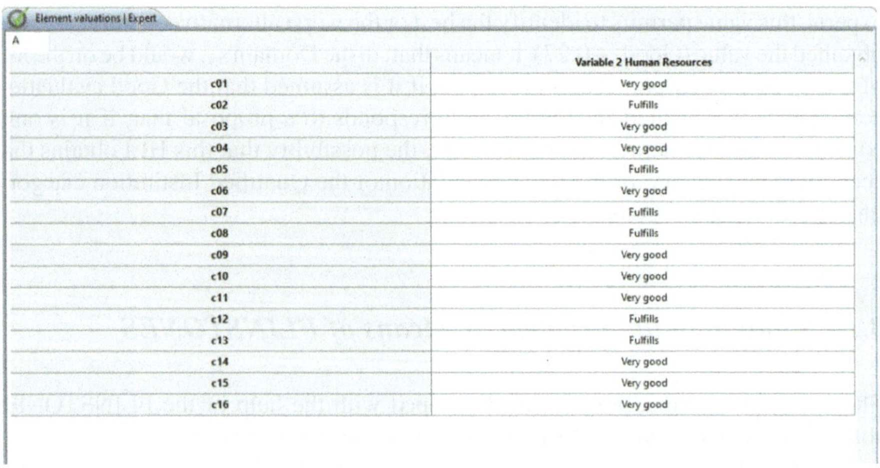

Fig. 3 Collecting data from experts for the studied example

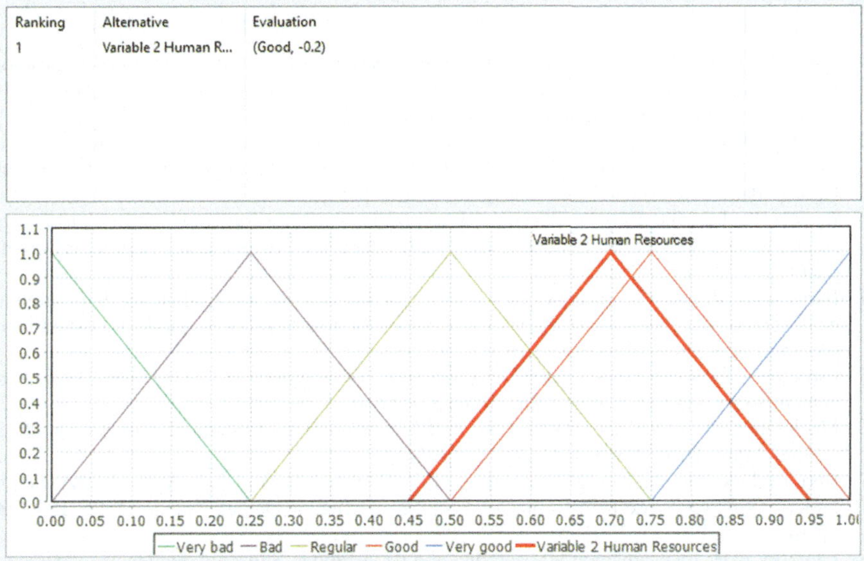

Fig. 4 Results in domain s5

Figure 4 shows the result obtained in domain s_5 where the unique alternative corresponding to Variable 2 (Human Resources) is evaluated as (Good, −0.27), slightly to the left of the Good value.

4 Conclusions

The accreditation of the quality of Higher Education institutions is a fundamental process in achieving the quality standards that today's society demands from universities.

The experts who participate in this evaluation do not always have the same degree of knowledge about the aspects to be analyzed, so it is necessary they be allowed to express their considerations in the domain of their preference. On the other hand, the different criteria do not have to be measured on the same scale.

Working the quality assessment process as a decision-making problem allows to take advantage of multiple existing alternatives for information management in uncertain environments. The results analyzed in this paper can be extended to all the evaluation variables and subjected to an aggregation as considered by the experts. This leads to a proposal that makes it possible to assign, on a more objective basis, the evaluations to the institution in question.

References

1. Varea, S., Coral, M.: Calidad en la educación superior, in Conferencia Nacional Educación Superior en el Ecuador. Investigaciones para una Nueva Agenda de Cambios Necesarios, pp. 1–30. Universidad Andina Simón Bolívar, Quito, Ecuador (2017)
2. Jiménez, H.R., Nossa, J.O.: Fundamentación teórica de evaluación para instituciones de educación superior. Master Master en Ciencias de la Educación, Facultad de Educación, Universidad de San Buenaventura, Bogotá, Colombia (2014)
3. Bonilla, L.: La calidad de la educación. Ideas para seguir transformando la educación venezolana, p. 70. Ministerio del Poder Popular para la Educación Universitaria, Miranda, Venezuela (2014)
4. Barrios, O., Roque, G.: Variables de calidad para una Guía de autoevaluación de programas de posgrado en ciencias médicas Quality variables for a self-evaluation guide of postgraduate programs in medical sciences (2010)
5. Farr, J.V., Bowman, B.A.: ABET accreditation of engineering management programs: contemporary and future issues. Eng. Manag. J. **11**(4), 7–13 (1999)
6. MacKinnon, R., Elder, K.L., Dyer, J.N.: ABET accreditation of IS and IT programs in 2013. J. High. Educ. Theor. Pract. **16**(1), 17–23 (2016)
7. Herrera, L.A., Gonzáles, E.M., Jaramillo, N.D., Simbaña, R.M., Ulloa, M.N.: La Educación Superior y el Proceso de Transformación Social en el Ecuador. Quipukamayoc **22**(42), 187–200 (2014)
8. Lamarra, N.F.: La evaluación y la acreditación de la calidad Situación, tendencias y perspectivas, Informe sobre la educación superior en América Latina y el Caribe 2000–2005: La metamorfosis de la educación superior (2006)
9. Couret, D.G.: Acreditación de programas de posgrado. Experiencias del ISPJAE. Congreso Univ. **2**(1) (2013)
10. Cruz, V., Martos, F.: Evaluación de la calidad de los programas de maestría y doctorado en Iberoamérica, Tema del mes (2010)
11. Telot, J.A., Addine, F., Estopiñán M.: Tendencias en la autoevaluación y la evaluación institucional en la educación superior en Cuba. Convención Internacional Universidad 2018, La Habana, Cuba (2018)
12. Addine, F.: Tras las huellas de algunos aprendizajes relacionados con el trabajo metodológico en las evaluaciones externas a las IES cubanas en los dos últimos cursos (2016)
13. MES: Resolución 150/2018. Reglamento del Sistema de Evaluación y Acreditación de la Educación Superior (SEAES). Ministerio de Educación Superior, Cuba (2018)
14. JAN: Sistema de Evaluación y Acreditación de las Instituciones de Educación Superior. Junta de Acreditación Nacional 24/14 (2014)
15. MES: Resolución Ministerial 62/2016. Ministerio de Educación Superior, Cuba (2016)
16. JAN: Junta de Acreditación Nacional de Cuba, Propuesta de modificaciones de indicadores (2020)
17. Telot, J.A., Estrada, A., Estopiñán, M.: Evaluación de universidades para un ranking académico mediante el modelo de representación 2 tupla lingüística. X Convención Científica Internacional de la Universidad de Matanzas, CIUM'2021 VII Taller Internacional de Ingenierías, Matanzas, Cuba, 23–25 de marzo de 2021 (2021) [Online]. Available: https://www.researchgate.net/publication/350524236. Accessed: 21-May-2021
18. Herrera, F., Martínez L., et.al.: A 2-tuple fuzzy linguistic representation model for computing with words. **8**(6), 746–752 (2000)
19. Peña, M., Rodríguez, C.R., Piñero, P.Y.: Computación con palabras para el análisis de factibilidad de proyectos de software. **20**(50), 69–84 (2016)
20. Gramajo, S.D., Gimenez, A.M.: Framework para toma de decisiones multi-experto multi-granular con información lingüística. XV Workshop de Investigadores en Ciencias de la Computación (2013)
21. Espinilla, M., Liu, J., Martínez, L.: An extended hierarchical linguistic model for decision-making problems. Comput. Intell. **27**(3), 489–512 (2011)

22. Estrella, F.J., Espinilla, M., Herrera, F., Martínez, L.: FLINTSTONES: a fuzzy linguistic decision tools enhancement suite based on the 2-tuple linguistic model and extensions. Inf. Sci. **280**, 152–170 (2014)

Chapter 16
Ontology-Based Management of the Scientific Activity in Software Development Projects

Eliani Varen Caballero and Nemury Silega Martínez

Abstract The management of the scientific activity in software development projects is essential for the success of the projects. This fact was confirmed in a survey applied to software project managers where the 93% of the surveyors point out that a deficient management of the scientific activity of the project may affect the quality of the developed products and provoke deviations in the project. To tackle these deficiencies, an ontology-based approach to support the management of the scientific activity in software development projects is introduced in this paper. Since ontologies are a formal language, with the adoption of this approach it is possible to achieve a semantic analysis of the represented information in order to detect inconsistencies as well as infer new knowledge. The ontology was created following a sound methodology and its evaluation confirmed that it is correctly designed, meets the specifications as a formal logical system and satisfies the requirements for which it was created. This ontology may be a useful instrument in order to enhance the efficiency and effectiveness of the management of the scientific activity.

Keywords Human resources · Ontology · Project · Scientific activity

1 Introduction

The Human Resources Management (HRM) is a crucial factor for the growth of a software development organization, taking into account that human resources are the key resources in the production [1].

HRM of the project includes the processes that organize, manage and lead the project team. It is composed by the stages of: HR Management Planning, Acquisition, Development and Leadership of the project team. Specifically, the HR development includes activities to improve the knowledge and skills of the team members and, improve feelings of trust and cohesion among team members [2].

E. Varen Caballero (✉) · N. Silega Martínez
Universidad de las Ciencias Informáticas, Carretera a San Antonio, Km 2½, Torrens, La Lisa, La Habana, Cuba

P. Y. Piñero Pérez et al. (eds.), *Artificial Intelligence in Project Management and Making Decisions*, Studies in Computational Intelligence 1035,
https://doi.org/10.1007/978-3-030-97269-1_16

A survey applied to managers of 13 software development projects that belong to a Cuban university reveals the existence of shortcomings in the process of planning, monitoring and controlling the scientific activity of the HR of the project teams. Some of the main issues are analyzed below:

- The research priorities of the project are not identified properly; causing conflict in the systematization of research topics and avoiding a common vision of the results in software production. The 85% of the respondents affirmed that the topics of research does not respond to research interests and priorities of the project.
- The processes of planning, monitoring and control of the scientific activity are affected due to the limitations of the support tools. According to 81% of respondents, these causes leads to insufficiencies in the fulfillment of scientific commitments and inaccuracies in the planning of STI objectives.
- Inadequate selection of specialists to develop a specific role within the project; caused by the lack of knowledge of the competencies of HR.
- The 78% of those surveyed affirmed that dependence of the information regarding scientific activity of each team member causes inaccuracy in STI projections. This occurs because the information is not centralized; causing information to be inconsistent.

The problems in planning, monitoring and controlling the scientific activity can affect several project indicators such as: Performance Index of Human Resources (IHR), Performance Index of the Project (PIP). 75% of those surveyed affirm that these indicators are significantly affected. Since the projects belong to an educational institution in addition to the production indicators, they also have an indicator focused on the scientific activity. According to the 100% respondents, this STI indicator is affected with the problems aforementioned.

For Dalkir [3], the man and his knowledge are the most important value that an organization has. This precept is recognized and used by the knowledge management. The knowledge management deals with techniques for acquiring, using and improving the necessary knowledge for the organization, creating an environment that allows to share and transfer them among all the members.

Since their characteristics, ontologies are a suitable mean of representing knowledge. Ontologies have been implemented in several research fields, such as: to calculate the percentage adherence of university professors [4]; creating recommendations of the best practice based on business for software process improvement [5]; generating and validating business process models [6]; calculating similarity between thesis [7]; representing the scientific production of a researcher [8] and establishing relationships among competency requirements of occupations in the market and the profile of higher education students [9].

Because of the benefits of the ontologies, we have developed an ontology-based approach to address the problems listed above. Therefore, this article describes our approach for enhancing the management of the scientist activity in software development projects.

The article is structured as follows: in Sect. 1 we discuss about the identified problems and related issues. Section 2 shows the tools and technologies to develop ontologies. In Sect. 3 the ontology to support the management of the scientist activity is introduced. In Sect. 4 the validation of the ontology is described. Finally, in Sect. V the conclusions and future work are presented.

2 Technologies and Tools

An ontology is a formal and explicit specification of a shared conceptualization [10] using concepts in a domain of discourse, properties of each concept describing its characteristics and attributes, as well as restrictions on the properties [11]. An ontology contains elements such as: individuals, properties, classes, relations and axioms [12].

Among the languages to specify ontologies, the following stand out: Ontolingua, XML Schema, Resource Description Framework (RDF), RDF Schema (or RDF-S) and Ontology Web Language (OWL). OWL is based on a logic model that makes it possible to define and describe the concepts.

The logic model allows the use of a reasoner that can verify whether all statements and definitions in the ontology are mutually consistent and can also recognize which concepts conform to which definitions [13, 14]. These advantages determined that in this research, OWL 2 is assumed as the language for the representation of ontologies.

The ontology to support the management of scientific activity is named OntoGAF. For its development, the Protégé tool is used, which is an open source editor used to build ontologies and as a general framework to represent knowledge, it has a flexible and extensible architecture [14].

Protégé is one of the most commonly used tools for ontological engineering [12, 14–16]. The ontologies created by means of Protégé can be exported in different syntaxes, such as: RDF/XML and OWL/XML. FaCT++, Hermit and Pellet reasoners [16] can be used in the inference process.

3 Ontology for the Management of Scientific Activity

To achieve satisfactory results developing an ontology, a methodology must be used to guide the construction process. Therefore, several methodologies were analyzed [11, 17–19]. The ontology development process presented in this research was guided by the methodology proposed by Noy and McGuinness [11] which includes the following steps:

Step 1 ***Determine the domain and scope of the ontology***

The ontology is focused on the domain of the management of the scientific activity of HR in project-oriented entities and contributes to strengthening the scientific development of software development projects. Ten questions of competence (CQ) that allow to specify the objective of the ontology were formulated:

CQ1. ¿What are the publications related to a determined research topic?
CQ 2. ¿What are the publications with a specific level that a researcher has made?
CQ 3. ¿What publications in journals with a specific level did the project staff produce?
CQ 4. ¿What are the research activities to be carried out by the project in a period?
CQ 5. ¿Who are the outstanding researchers in a given period?
CQ 6. ¿Who are the people who have experience in a subject?
CQ 7. ¿Who are the people who have experience in a type of project?
CQ 8. ¿Who are the people who must complete a master's or doctorate?
CQ 9. ¿Who are the appropriate specialists to deliver a certain training?
CQ10. ¿Which specialists meet the necessary requirements to perform a specific role?
These requirements are the main reference to check the quality of the ontology. As a non-functional requirement, it is specified that the ontology will be implemented in English.

Step 2 ***Reuse existing ontologies***

During the development of an ontology, it is convenient to check if concepts from existing ones can be reused. Other works of interest on the development of ontologies can be found in various ontology repositories [19]. However, the lack of formal and explicit representations for project management is striking, regardless of the domain or area of knowledge [20]. The DAML Ontology Library and DBpedia repositories in which ontologies from multiple domains are published were also reviewed. In this investigation, concepts from the domain of the HERO_ONTOLOGY ontology were used.

Step 3 ***List the relevant terms of the ontology.***

The main terms of the ontology are: publication, journal, event, people, roles, project, university, course, topic and projection. The terms that were reused from the HERO_ONTOLOGY ontology are: book, research report and technical report.

Step 4 *Define the classes and class hierarchy.*

This step describes the steps developed in the Protégé tool to implement the ontology.

Classes Definition
A total of 140 entities were defined. These were identified according to the domain terms described in Step 3 of the methodology used. In OWL there are two types of classes: Primitive and Defined. Primitive classes contain only sets of necessary conditions. On the other hand, classes defined will establish necessary and sufficient conditions, allowing the reasoner automatically infer individuals belonging to these classes.

Additionally, 13 defined classes were specified in OntoGAF: *Professor*, *Specialist_ Production*, *Level1_journal_publication*, *Level2_ journal_publication*, *Level3_journal_publication*, *Level4_ journal_publication*,

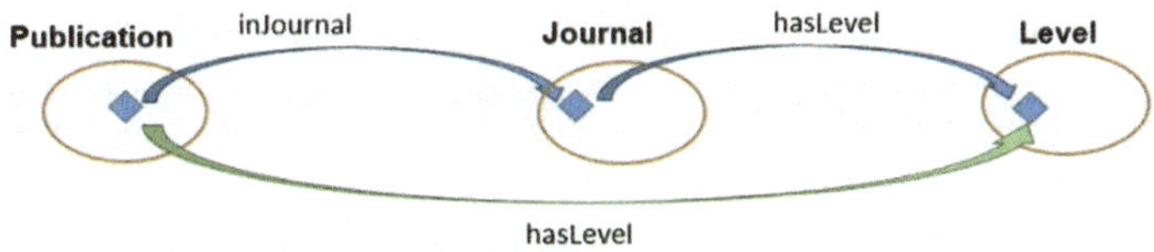

Fig. 1 Example of property chains representation in OntoGAF

Professor_change_Dcategory_assistant, Professor_change_Dcategory_associate Professor_change_Dategory_FullProfessor, Posgrade_professor, Professor_can_teach postgrade, Investigative_priority_topic.

Step 5 *Define the properties.*

They are two types of properties: object property or data property [13]. A total of 180 object properties were represented. These properties illustrate the expressive richness of OWL, allowing the types of relationships between individuals to be described in detail.

Each object property has a corresponding inverse property. In this way, if a property relates individual *a* to individual *b* then its inverse property relates individual *b* to individual *a*, an example is: a P*rofessor teachesPostgraduateCourse*, therefore the postgraduate course *isTaughtBy* that *Professor*. We specified the corresponding inverse relationship for all properties created in OntoGAF.

The property chains were incorporated into the OWL 2 language to express relationships that due to their complexity could not be expressed in previous versions of the language. Property chains allow to relate several individuals using chains of properties. An example of this property is: a P*ublication* is made *inJournal* and a journal *hasIndexation*, therefore the *Publication* also *hasIndexation*, in Fig. 1 this example is illustrated.

Due to the complexity of several properties, Semantic Web Rule Language (SWRL) was used to represent them. This is a language that complements OWL in order to improve its expressiveness [21, 22]. To verify the experience of a person in a subject, the properties were used: the properties *hasJournalPublication* and *hasTopic* are related through the property *hasTopicExperience*. In this way, it is possible to know the topic of the publications that a person has. Figure 2a shows in the red box the individual *Miguel_Angel_Sanchez* created as an instance of the class *Professor*. The green box in Fig. 2b shows the result of the reasoner which inferred that the individual has experience in the topic *Ontology*, because this individual *hasJournaRCCIPublication* which is related to this topic, as illustrated in the blue box. This example is aligned with CQ 6 of the ontology.

Attributes Definition

Data properties are the other type of property defined in OWL. Data properties allow to describe attributes of individuals. Figure 3 shows a representation of the use of the ProjectionYear attribute, which defines the year corresponding to the STI projection, as shown in the red box. The blue boxes contain the individuals

Fig. 2 (**a**) Example of creation of the individual *Miguel_Angel_Sanchez*, **b** result of the reasoner when executing the property *hasTopicExperience*

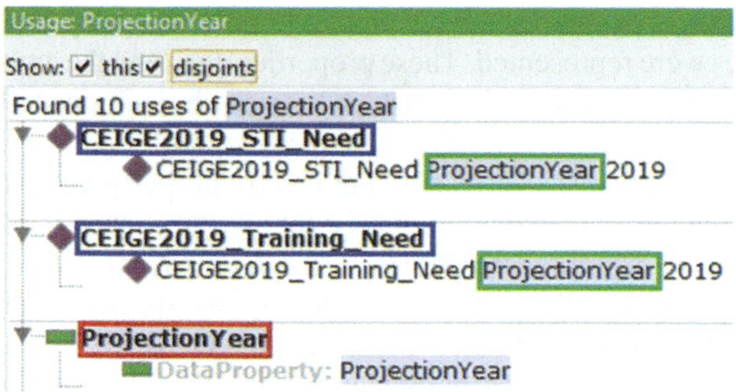

Fig. 3 Example of use of the *ProjectionYear* attribute

CEIGE2019_Training_Need and *CEIGE2019_STI_Need*, which have this attribute, while the green boxes show the value for this attribute.

Step 6 ***Define facets and / or restrictions on the slots or relationships***

In this step, a set of existential, universal and cardinality constraints were defined. As a necessary and sufficient condition for the class *Professor*, it was defined that it must have a teaching category. In this case, a universal restriction was defined, characterized in OWL by the expression ***some***, as shown in Fig. 4.

Fig. 4 Professor class definition

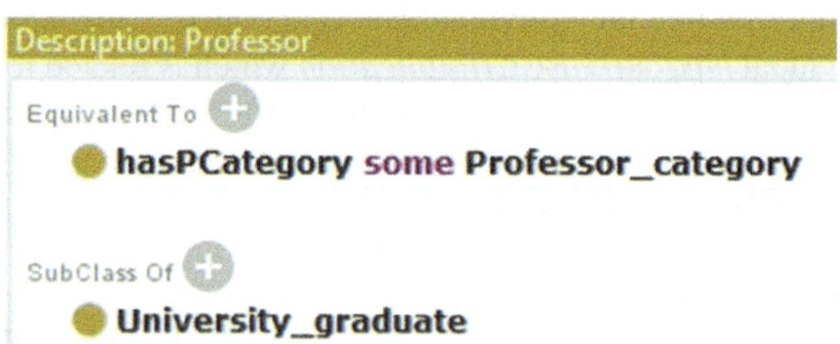

Additionally, the class *Professor_change_Pcategory_FullProfessor* must have a scientific degree, having the Associate professor category and have at least 10 professor evaluations.

Step 7 ***Define instances***

In this step, individuals are defined for the specified classes and the corresponding axioms are established. The result of the execution of this step can be verified in the next section, where the ontology evaluation process is presented.

4 Ontology Evaluation

According to Guerrero and Martínez [8] the ontology validation process is not a specific event for a specific moment. López [23] proposes the use of test cases, reasoning and a checklist with the most common errors detected during the development of ontologies to verify the design. In the design of an ontology, the following issues must be verified: the conditions and properties as a formal logical system, the structural design and the fulfillment of the defined requirements.

4.1 Checking the Formal Logic System

The possibility of inferring knowledge from what is already explicit is one of the benefits that ontologies provide. Silega [24] explains that according to several of the consulted works [7, 9, 24, 25] it is agreed that the best way to verify the formal logical properties of an ontology is through the use of reasoners. In this case, the Pellet reasoner v2.2.0 was adopted.

Figure 5a depicts the individual *Nemury_Silega_Martinez* which is an instance of the class *Person* as indicated by the red box. By assigning the corresponding Object Properties shown in the green box in Fig. 6, the reasoner classifies the individual as *Postgraduate_professor* and *Professor_can_teach_posgraduate* as shown in the blue box in Fig. 5b. This reasoning is due to the *Postgraduate_professor* and *Professor_can_teach_posgraduate* classes are defined classes as we presented in the previous section.

4.2 Checking the Design

To verify the design of the ontology, the proposal of López [23] was used as a tool, where a checklist is established with the errors that commonly appear in this activity.

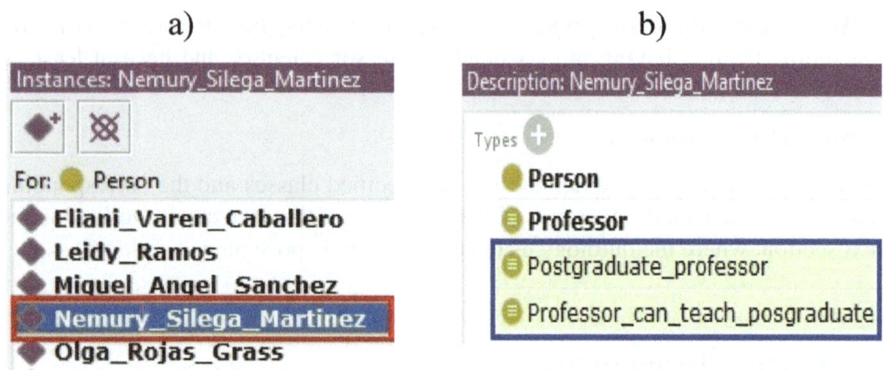

Fig. 5 Example of **a** creation of the individual Nemury_Silega_Martinez **b** classification of the individual Nemury_Silega_Martinez

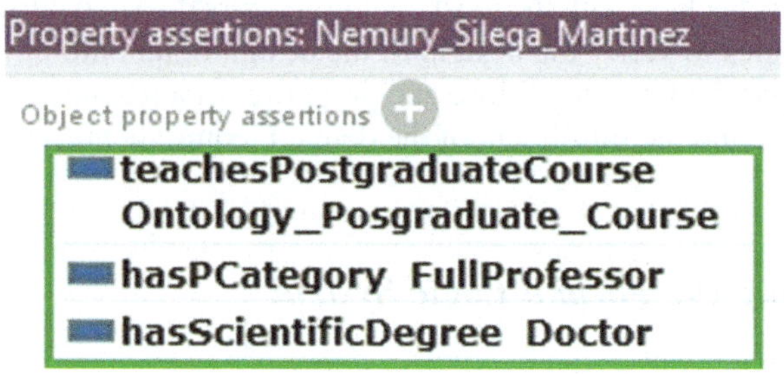

Fig. 6 Property value for the individual Nemury_Silega_Martinez

Type II and VI errors were obtained in the different evaluations carried out. All cases were corrected. Table 1 shows the result related to the application of the checklist.

Verifying the Requirements

Functional requirements or competency questions can be verified through the ontology. Test cases were used to verify how the ontology answers the competency questions. A test case for CQ 3 is illustrated below.

Test case 1

Table 1 Actions carried out after the application of the checklist

Verified item	Applied solution
II: Excessive use of the is-a relationship	The number of *is-a* relationships was identified, analyzed and reduced as much as possible
VI: Lack of completeness. Subclasses are declared regardless of the complete division of concepts into parts	A more detailed description of each concept was made. These elements were useful for defining properties between classes

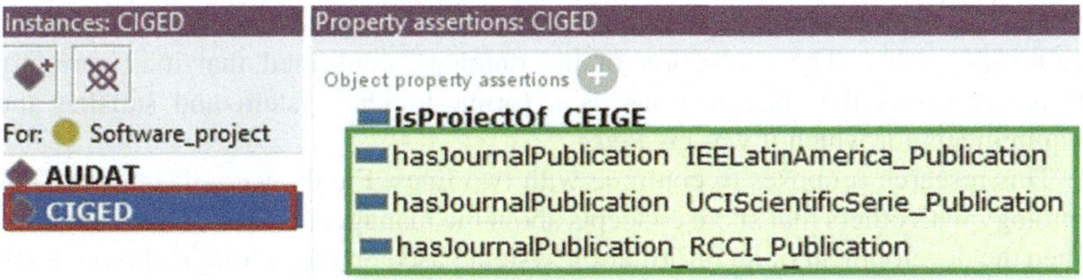

Fig. 7 Result shown by the reasoner for test case 1 (CQ 3)

CQ. ¿What publications in journals with a specific level did the project staff produce?

Expected result: When applying the Pellet reasoner, the defined restrictions are met and the reasoner inferred the publications that the people of the *CIGED* project have developed.

Result obtained: Satisfactory. Figure 7 shows the result inferred by the reasoner where the publications in magazines made by members of the project appear. In the red box appears the instance of the *Software_Project* class and in the green box appears the result of the reasoner where the publications in the level magazine that the project personnel have made are observed.

During the research, the domain concepts related to the HR scientific activity management in software development projects were represented. However, in spite of encompassing a broad vocabulary in this sense, it is proposed to enrich the developed ontology with others that describe specific aspects such as [9] and [24] as lines of continuity of the research.

On the other hand, although a set of steps have been defined to obtain relevant information, this ontological model can be consulted by different software agents that support OWL. Furthermore, the model can be exploited to perform queries from different languages, such as SPARQL [26].

5 Conclusions

In this work, an ontology to support the management of scientific activity in soft-ware development projects was described. This approach exploits the expressive richness of OWL to represent research results of an individual as well as a project team. With this ontological model it is possible to carry out automated analyses of the information represented and get useful inferences for monitoring and controlling the scientific work of the project. Likewise, this approach helps to verify the consistency of the information and check the status of planned scientific results. Besides, this proposal provides inputs to assign research responsibilities according to the needs of the project.

The ontology was built following a solid methodology that contributed to guaranteeing its quality. The evaluation of the ontology confirmed that it is correctly designed, meets the specifications as a formal-logical system and satisfies the requirements for which it was created.

This research proposes to continue with two lines. Firstly, we will integrate our ontology with others that share concepts about the management of scientific activity. Then the development of an information system based on the ontological model will be proposed. Although the motivation of the research is focused on the representation and analysis of the scientific information of the project, the developed ontology can also be used to represent and analyze the results of the HR individually.

References

1. Villegas, S.A.: Administración de Recursos Humanos, vol. 1. Perú (2004)
2. PMI.: The standard for project management and a guide to the project management body of knowledge seventh edition. Ed.: Project Management Institute. ISBN: 978-1-62825-664-2, pp. 370 (2021)
3. Dalkir, K.: Knowledge management in theory and practice. Massachussetts Institute of Technology (2017).
4. Vasconcelos, I.F.: IBM: o desafio da mudança. Revista de Administração de Empresas **33**(3), 84–97 (1993)
5. Mata, E.M., Mata, M.A., Miranda, J.M.: Knowledge management in process improvement and best practices sharing. IEEE **12**(3), 469–474 (2014)
6. Silega, N., García, M.N.: Applying an MDA-based approach for enhancing the validation of business process models. Elsevier Comput. Sci. **184**, 761–766 (2021). https://doi.org/10.1016/j.procs.2021.03.094
7. Nava, R.V., Menéndez-Domínguez, V.H., Montalvo, J.R.: A document recommendation system using a document similarity ontology IEEE Lat. Am. Trans. **14**(7), 3329–3334 (2016)
8. Guerrero-Proenza, R.S., Martínez, A.G.: Ontología para la representación de las preferencias del estudiante en la actividad de aprendizaje en entornos virtuales. Revista Cubana de Educacion Medica Superior (2013)
9. Gasmi, H., Bouras, A.: Ontology-based education/industry collaboration system. IEEE Access **6**, 1362–1371 (2017). https://doi.org/10.1109/ACCESS.2017.2778879
10. Gruber, T.R.: A translation approach to portable ontology specifications. Elsevier Sci. Direct **5**(2), 199–220 (1993). https://doi.org/10.1006/knac.1993.1008
11. Noy, N.F., McGuinness, D.L.: Ontology development 101: a guide to creating your first ontology. Stanford knowledge systems laboratory technical report KSL-01–05 and Stanford medical informatics technical report SMI-20010880 (2001)
12. Choi, C., Choi, J.: Ontology-based security context reasoning for power IoT-cloud security service. IEEE Access **7**, 110510–110517 (2019)
13. Horridge, M., Knublauch, H., Rector, A., Stevens, R., Wroe, C.: A practical guide to building OWL ontologies using the protégé-OWL plugin and CO-ODE tools edition 1.0. The University of Manchester (2011)
14. Informatics, S.M.: Protégé Wiki: Protégé Ontology Library (2019). [Online] Available on: http://protegewiki.stanford.edu/index.php/Protege_Ontology_Library. Accessed 21 May 2021
15. Alsanad, A.A., Chikh, A., Mirza, A.: A domain ontology for software requirements change management in global software development ENVIRONMENT. IEEE **7**, 49352–49361 (2019)
16. Rhayem, A., Mhiri, M.B., Salah, M.B., Gargouri, F.: Ontology-based system for patient monitoring with connected objects. In: International Conference on Knowledge Based and Intelligent Information and Engineering Systems, Marseille, France, pp. 683–692 (2017)

17. Alvarado, R.D.: Metodología para el desarrollo de ontologías. PhD Thesis (2010).
18. Figueroa, B.S.: NeOn Methodology for Building Ontology Networks: Specification, Scheduling and Reuse. (P. Thesis, Ed.) Universidad Polit, España: Departamento Inteligencia Artificial (2010)
19. Pagels, M.: The DARPA agent markup language ontology library (2019)
20. Bertol, F.J., Dolado, J.: Una ontología para la gestión del conocimiento de proyectos software. Revista Española de Innovación, Calidad e Ingeniería del Software, vol. 4 (2008)
21. Caballero, E.V., Martínez, N.S., Castro, G.F., Alcivar, I.A. Método Basado en Ontologías para la Gestión de la Actividad Científica en Proyectos de Desarrollo de Software. Experiencias iberoamericanas en ingeniería de proyectos. Guayaquil, Ecuador (2020)
22. Horrocks, I., Patel-Scheinede, P.F., Boley, H., Tabet, S., Grosof, B.: SWRL: A Semantic Web Rule Language Combining OWL and RuleML. W3C Member Submission (2004)
23. Rodríguez, Y.A.: Método Para la Integración de Ontologías en Sistemas Relacionales Para la Evaluación de Créditos. Maestría, La Habana, Cuba (2017)
24. Martínez, N.S., García, M.N., Macias, D.: Ontology-based transformation from CIM to PIM. IEEE America Trans. **14**(9), 4156–4165 (2016)
25. Sosa, J.D., Domínguez, V.H., Bolaños, M.E., Montalvo, J.R.: Use of an ontological model to assess the relevance of scientific production. IEEE Lat. Am. Trans. **17**(9), 1424–1431 (2019)
26. López, A., Delgado, Y.H., Martínez, N.S.: Un Método Práctico para la Evaluación de Ontologías. Revista Cubana de Ciencias Informáticas, pp. 76–85 (2018)

Chapter 17
Project Management Repository for Decision-Making Researches

Pedro Y. Piñero Pérez, Iliana Pérez Pupo, Pedro E. Piñero Ramírez, Yosvani Marquez Ruiz, and Yulia Fustiel Alvarez

Abstract Data repositories facilitate basic research and applied research in multiple fields of knowledge. They contribute to the validation of research and the development of new algorithms and artificial intelligence techniques, aimed at the discovery of knowledge. In particular, they are useful for the development of research associated with the development of new methods to aid decision-making. This article proposes a new repository for the development of research in project management. The repository is made up of twenty databases that cover different areas of knowledge in project management. The data reflects the behavior of software development projects. In most of the databases presented, heterogeneous data can be found that reflect different processes of project management. In addition, a set of technical processes for managing the repository are presented. In the results and discussion section, the authors reference a set of doctoral theses, master's theses and articles that used the repository data for different investigations. The authors evaluate the perceived quality of the researchers by applying a questionnaire to them that assesses the variable's relevance and usability.

Keywords Artificial intelligence · Project management · Research repository

P. Y. Piñero Pérez (✉) · I. Pérez Pupo · Y. Marquez Ruiz · Y. Fustiel Alvarez
Grupo de Investigaciones en Gestión de Proyectos, Universidad de las Ciencias Informáticas, 54830 La Habana, Cuba
e-mail: ppp@uci.cu

I. Pérez Pupo
e-mail: iperez@uci.cu

P. E. Piñero Ramírez
Intelligent Software Ecosystems Group, Business Redmine Corporation, Mississauga, Canada

P. Y. Piñero Pérez et al. (eds.), *Artificial Intelligence in Project Management and Making Decisions*, Studies in Computational Intelligence 1035,
https://doi.org/10.1007/978-3-030-97269-1_17

1 Introduction

Software engineering is one of the fastest growing emerging industries in recent years. There are numerous guides and standards for the development of this industry, among which the following stand out: PMBOK [1], ISO 21500 [2] and CMMI [3].

But despite the large number of guides, difficulties persist, as reflected in the Standish Group's CHAOS Reports [4]. In recent years, reports indicate that medium-sized projects: approximately 31% of projects fail, that around 26% of projects are renegotiated due to variations in scope, cost or quality, and that only 29% of the projects were successful [4, 5]. In this scenario, learning from the experience of previous projects would help reduce the high number of canceled or renegotiated projects. Starting with version 6 of the PMBoK and in the most current versions [1], the introduction of data mining and artificial intelligence techniques is strongly recommended. In this context, authors formalize data mining and artificial intelligence as shown below:

- Artificial Intelligence as the set of Software and Hardware techniques that simulate different human reasoning methods and the collective intelligence of natural ecosystems, to solve complex problems where the algorithm for its solution has non-polynomial complexity, or there are no algorithms for its solution.
- Data mining as an interdisciplinary field where statistical techniques, mathematics, artificial intelligence, pattern recognition and databases converge for the discovery of non-trivial knowledge that resides implicitly in the data [6].

But to guarantee the development of investigations using these techniques it is necessary to have sufficient data. In the particular field of project management, the number of accessible databases and repositories is still insufficient. The data that is generated in the management of software projects generally comes from information systems that are not public, making it difficult to access the information.

There are different repositories of international databases for research development. One of the most recognized repositories is the UCI Repository [7] for machine learning that includes numerous databases of different fields of application. The PSPLIB [8] for its part is specialized in databases for resource allocation, useful in testing optimization techniques. In [9] it is exposed a comprehensive list of data repositories for every type of problem and industries, but not about project management; besides, most of those repositories are not-free. In Google Dataset Search, a search engine for researchers to locate online data, we can fain some repositories about project management, but the majority of them are not-free.

This paper proposes a repository for the development of research in project management that includes twenty databases associated with software projects. In the second section of the paper, a set of processes associated with the construction and management of the repository is proposed. The third section explains the structure of the repository and the databases that comprise it. Four section presents the analysis of results of the proposed repository through a set of investigations that have used the repository. The last section presents de conclusions.

2 Repository Construction Process

The repository that is proposed in its current stage is a repository focused on data associated with decision-making in project management. Its development has followed four fundamental stages that are listed below:

1. Stage of identification and selection of information sources and information systems that provide information to the repository.
2. Data pre-processing and cleaning stage, to support the construction of the repository.
3. Stage of publication and management of the repository.
4. Stage of exploitation of the repository.

Figure 1 describes the stages in the construction of the repository and the main results in each stage. These stages were proposed by Iliana Pérez in [10].

As part of the first stage, the information systems BusinessRedmine [11] and GESPRO [12] were identified. These systems collect information from real software projects since 2010. These systems constitute the fundamental sources of data feed for the repository. In addition, other data sources and expert information providers are identified.

In the database construction stage, the datasets to be constructed are defined from the nature of the re-collected data. The data is then pre-processed using data cleansing techniques, including:

- Elimination of records with a high degree of incompleteness.

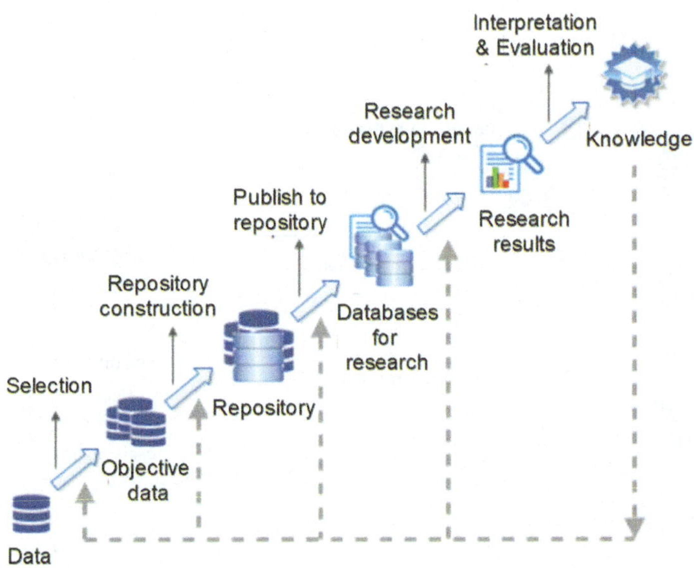

Fig. 1 Stages in repository creation

- Anonymization and pseudonymisation of data.
- Imputation in incomplete data in some databases that require it. Maintaining that some databases also have missing data.
- Data verification.
- Standardization of data in cases where the standardization process is of interest.
- Attribute selection processes are applied.
- The datasets are formed by eliminating normal forms established in the original databases and promoting the rapid retrieval of the records.

In addition, improvements are made to the information systems that contribute data to the repository. The improvements that are introduced are aimed at reducing noise and incompleteness in the data. In this sense, the following actions are carried out on the system:

- In the forms, list-type fields are introduced that facilitate the standardization of the information.
- Validations are established that regulate the way of writing the information through the use of regular expressions.
- Mandatory fields are established in the forms to avoid the incompleteness of the data.
- Management indicators are calculated whenever possible, reducing subjectivity.
- Information systems are updated to capture new necessary data.

In the stage of publication and management of the repository, the publication of the databases in the repository is verified and approved, see Fig. 2. As part of this stage, the following set of processes are established that facilitate the management of the repository:

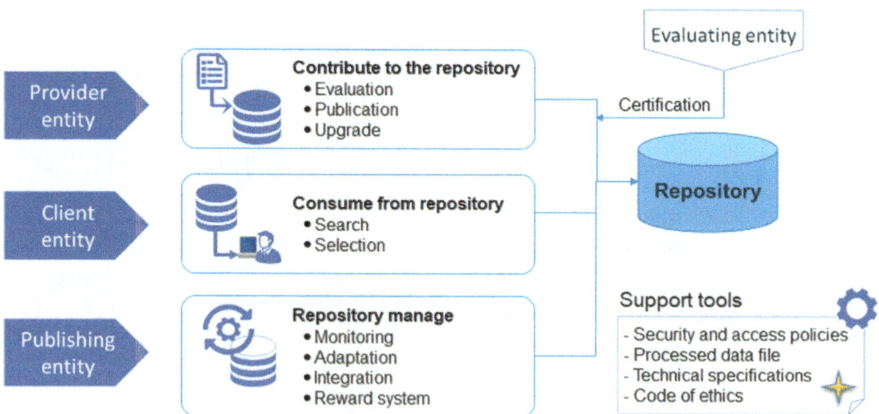

Fig. 2 View of the processes for managing the repository

- Process to contribute information to the repository: where the policies are established so that different stakeholders can contribute new datasets to the repository. This process is subdivided into the following threads:
  - Receipt of contributions from stakeholders.
  - Updating of existing datasets.
  - Evaluation of the datasets.
  - Publication of datasets
- Process of consuming and using the repository, where the policies for the use of datasets by researchers and other stakeholders are established. As part of this process, sub-processes are established to facilitate: information search and data selection.
- Repository management process where the following threads are established:
  - Monitoring the use of the repository.
  - Adaptation and maintenance of the repository: which includes corrective maintenance actions, adaptive maintenance and perfective maintenance.
  - Data integrity actions.
  - System of rewards and penalties for the use of the repository.

In the management of this stage, numerous techniques and tools are used to support the management of the installed technological resources and visualization of the repository.

In the repository exploitation stage, the use of datasets in research development is promoted. In addition, the correct referencing of the repository and of each dataset is encouraged.

Then, in the improvement and evaluation stage, the impact of the repository is analyzed in different investigations. Lessons learned are collected, and improvement processes are carried out.

For the management and publication of the repository, the following roles are important:

- Repository manager: personnel in charge of maintaining the order and cleanliness of the data in the repository.
- Researcher: is anyone who will use the data for the development and validation of their research.

3 Estructure of Project Management Research Repository

In the repository, each database is represented by a ´.name´ and ´.data´ files, similar to the structure of the UCI Repository [7]:

- The ´.name´ file represents the structure of the database and must contain the following information:

- Title: name of the database, it must be representative of the area of knowledge of project management corresponding to the data it contains.
- Source: entity or institution that provides the data. Here, the largest possible amount of contact and identification data should be offered; For example, address, telephone numbers, post office, etc.
- Cited: the way in which the database should be cited, the name or the date on which the data was provided should not be missing.
- Usage history: all the investigations carried out with that database are listed, referencing each one of the publications.
- Relevant information: this is where the attributes that make up the database are described.
- Number of records in the database.
- Number of attributes.
- Missing attribute values: the term or value to be used in the cases of missing values is indicated.
- Information on anomalous data: attributes that is necessary, the value from which it is considered that it is an anomalous data or out of range is specified.

- The ´.data´ file is a text file in ´csv´ format that contains the data of each of the records and the attributes in the same order as explained in the ´.name´ file.

The current version of the repository is made up of the following databases:

Associated with the subject of project and program evaluation:

1. Project management database: contains general information on projects that can be used to estimate time and cost of projects.
2. Project evaluation database: contains information on the indicators in the different evaluative cuts of the projects with mixed data.
3. Hard project evaluation database: contains information on the indicators in the different evaluative cuts of the projects with hard data. It is based on indicators with nominal values that evaluate projects.
4. Database of fuzzy evaluation of projects: contains information on the indicators in the different evaluative cuts of the projects with fuzzy data.
5. Indicator management database: contains information on general project indicators.
6. Client indicator management database: contains information on general indicators of project programs associated with clients. Relates customer-oriented indicators.
7. Organization indicator management database: contains information on general control indicators of project-oriented organizations. Relates indicators oriented to the organization.
8. Program indicator management database: contains information on general indicators of project programs associated with themes. It relates program-oriented indicators.

Associated with the issue of allocation to human and non-human resources projects:

9. Task management database: contains general information about tasks and their execution.
10. Database assignment of resources to tasks: contains general information about the assignment of resources to tasks.
11. Human and material resource management database: contains general information on the allocation of human and non-human resources.

Associated with the scope management theme:

12. Requirements management database: contains general information about requirements management.
13. Change management database - Contains general information about change management.

Associated with the risk management theme:

14. Risk management database: contains general information on risk management.
15. Deviation management database: contains general information about deviation management.

Associated with the subject of human resources management:

16. User competencies management database: lists the users and the competencies they present. It is based on 4 indicators that yield an evaluation of the user, which is stored with linguistic terms [13].
17. Management database of project human resources indicators: contains general information on the management of project human resources indicators. It relates indicators that yield a qualitative and quantitative evaluation of human resources by projects.
18. Management database of the organization's human resources indicators: it contains general information on the management of the organization's human resources indicators. It relates indicators that yield a qualitative and quantitative evaluation of the human resources it stores.

Next, the general characteristics of the databases that make up the repository are exposed, see Table 1.

Table 2 hows a brief analysis that shows the relationship of the repository databases with 9 knowledge areas of project management proposed by the PMBOK and ISO 21500.

4 Results Analysis

This section presents the experiences in using the repository. This section is organized considering the following sets of indicators:

- Indicators associated with the level of exploitation of the repository in doctoral and master's research.

Table 1 General characteristics of the repository databases

Database	Heterogeneous data	Attributes	Records	Projects
Project management	Yes	49	136	134
Project evaluation	Yes	84	8 430	667
Evaluation of hard projects	Yes	19	8 430	667
Fuzzy project evaluation	Yes	19	8 430	667
Indicator management	Yes	93	34 810	996
Client indicator management	Yes	16	387	0
Management of organization indicators	Yes	16	87	0
Management of program indicators	Yes	16	87	0
Task management	Yes	26	12 563	103
Assignment of resources and tasks	Yes	38	9 569	88
Management of human resources and materials	Yes	28	612	35
Requirements management	Yes	17	1 652	36
Change management	Yes	20	593	44
Risk management	Yes	19	2 379	345
Deviation management	Yes	10	2 506	330
User competency management	Yes	11	10 452	0
Management of the project human resources indicators	Yes	9	6 539	84
Management of the organization's human resources indicators	Yes	8	3 149	0

Table 2 Relationship between the repository databases and the project management knowledge areas

Knowledge areas	Databases
Scope	1, 5, 9,10,11,12,13 y 15
Costs	1,2,3,4,5,6,7,8,9,10,11, 16
Time	1,2,3,4,5,9,10,11,12,13,14 y 15
Quality	3,5,7,8,12,13
Human resources	1,2,4,5,6,7,8,9,11,14,16,17,18
Risks	14,15
Acquisitions	5,10,11
Stakeholder	1,2,6
Integration	3,4,5,7,8,10,11,14,17,18

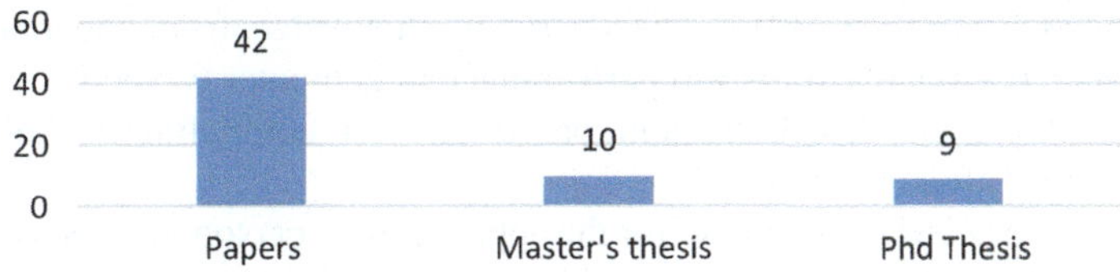

Fig. 3 Research conducted using the project management repository

- Amount of master's research uses the repository.
- Number of doctoral researches uses the repository.
- Number of scientific articles that use the repository.

- Indicators associated with the perceived quality of end users of the repository.

The repository has allowed an increase in research associated with project management, see Fig. 3. Research carried out that used the repository in the validation of results is listed below.

Completed PhD research that has used the repository in the validation:

1. Competency assessment model based on evidence during project management [13].
2. Project control model based on geo-referential analysis to aid decision-making [14].
3. Model for the assurance of income in organizations oriented to projects based on anomalous data mining [15].
4. Model for the feasibility analysis of software projects in uncertain environments [16]
5. Model for the control of project execution based on soft computing [17].
6. Ontological model of information retrieval for decision-making in project management [18].
7. Algoritmos de sumarización lingüística de datos para la ayuda a la toma de decisiones [19]
8. Nuevas extensiones de mapas cognitivos difusos para problemas de toma de decisiones secuenciales multietapa [20].
9. Algoritmos de estimación de distribuciones con tratamiento de restricciones para la construcción de cronogramas de proyectos.

Master's research that has used information from the institutional 9repository:

1. Marieta Peña Abreu, 2012, Model for feasibility analysis in the evaluation of Rosa
2. Rodríguez Stiven, 2012, Model for evaluating the composition of computer project teams.
3. Perdomo Alonso, 2013, Model for the evaluation of competences as multi-label problems from systems based on cases.
4. Erick Piñera Trinchet, 2013, Linguistic summarization algorithm to support decision-making in project management.

5. Pacheco Rodríguez, 2014, Discovery of knowledge from the relationship between personality traits-job performance in computer projects.
6. Bermudez Peña, 2015, System based on soft computing techniques for evaluating project execution.
7. Pastor López 2018, Procedure for the application of personality tests as support to the management of human resources in computer projects.
8. Maria Rente, 2020, Evaluación del impacto de los cambios durante la gestión de proyectos software projects.
9. Marbelis Rojas Rodríguez, 2021, Objetos de aprendizaje para la enseñanza de la Dirección Integrada de Proyectos.
10. Bárbara Bron Fonseca, 2021, Sistema para la generación de recomendaciones sobre la evaluación de proyectos de desarrollo de software.

Regarding the number of scientific articles that use the repository, more than 36 publications were generated some of them are shown below:

- Bermúdez Peña, J.A. Lugo García, et al, "An adaptive-network-based fuzzy inference system for project evaluation", Ing. Unv, vol. 19, no. 2, pp. 299-313, 2015. http://doi.org/10.1114/javeriana.iyu19-2.sdib ISSN: 0123-2126.
- Bermúdez Peña A., J. A. L. García, et al, "Sistema neuro-borroso de apoyo al control de la ejecución de proyectos," Revista Cubana de Ingeniería, vol. 5, no. 2, pp. 41–51, 2014.
- Lugo García J. Alejandro, Torres López, Surayne et al. Automatic calculation of indicators as a decision support technique in the National Health System.Revista Cubana de Información en Ciencias de la Salud (ACIMED), Medigraphic, Vol 24 No (1) pp 472-481, 2013.
- Lugo García, J. Alejandro, Torres Surayne, et al, "Control de la ejecución de proyectos basado en indicadores y lógica borrosa, Iberoamerican Journal of Project Management, Vol 4 No.1, 2013, ISSN: 2346 -9161, URL: http://journal.riipro.org/index. php/IJOPM/article/view /118
- Castro Aguilar G. F., Pérez Pupo I., et al. "Platform for Project Evaluation Based on Soft-Computing Techniques," II Congreso Internacional de Tecnologías e Innovación CITI2016, Springer, International Publishing, vol 658, pp 226-240, octubre 2016.
- Castro Aguilar G. F., Pérez Pupo I., et al, "Aplicación de la minería de datos anómalos en organizaciones orientadas a proyectos," Revista Cubana de Ciencias Informáticas, vol. 10, pp. 195–209, 2016.
- Castro Aguilar G. F., Pérez Pupo I, et al, "Método para el aseguramiento de ingresos en entornos de desarrollo de software," Revista Cubana de Ciencias Informáticas, vol. 10, pp. 43–57, 2016.
- Aguilar G. F. C., Pérez Pupo I, et al, "PRODANALYSIS, un Sistema para el Aseguramiento de Ingresos Basado en Minería de Outliers," INNOVA Research Journal, vol. 1, no. 7, pp. 18–36, ISSN 2477-9024, 2016.
- Villavicencio N., Peña Abreu M, et al. Experiencias en la integración de procesos en las organizaciones orientadas a proyectos de software. Revista Cubana de Ciencias Informáticas, Vol. 10, 2016.

- Sosa González R. Pérez Pupo I, et al, "Ecosistema de Software GESPRO-16.05 para la Gestión de Proyectos," Revista Cubana de Ciencias Informáticas, vol. 10, pp. 239–251, 2016.
- García Vacacela R., Pérez Pupo I., et al, "Experiencias usando algoritmos genéticos en la planificación de proyectos," Revista Cubana de Ciencias Informáticas, vol. 10, pp. 71–86, 2016.
- S. Torres, Piñero, P.Y. et al. Red neuronal multicapa para la evaluación de competencias laborales. Revista Cubana de Ciencias Informáticas, Vol. 10, 2016.
- Torres López, S, Castro Aguilar G. F., et al, "Rough Sets for Human Resource Competence Evaluation and Experiences," Applied Mathematics, vol. 7, no. 12, pp. 1317–1325, ISSN 2152-7393, 2016, (https://doi.org/10.4236/am.2016.712116).
- Guerrero A., Pérez I., et al. "Evaluación de proyectos usando sistemas basados en algoritmos genéticos de aprendizaje de reglas". Revista Cubana de Ciencias Informáticas, vol. 11, no. 4, pp. 39–56, 2017.
- Rodríguez C. R., Peña M., et al, "Sistema clasificador borroso basado en algoritmos genéticos para evaluar el estado de ejecución de proyectos". Revista Cubana de Ciencias Informáticas, vol. 11, no. 3, pp. 174–189, 2017.
- Jimenes Moya G. E., Lugo García J. A. et al, "Control de proyectos: hacia un modelo de análisis geo-referencial basado en softcomputing," Ciencias de la Información, vol. 48, no. 3, pp. 18–27, 2017.
- Lugo García J. A., Bermudez. Peña A, et al, "Project Control and Computational Intelligence: Trends and Challenges," International Journal of Computational Intelligence Systems, ISSN: 1875-6891, vol. 10, no. 1, pp. 320–335, Jan. 2017, (https://doi.org/10.2991/ijcis.2017.10.1.22)
- Cabrera Rodríguez L., García Vacacela R., el al, "Factores que influyen en el aseguramiento de ingresos en pequeñas y medianas empresas orientadas a proyectos de software," in Global Conference on Business and Finance Proceedings, vol. 12, Las Vegas, Nevada: The Institute for Business and Finance Research, 2017, pp. 863–876, ISSN: 1941-9589.
- Peña Abreu M, Rodríguez Rodríguez C. R., et cal, "Computación con palabras para el análisis de factibilidad de proyectos de software," Tecnura, vol. 20, no. 50, pp. 69–84, Dec. 2016, (https://doi.org/10.14483/udistrital.jour.tecnura.2016.3.a05).
- Al-Subhi, S. H. S. Mahdi G.S., et al, "Operador media potencia pesada lingüística y su aplicación en la toma de decisiones," International Journal of Innovation and Applied Studies, vol. 22, no. 1, pp. 38–43, 2017.
- Pérez Pupo I. P., López Gómez P. et al "Construcción de resúmenes lingüísticos a partir de rasgos de la personalidad y el desempeño en el desarrollo de software," Rev. Cuba. Cienc. Informáticas, ISSN: 2227-1899, vol. 12, pp. 135–150, 2018.
- Pérez. Pupo I, Acosta O. S., et al, "Descubrimiento de resúmenes lingüísticos para ayuda a la toma decisiones en gestión de proyectos," Rev. Cuba. Cienc. Informáticas, ISSN: 2227-1899, vol. 12, pp. 163–175, 2018.
- Pérez Pupo I., Piñero Pérez P. el at "Extensions to Linguistic Summaries Indicators based on Neutrosophic Theory, Applications in Project Management Decisions,"

Neutrosophic Sets and Systems, University of New Mexico, Book Series ISBN: 978-1-59973-587-0, vol. 22, pp. 87–100, 2018.

- Peña Abreu M., Rodríguez R., C.R, et al. "Economic Feasibility of Projects Using Triangular Fuzzy Numbers". 6th International Workshop, IWAIPR 2018, Havana, Cuba, September 24–26, 2018, Proceedings in Progress in Artificial Intelligence and Pattern Recognition, 2018, pp. 288–298, (https://doi.org/10.1007/978-3-030-01132-1_33). https://www.springerprofessional.de/economic-feasibility-of-projects-using-triangular-fuzzy-numbers/16143680.
- Al-Subhi S., Pérez I., el at, "A New Neutrosophic Cognitive Map with Neutrosophic Sets on Connections, Application in Project Management," Neutrosophic Sets and Systems, University of New Mexico, Book Series ISBN: 978-1-59973-587-0, vol. 22, pp. 63–75, 2018.
- Torres, S., Cuesta-Santos, A. et al, "Evaluación de competencias laborales a partir de evidencias," Ingeniería Industrial, vol. 39, no. 2, pp. 124–134, 2018.
- Mahdi G.S., Al-Subhi. S., et al, "Algoritmo de estimación de distribución con tratamiento de restricciones en el modelo probabilístico en problemas de scheduling," Revista de Investigación Operacional, vol. 40, no. 5, pp. 674–686, ISSN 0257-4306, 2019.
- Pérez, I., Acosta O., et al, "Algoritmos para la sumarización lingüística de datos, ayuda a la toma de decisiones en organizaciones orientadas a proyectos," in Proceedings of the XXII Iberoamerican Conference on Software Engineering, CIbSE 2019, La Habana, Cuba, April 22-26, 2019, 2019, pp. 633–640.
- Al-Subhi S.H., Mahdi G.S., el al, "Neutrosophic Cognitive Map for a Multi-Objective Decision Making in Software Projects," in Proceedings of the XXII Iberoamerican Conference on Software Engineering, CIbSE 2019, La Habana, Cuba, April 22-26, pp. 58–71, 2019.
- Pérez Pupo I., Piñero Pérez P.Y., el at, "Linguistic Summaries Generation with Hybridization Method based on Rough and Fuzzy Sets". In International Joint Conference on Rough Sets (IJCRS 2020) 2020.
- Pérez Pupo I, Piñero Pérez P.Y, et al, "Discovering fails in software projects planning based on linguistic summaries". In International Joint Conference on Rough Sets (IJCRS 2020) 2020.
- Pérez, I., García-Vacacela, R. et al, "Experiencias en el uso de técnicas de soft-computing en la evaluación de proyectos de software," Revista Investigación Operacional, **41**(1), 106–117, 2020.
- Mahdi G.S., Madera Quintana J., el at Estimation of Distribution Algorithm for solving the Multi-mode Resource Constrained Project Scheduling Problem, EAI Endorsed Transactions on Energy Web (ISSN: 2032-944X). Vol 7 (27), 2020, https://doi.org/10.4108/cai.13-7-2018.164111.
- Al-subhi S. Hassan, Piñero Pérez P. Y., et al, Sistema de apoyo a la toma de decisiones durante la gestión de proyectos basado en mapas cognitivos neutrosóficos, Revista investigacion operacional, vol. 41, no. 5, 768-779, 2020
- S. Hasan Al-subhi, Román Rubio, P. A., et al. A new neutrosophic clinical decision support model for the treatment of pregnant women with heart diseases, Revista Investigacion Operacional vol. 41, no. 5, 780-790, 2020.

- Al-Subhi S. H., Papageorgiou E. I., et al. "Triangular neutrosophic cognitive map for multistage sequential decision-making problems", International Journal of Fuzzy Systems, vol. 23, no. 3, 657-679. https://doi.org/10.1007/s40815-020-010 14-5. (2021)

The study carried out on the perceived quality of the researchers who have used data from the repository is shown below. The researchers involved add up to 57 people, including master's students, doctoral students, and researchers. A survey associated with the use of the repository is designed that evaluates the following variables: relevance of the repository and usability. The questions are formulated in such a way that the answers correspond to one of the following linguistic terms *LBTL = {none, very low, low, mean, high, very high, perfect}*. The researchers' responses to each question are aggregated and computed using the 2-tuple word computation technique (Herrera & Martínez, 2000).

The obtained results are showed next tables, see Tables 3 and 4.

In general, a high level of satisfaction is manifested in the use of the repository. Regarding the relevance variable, the criterion with the lowest score was the level of exclusivity of the element repository, which is due to the existence of the PSPLIB repository for the development of research associated with solving optimization problems. But the authors of this work have not really found other public repositories for the development of research in software project management. Regarding the usability

Table 3 Analysis of criteria regarding the relevance variable

	Result
To what degree was your research related to the data in the repository?	High, 0.02
Evaluate the level of coverage of knowledge areas in project management	High, – 0.43
To what extent did the repository facilitate the validation of your research results?	High, 0.04
Indicate the impact the repository may have on collaborative networks	High, 0.01
Indicate the level of exclusivity of the repository, considering the existence of other similar public repositories for its use	Mean, – 0.02
To what degree do you consider the level of impact of the repository?	High, – 0.03

Table 4 Criteria analysis regarding usability

Is the structure of the repository adequate to facilitate its exploitation?	Very high, – 0.02
Evaluate the level of accessibility of the repository on the network	Mean, – 0.43
Assess the level of complexity in working with the repository	Low, 0.36
Indicate the level of ease for incorporating new databases to the repository	High, 0.3
Evaluate the level of satisfaction with the help system and / or description of the repository	Mean, 0.29
Assess the level of data quality in terms of the number of missing values	Mean, 0.4
Assess the level of consistency of data records	High, – 0.3

variable, the criteria with the lowest evaluation were: the level of accessibility of the repository, the quality of the data regarding the amount of missing values and satisfaction regarding the help in the use of the repository. These are elements that must continue to be worked on by the authors to guarantee higher levels of use.

5 Conclusions

The construction of the repository of research in project management enables the development of dissimilar research in this thematic area. In particular, the repository presented in this research contains 18 databases associated with software projects, and they cover most of the project management knowledge areas as proposed by the PMBOK and ISO 21500. In general, they are They have developed more than 20 publications that use the data from the repository, as well as doctoral and master's theses. Researchers who benefit from the use of the repository show a high degree of satisfaction with it, considering both relevance and usability. The elements that the authors consider that they should continue working to achieve better results are: the level of accessibility of the repository, the quality of the data regarding the amount of missing values and satisfaction regarding the help in the use of the repository.

References

1. PMI.: The standard for project management and a guide to the project management body of knowledge seventh edition. Ed.: Project management institute. ISBN: 978-1-62825-664-2, pp. 370 (2021)
2. SEI.: CMMI for development v1.3, SEI CMMI production, CMU/SEI-2010-TR-034, Technical report. Software Engineering Institute (2010)
3. Stellingwerf, R., Zandhuis, A.: ISO 21500 Guidance on Project Management–A Pocket Guide. Van Haren (2013)
4. Johnson, J.: CHAOS Report: Decision Latency Theory: It Is All About the Interval. The Standish Group (2018).
5. Hastie, S., Wojewoda, S.: Standish Group 2015 Chaos Report-q&a with Jennifer Lynch. Retrieved **1**(15) (2015). [Online] Available on: https://www.infoq.com/articles/standish-chaos-2015. Accessed 2 May 2021
6. Gorunescu, F.: Data Mining: Concepts, models and techniques. Springer Science & Business Media, vol. 12 (2011)
7. Dua, D., Graff, C.: UCI Machine Learning Repository. University of California, Irvine, School of Information and Computer Sciences (2021). [Online] Available on: http://archive.ics.uci.edu/ml. Accessed: 12 May 2021
8. Kolisch, R., Sprecher, A.: PSPLIB-a project scheduling problem library: OR software-ORSEP operations research software exchange program. Eur. J. Oper. Res. **96**(1), 205–216 (1997)
9. Tyagi, H.: Data Repositories for almost Every Type of Data Science Project. Medium. (2020). [Online] Available on: https://towardsdatascience.com/data-repositories-for-almost-every-type-of-data-science-project-7aa2f98128b. Accessed 12 May 2021

10. Pérez, I.: Propuesta de metodología para el diseño e implantación de repositorios de activos de software reutilizables. Tesis de Maestría en Gestión de Proyectos, Departamento de Investigaciones en Gestión de Proyectos, Universidad de las Ciencias Informáticas (2011)
11. Piñero, P.Y., et al.: BusinessRedmine Sistema para la toma de decisiones empresariales y la gestión de proyectos. Número de Registro de Propiedad Intelectual, Cuba. CENDA 4076-12-2019 (2019)
12. Piñero, P.Y., Lugo, J.A, Menéndez, J., et al.: Software application XEDRO GESPRO v13.05. National Office of Copyrights Registrations of Cuba, CENDA Registration Number: 2336-06-2015. La Habana, Cuba, DCN-002/2016 (2015)
13. Torres, S.: Modelo de evaluación de competencias a partir de evidencias durante la gestión de proyectos. PhD Thesis. Departamento de Investigaciones en Gestión de Proyectos, Universidad de las Ciencias Informáticas, La Habana, Cuba (2015)
14. Jiménez, G.E.: Modelo de control de proyectos basado en análisis geo–referencial para la ayuda a la toma de decisiones. PhD Thesis. Departamento de Investigaciones en Gestión de Proyectos, Universidad Católica de Santiago de Guayaquil, La Habana, Cuba (2017)
15. Castro, G.F.: Modelo para el aseguramiento de ingresos en organizaciones orientadas a proyectos basado en minería de datos anómalos. PhD Thesis. Departamento de Investigaciones en Gestión de Proyectos, Universidad de las Ciencias Informáticas, La Habana, Cuba (2017)
16. Peña, M.: Modelo para el análisis de factibilidad de proyectos de software en entornos de incertidumbre. Phd Thesis. Departamento de Investigaciones en Gestión de Proyectos, Universidad de las Ciencias Informáticas, La Habana, Cuba (2017)
17. Lugo, J.A.: Modelo para el control de la ejecución de proyectos basado en soft computing. PhD Thesis. Departamento de Investigaciones en Gestión de Proyectos, Universidad de las Ciencias Informáticas, La Habana, Cuba (2015)
18. Fernández, A.: Modelo ontológico de recuperación de información para la toma de decisiones en gestión de proyectos. Universidad de Granada, Ciudad de Granada, España (2016)
19. Pérez, I.: Algoritmos de sumarización lingüística de datos para la ayuda a la toma de decisiones. PhD Thesis. Departamento de Investigaciones en Gestión de Proyectos, Universidad de las Ciencias Informáticas, La Habana, Cuba (2021)
20. Al-Subhi, S.: Nuevas extensiones de mapas cognitivos difusos para problemas de toma de decisiones secuenciales multietapa. PhD Thesis. Departamento de Investigaciones en Gestión de Proyectos, Universidad de las Ciencias Informáticas, La Habana, Cuba (2021)
21. Mahdi G.S.: Algoritmos de estimación de distribuciones con tratamiento de restricciones para la construcción de cronogramas de proyectos. PhD Thesis. Departamento de Investigaciones en Gestión de Proyectos, Universidad de las Ciencias Informáticas, La Habana, Cuba (2021)

Part IV
BIM and Other Emerging Technologies in Project Management

Chapter 18
Relevance of the National Program of Automatic, Robotics and Artificial Intelligence Projects in Applications

Armando Plasencia Salgueiro, Ileana Dopico Mateo, Arlety García García, Ystria Rivero Ripoll, and Ileana Suárez Blanco

Abstract The use of Automation Robotics and Artificial Intelligence is one of the ways for the effective use of resources, since it guarantees a constant quality and contributes to the technological discipline and in essence a driving force behind the development of the economy and the society in general. The Ministry of Science, Technology and Environment of Cuba, created a National Program of Projects to promote these scientific disciplines and the application and generalization of their results. The work highlights the importance of the interdisciplinary nature of the Program and demonstrates the need, for the successful execution of projects, of the application of correct methologies as a guide to design and development work. It shows, through the converge of Control and Machine Learning Theory in the Iterative Learning Control, how addressing these issues together helps to promote them. The operation of deep reinforcement algorithms is explained and how their applications in robotic allows themto accelerate their development and make their behavior more flexible in an autonomous way. Finally, through three projects that are developed in the program, the relevance of the interdisciplinary conception of the program is demonstrated to carry out the doctoral research topics corresponding to each project.

Keywords Artificial intelligence · Automation · Deep reinforcement learning · Project management · Project schedule · Robotics

A. P. Salgueiro (✉) · I. S. Blanco
Institute of Cybernetic, Mathematic and Physic, 10400 Havana, Cuba
e-mail: armando@icimaf.cu

I. D. Mateo
Havana University, 10400 Havana, Cuba

A. G. García · Y. R. Ripoll
Universidad de la Isla de la Juventud "Jesús Montané Oropesa", Nueva Gerona, Cuba

P. Y. Piñero Pérez et al. (eds.), *Artificial Intelligence in Project Management and Making Decisions*, Studies in Computational Intelligence 1035,
https://doi.org/10.1007/978-3-030-97269-1_18

1 Introduction

Humanity advances in science, technology, communications, computerization, industry, food production, transportation, health and education, among other spheres, as well as the experiences of different countries, provide significant opportunities for our development.

To the extent that the economy manages to generate repeated positive effects in areas with these characteristics, a sustained increase in productivity is experienced, traditional branches are transformed technologically, new growth poles emerge characterized by more complex production processes due to greater technological intensity and use of knowledge, productivity differentials between activities are reduced, the sectoral base of the economy is diversified, consolidating its internal integration and with it, the increase in activity levels and productive capacities of other sectors, branches and industries.

The use of Automation, Robotics and Artificial Intelligence is one of the ways for the effective use of resources since it guarantees constant quality and contributes to technological discipline. Through intelligent monitoring and control, it is possible to know, in real-time, the behavior of the process of transformation of raw materials into products, the use of parts and pieces used to create equipment and contribute to the rational use of energy, among others, incorporating the elements of robotic production with greater efficiency in the realization of production processes and the incorporation of the self-learning capacity of the technological elements themselves. On the other hand, Artificial Intelligence techniques allow the efficient processing of the large amount of data and information that today accumulates, as well as providing decision-makers and computer systems with advanced capabilities for analysis, inference, simulation and prediction.

The 2030 Agenda for Sustainable Development (approved at the Sustainable Development Summit in the framework of the 70th session of the UN General Assembly in September 2015) identifies 17 Sustainable Development Goals (SDG) and 169 targets, which means a commitment of the countries to achieve a new paradigm of development and well-being to be fulfilled in 15 years. The 17 points propose topics ranging from the eradication of poverty and hunger, health and education for all, gender equality, taking action for the climate, sustainable development, economic growth, industrialization, innovation, access to water and energy, people's safety and peace. To achieve these objectives, undeniably, innovation and the use of advanced technologies in a transversal or direct way mediate.

The Cuban State abides by the 2030 Agenda and in [1] it is declared that in this context, the process to form a development strategy proposal, known as the National Plan for Economic and Social Development until 2030 (PNDES 2030), converged in Cuba, with a systemic, comprehensive and sustainable approach that responds to a strategic and consensual vision in the medium and long term, consistent with what is proposed in the Conceptualization of the Cuban Economic and Social Model and the Guidelines of the Economic and Social Policy of the Party and the Revolution.

In the PNDES 2030, Six Strategic Axes were also identified at the country level that is interrelated with the Development Goals (SDGs) of the 2030 Agenda, which are designed under a systemic approach, articulating the proposal for economic and social development of Cuba until 2030. From them, 22 general objectives and 106 specific objectives are defined.

In more specific documents [2] it is declared that fulfilling the Vision of the Nation requires a high economic and social development, with sustained growth of the economy, which demands a deep and systematic transformation and technological modernization of the production plant, which is fundamentally based on the development and application of science, technology, innovation and the increase in the effectiveness, efficiency and competitiveness of the country's economic activity, primarily in the export sector. For this, the objectives outlined in the Human Potential Strategic Axis, science, technology and innovation must be met, which are:

1. Ensure the care, protection, motivation and stability of highly qualified, more experienced and difficult-to-replace personnel under their professional and scientific profile, as well as the development of young talents in their working lives.
2. Increase the impact of science, technology and innovation in economic and social development based on an improvement of the institutional framework focused on the introduction of results.
3. Promote the development, capture and assimilation of advanced technologies and promote, at the same time, technological sovereignty.

Cuba has outlined its Science, Technology, Innovation and Environment Policy from the Constitution of the Republic itself (endorsed in 2019), which establishes in its Article 21: "the State promotes the advancement of science, technology and innovation as essential elements for economic and social development; implements forms of organization, financing and management of scientific activity; it encourages the systematic and accelerated introduction of its results in the production and service processes, through the corresponding institutional and regulatory framework" [3].

The System currently has fourteen National Programs of Science, Technology and Innovation (CTI), elaborated from previous experiences, also in consultation with the Cuban Academy of Sciences, the Central State Administration Organizations (OACE), the Organizations Superiors of Economic Management (OSDEs) and national entities and territories. These national programs receive government funding for their development.

The progressive digitization of human activity is generating massive data sets, whose exploitation through technologies for their treatment and Artificial Intelligence (AI) is allowing the development of new business models, products and services, impacting the improvement of society and economics by allowing the optimization of processes and the identification of patterns at scales not previously observed.

The National Program for Automation, Robotics and AI (ARIA), managed by the Ministry of Science, Technology and Environment of Cuba (CITMA), arises to respond to growing demand and expectations in these domains. It is therefore an environment that implies great technological, legal and ethical challenges, together with the need to establish multidisciplinary teams in cooperation to cover the wide range of applicable technologies (Artificial Intelligence, HPC, Cloud, security, semantics, interfaces, IoT, eServices, etc.), transversal to various sectors (energy, food, industry, health, etc.).

However, for the successful development of the ARIA´s Program is necessary to carry out three fundamental aspects, first, a deep understanding of definitions and the scope of each of its branches (automation, robotics and artificial intelligence), second, understand and apply appropriate interdisciplinarity in each project and third, follow a methodology that guarantees the successful projects ending.

The article is divided into Introduction, Methods, were exposed the concepts and definitions, the interdisciplinarity fundaments and the proposed methodologies for the realization of the projects; in Results, is showing the actual interdisciplinarity obtained by the program and some projects examples that highlight the application of different knowledge branches in one project and the methodology application and Conclusions.

2 Methods

One of the fundamentals characteristics of the National Program for Automation, Robotics and AI is the development and interrelation of three knowledge branches in one program. For the better comprehension of the necessity of this interrelation, will began with definitions of these branches.

2.1 Concepts and Definitions

2.1.1 Automation

According to "Definición de automática"—RAE: https://dle.rae.es/automático, automation is the Science that tries to replace the human operator in a process with mechanical or electronic devices. According to Wang and Siau in [4], automation is a technology that "actively selects data, transforms information, makes decisions, or controls processes", as well as exhibiting the potential to extend human performance and improve safety. Although automation has been defined in different ways, the common theme is that automation frees humans from time-consuming and repetitive tasks, whether the tasks involve physical functions and/or cognitive functions [4]

The contrast between the two previous definitions is the one that is the first, that of the DRAE, identifies automation as a science, while in the second, it speaks of technology. However, the two complement each other in that the first definition gives the reason for automation: "replace the human operator" with an artifact and in the second, it emphasizes how to perform that "replacement", giving importance to the management of information in the processes.

2.1.2 Artificial Intelligence (AI)

Artificial intelligence is a scientific discipline that deals with creating computer programs that perform operations comparable to those performed by the human mind, such as learning or logical reasoning (RAE: https://dle.rae.es/inteligencia), AI is an umbrella concept that influences and is influenced by many disciplines, such as computer science, engineering, biology, psychology, mathematics, statistics, logic, philosophy, business and linguistics. Generally, AI can be classified into weak AI and strong AI. Weak AI, also known as narrow AI, excels in specific tasks. Most advancements in AI, that have been achieved to date, can be classified as weak AI, such as Google Assistance and Alpha Go.

Researchers from different domains are, however, competing to create a strong AI (also called human-level artificial general intelligence or artificial superintelligence), which will process multiple tasks proficiently. Superintelligence can be loosely defined as "any intellect that greatly exceeds the cognitive performance of humans in virtually all domains of interest". In other words, a strong AI would be able to outperform humans in nearly every cognitive task.

In the case of DRAE, also identifies AI as a scientific discipline in which various disciplines converge, especially those related to or studying cognitive processes in humans. In turn, within Artificial Intelligence, Machine learning and Deep learning can be identified as fields of study (Fig. 1).

Arthur Samuel in 2000 coined "machine learning" in 1959 and defined it as a field of study that enables computers to learn without being explicitly programmed.

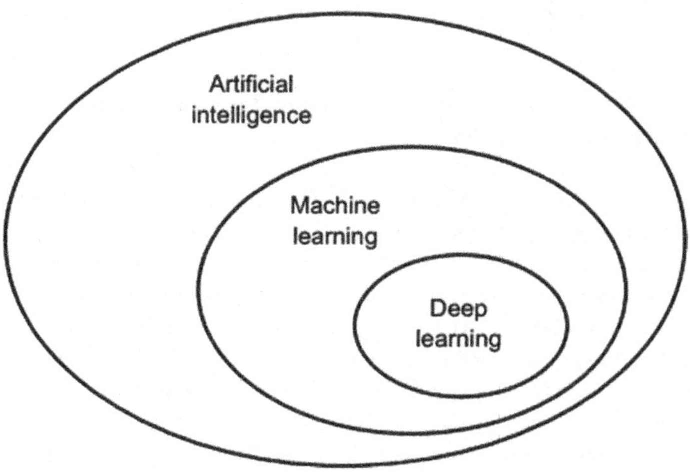

Fig. 1 Interrelations of artificial intelligence

Although this definition is rather vague, it indicates a significant feature of machine learning—it does not follow pre-programmed "rules". In general, machine learning is an automated process that enables machines to analyze a huge data set, recognize patterns and learn from the data to provide support for predictions and decision-making.

An example of this process is realized by "Reinforcement machine learning" algorithms. Learned by adjusting actions, based on continuous feedback and Reinforcement machine learning can also bypass human biases that may reside in data when using big data for training. Machine learning can be regarded as the automation of cognitive functions, or the automation of knowledge work. The drawback in machine learning, at the moment, is that the inner working of these self-learning machines is a black box, which makes it difficult to understand and explain the reasoning process and to justify the recommendations. As a result, trusting these machines is a challenge. People tend to only trust a system when the reasoning process is known and interpretable [4].

2.1.3 Robotics

Robotics is a branch of mechanical engineering, electrical engineering, electronic engineering, biomedical engineering and computer science, which deals with the design, construction, operation, structure, manufacture and application of robots (RAE: http://lema.rae.es/drae/?val=rob%C3%B3tica). Robotics is the technology used to develop machines, called robots, that can replicate human actions.

They are not necessarily like humans, since many are designed to carry out repetitive and dangerous tasks, rather than to perform high-precision and creative activities. Artificial intelligent robots are also expected to learn by themselves, behave like humans and even have the potential to surpass humans. In this case, a robot is automation of physical functions [4] and is directed by machine learning.

2.1.4 The Converge of Control and Machine Learning Theory. The Iterative Learning Controls

The key idea of Iterative learning control (ILC) is that as the system is usually working repetitively (e.g. robotic arms on car assembly lines working repetitively for each car), one can learn the best input action to track a given reference by learning from the data (input, tracking error) collected from previous attempts of performing the task. These data contain (implicitly) important information about the system dynamics. As a result, an (explicit) highly accurate system model is not required. ILC has achieved high tracking performance in a wide range of areas [5]. Several ILC design methods have been proposed. Crudely speaking, they can be classified into two categories, namely, model-free ILC design and model-based ILC design. Model-free ILC design learns the input using only previous input and error, examples of which include Proportional-Integral-Derivative (PID) type ILC, phase lead ILC [6], adaptive ILC [7] and most recently a data-driven ILC [8], Model-based ILC design, on the other hand, using some explicit (not necessarily accurate) model information. Some examples of model-based ILC design include inverse ILC [9], gradient ILC [10] and optimization-based ILC [11]. In general, due to the use of model information, model-based ILC tends to have improved convergence performance compared to model-free design.

The idea of learning from past information and adjusting the action is also the key idea of reinforcement learning (RL), a machine learning design method. RL aims to learn an optimal action policy such that some performance of interest (called return) is maximized [12]. To do so, it continuously interacts with the environment (system) by applying a policy, receiving a reward (to reflect how good the action is) and then improving the action policy. RL has achieved great success in some areas and particularly its recent achievement in Alpha Zero has attracted significant interest [13]. RL is particularly useful when a model of the system is either very complex where even a solution based on accurate model information is not possible to find (e.g. in the problem of Go, the optimal action policy is difficult to obtain due to the extremely large action space), or a model is not available at all (e.g. when the environment is time-varying). Similar to ILC, there are two main classes of RL algorithms. When a model is presented, the best action policy can be obtained by solving the Bellman Equation, using policy iteration, value iteration, or policy search. When the model information is not available, various model-free RL methods have been proposed, e.g. Q-learning, SASAR, Monte Carlo methods, or policy gradient. For more information on RL, please refer to [12].

The results of a comparison of their performance, both analytically and numerically on a high-performance tracking problem show that both methods have the same performance when accurate model information is available. Their performance, however, differs significantly when no model information but only data are given: ILC appears to learn information from the data in a 'clever way than RL such that it requires much fewer experiments (interactions with the environment/system) than RL to achieve a given accuracy; the mechanisms behind this need to be further investigated and might potentially be very useful to improve the convergence per-performance of RL algorithms which usually require a large number of experiments to converge (to optimal solutions). On the other hand, RL can be applied to a broader class of problems, e.g. with non-fixed task duration, non-identical initial conditions and therefore ideas from RL can be used to solve some emerging ILC applications, e.g. varying trial length ILC design, trial varying tracking tasks.

By the other hand, Deep Reinforcement Learning (DRL) uses deep networks to represent the value function, the Policy and the model. Also, optimizes the loss function using the stochastic descending gradient, the value function, the policy and the model.

The comparison of the current state-of-the-art robotics with applications of DRL in robotics can be seen in Fig. 2.

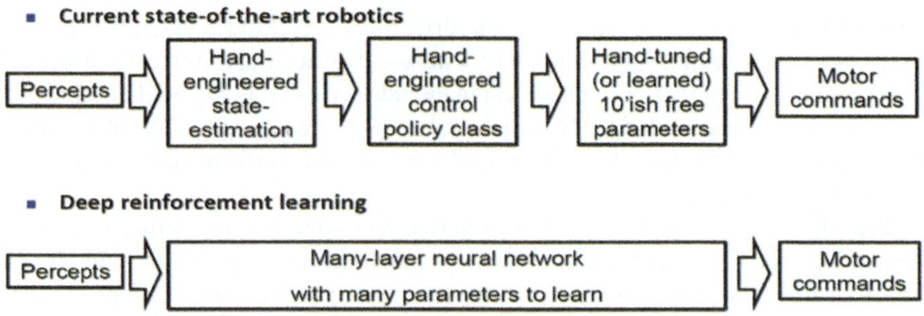

Fig. 2 Conceptual difference from the traditional independent block process of robotics conception with deep reinforcement learning algorithms applied to robotics

2.1.5 Uncertainty

What kind of uncertainty can be used in the model? [14]. There are two kinds of uncertainty:

1. Epistemic—uncertainty caused by data generation—parameter/model uncertainty. The origin of this kind of uncertainty are the errors in the measures and can be explained away by unlimited data, and
2. Aleatoric uncertainty—It is the inherent uncertainty of the system (exogenous uncertainty). The origin of this kind of uncertainty is the measures that can't be understood from the data and can be explained away by unlimited sensing.

Modeling Epistemic Uncertainty with Bayesian Deep Learning

Epistemic uncertainty can be model in deep learning models using Monte Carlo dropout sampling at test time. Dropout sampling can be interpreted as sampling from a distribution over models (Fig. 3).

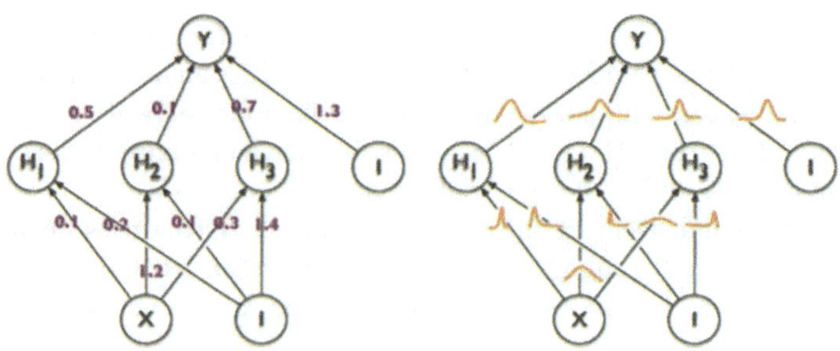

Fig. 3 Bayesian deep learning epistemic uncertainty. Left: each weight has a certain value. As stated in classical backpropagation. Right: each weight is assigned a Bayesian distribution

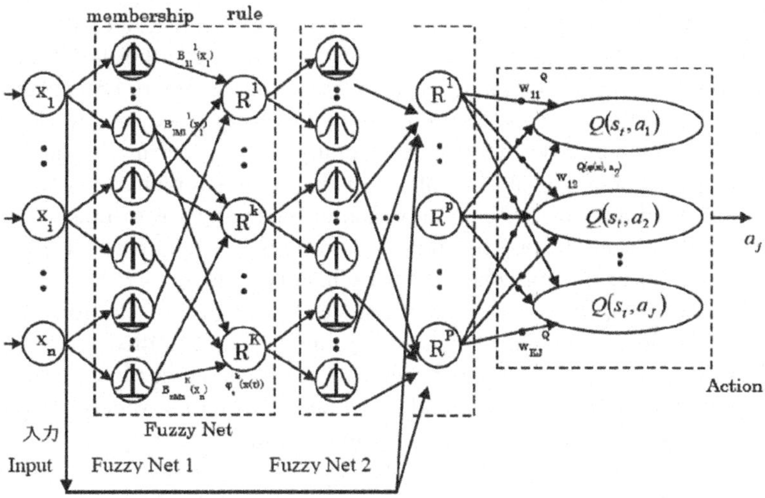

Fig. 4 Reinforcement learning system with multi-layered fuzzy neural network [15]

Modeling Epistemic Uncertainty with Fuzzy Deep Reinforcement Learning

The structure of Reinforcement Learning with a fuzzy self-organizing multi-layer fuzzy inference neural network [15] (MLFNN) is given in Fig. 4.

The MLFNN structure can be made up of a fuzzy network and a Temporal difference (TD) time difference of the learning algorithm. E.g. Q-learning and Sarsa.

TD methods start from the assumption that temporally consecutive states should have closer values.

The main advantage of TD methods over other RL methods is that they update the value of the function with each transition state, not just after the transitions resulting from a direct reward.

2.2 Theory Fundamentals

2.2.1 Interdisciplinarity

The convergence of these three disciplines in the same National Program responds to the precepts of interdisciplinarity, arising as a result of the interrelation of the sciences to join collaborative efforts in solving scientific problems in the modern world.

Lage states in his book [16] that in the twentieth century it became clear that knowledge in each specialty accumulated at a speed that exceeded the speed of interpretation of the impact of these in other fields of knowledge and even within from the same field. It was then that "interdisciplinary or multidisciplinary projects" began to be spoken of as a way of adapting to the new reality, but still preserving the traditional structure based on scientific disciplines. The author himself clarifies that later on, the concept of convergence sciences such as bioinformatics emerged, in which there is the intelligent recombination of existing pieces of knowledge.

For his part, Núñez Jover in the book Thinking Science, Technology and Society [17] clarifies that the practical and complex nature of the problems to be undertaken requires overcoming the disciplinary approach and opens the way to transdisciplinary, which is privileged and attribute of "mode 2" knowledge. Absolutism and disciplinary arrogance give way to a more open and participatory dialogue.

For Núñez Jover, the knowledge of "mode 2" distances itself from the traditional way of producing knowledge called "mode 1", identified with academic science, primarily cognitive purposes, search for truth and objectivity, organization of disciplinary-based knowledge produced mainly at the University. Meanwhile, "mode 2" demonstrates attributes such as being a knowledge produced in the context of an application, which from its conception has the intention of satisfying a practical interest, while the relevant science is produced in the same course which seeks to provide solutions to problems defined in the context of the application.

The definition of the Interdisciplinarity by the OECD, the Organization for Economic Cooperation and Development [18], given in the report by de Boer, reads as follows:

Interdisciplinarity is a noun describing the interaction of two or more different disciplines. This interaction may range from simple communication of ideas to the mutual integration of organizing concepts, methodology, procedures, epistemology, terminology, data and terms organized into a common effort on a common problem with continuous intercommunication among the participants from the different disciplines [19].

Employing differentiation, specialization and professionalization, scientific disciplines have made an enormous contribution to scientific progress, from which society has benefited. And that is still happening, unabatedly. The value of disciplines lies in the fact that they can account for their conditions of existence and how they gather their knowledge. At the same time, it can be said that progress within individual disciplines will sometimes remain more limited than progress in collaboration between disciplines.

The risk that monodisciplinary approaches fall short is not negligible. In discussions, it is sometimes pointed out that mono-disciplines wrongly claim to have the analysis and solution for major problems, such as the major economic crises or climate change and that, therefore, these problems are, in part, maintained. As already mentioned, interdisciplinarity contains a great promise of added value through collaboration. This promise means that it is possible to broaden and enrich knowledge and to move beyond individual disciplines.

Is important to remark, that interdisciplinarity is not against the disciplines but a driver of progress in the creation of knowledge. The League of European Research Universities [20] also sees no contradiction between discipline and interdisciplinarity, but rather complementarity.

So, disciplines should remain a central element of the academic system, as they have an unrivaled power to structure and understand the world. However, as the constant tensioning of ideas is central to the creation of knowledge, the collaboration between disciplines is a way of questioning the potential, limits and margin of progression of the disciplines.

For interdisciplinary research, it is, even more, the case that this is done or should be done in teams [21], interdisciplinary research is team research, that is, research conducted by a team—it is team science because it is infeasible to conduct interdisciplinary research independently.

According to [22], there is no scheme for the integrative process of interdisciplinary research, but the following steps are important:

Step 1 To define the problem

a. Define the problem (question, topic, issue);
b. Determine all knowledge needs, including the appropriate disciplinary representatives and advisors, as well as relevant models, traditions and literature;

Step 2 Research management

a. Specifying particular studies to be undertaken.
b. Engaging in "role negotiation" in teamwork.
c. Gathering all knowledge and new information.
d. Resolving disciplinary conflicts by working towards a common vocabulary (and focusing on reciprocal learning in the teamwork)
e. Building and maintaining communication through integrative techniques.

Step 3 Impact evaluation

a. Collocating all contributions and evaluating appropriateness, relevancy and adaptability
b. Integrating the individual pieces to determine a pattern of mutual relatedness and relevancy
c. Confirming or disconfirming the proposed solution (answer)
d. Deciding about future management or disposition of the task/project/patient/curriculum.

2.3 Methodology

The advantage of following a methodology is that a logical sequence is established to follow to develop a project in the best possible way. For the development of a research project, particularly a research project in the Program ARIA is very important to follow a methodology, but at this moment don't exist. The proposed solution is to make a mixture of methodologies that can answer this problem.

One of the first developed methodologies was CRISP-DM. It is one of these process models that can be used as a reference for knowledge extraction. CRISP-DM is composed of a series of stages that are, understanding the business, understanding the data, preparing the data, modeling, evaluation and deployment.

Unfortunately, even with this process model in place, many of the Knowledge Data Mining Projects (KDDM) currently underway fail. An important factor that determines the success or failure of these projects is the way they are executed during their life cycle [23].

The greatest weakness of CRISP-DM is that it is at the same time a process model, a methodology and a life cycle. In this way, it weighs down the definitions and details of it. Also, the data mining process is completely decoupled from the software engineering process, even though its results have a concrete impact on this process. This methodological deficiency is one of the fundamental reasons why many data mining projects are not completed or if they did, they fail to meet the expectations of the customer who does not use it. To mitigate this problem Segovia [24] proposes:

1. Define and integrate the data mining process with a software process, examining what tasks they have in common, their respective inputs and outputs and the weaknesses of CRISP-DM.
2. Develop an instance of the process tailored to CRISP-DM and unify that instance with a Unified Process (RUP) and
3. Validate and transfer the technology to real cases.

So, in [24] was proposed to establish a common development process framework based on two software standards, IEEE Std. 1074 and ISO 12207 to accommodate the tasks and processes established by CRISP-DM. The Fig. 5 below summarizes the result obtained.

For the industrial sector related to control, supervision and diagnosis, the application of a methodology presents peculiarities of this environment. Among these particularities we have how noisy industrial data is, having few reference elements when it comes to filtering and processing, the large volume of data generated in a very short time in a completely automatic way, the little informational culture of the personnel that works in the industry, the diversity of standards of measurement means that make the task of acquiring data from various sources and protocols a real technological challenge, to name the most significant in our understanding.

Under these criteria, the methodology called CONTROL-DM was proposed in [23], which is nothing more than the adaptation of the CRISP-DM methodology for the application of data mining to control processes (supervision and diagnosis) in organizations, following two guidelines:

1. Replace any business approach as an activity and as a process with the Control approach, which encompasses the actual Control activities and that of supervision and diagnosis. With this, we gain clarity of understanding and specificity of the methodology to the control process, with the consequent increase in efficiency in the introduction of it.

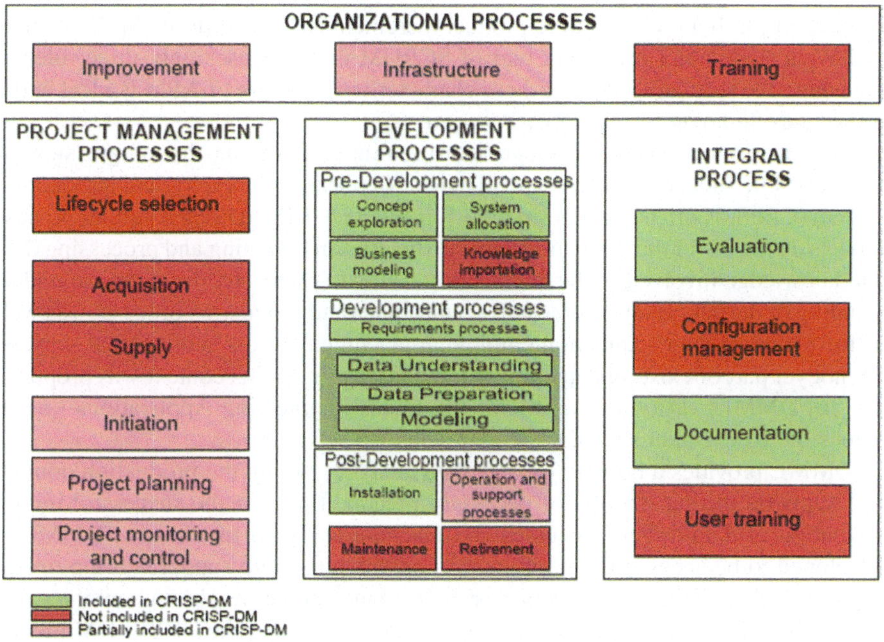

Fig. 5 Common framework of the development process

2. Apply the fundamentals and methodology of the CASE tools common to software development to the development of KDDMs to gain consistency, agility and understanding in the selected Data Mining project.

2.3.1 Agile Project Management

Agile Project Management (APM), which is commonly known as *Agile* refers to an iterative method for managing, designing and developing activities in the software industry mainly in engineering, information technology and other businesses that provide new products or services developed interactively. The *Agile* approach is very useful, helping teams in overpassing or even excluding unpredictable behavior in the project management process by using a management model based on iterative work and heuristic feedback.

SCRUM is the best way of introducing the *Agile* approach to project management as it's simple and easy to handle by the whole team involved in the project. SCRUM represents a management framework for incremental product development using one or more cross-functional, self-organizing teams of about seven people each. It uses heuristic feedback, team self-management and aims to build properly tested product increments within short iterations.

The short iterations are called Sprints and they have a fixed length not exceeding 30 days but *are* usually kept as short as possible. The idea of a Sprint is for the teams to try to build a releasable product increment at the end of the Sprint (James and Walter). Another important part of SCRUM is that it uses only three roles in managing the software [25]. For these characteristics, the use of APM is ideal in the project development inside ARIA Program.

The increasing output in manufacturing-related data for the progression of Industry 4.0, Industrial Internet of Things (IIoT) and Cyberphysical production system (CPPS) is often not utilized to its full extend. CRISP-DM and similar process models do not cover the domain-specific difficulties in obtaining and processing data and knowledge involved in the engineering context. Therefore, the technical understanding of the data analysis objects, e.g. machines and production processes, as well as the technical data acquisition employing machine controls and external sensors, are not yet part of existing data mining methodologies. In this context was proposed by [26] DMME—Data Mining Methodology for Engineering Applications—as an extension to the established CRISP-DM standard.

DMME provides a holistic view of data analytics within the production domain that supports the satisfaction of necessary preconditions for the successful implementation of data-driven processes and machine analysis. In addition, DMME is developed to be accessible for engineers implementing it to support startup operations as well as optimize production or maintenance processes throughout the value chain. DMME provides three novel steps as an extension to the CRISP-DM model. These are the "Technical Understanding", "Technical Realization" and "Technical Implementation":

1. Technical Understanding. The goal of this task is to transform the business goals into measurable technical goals, gather existing expertise of the related physical and process effects as well as develop an experiment plan.
2. Technical Realization. The goal of this task is to test and select the measuring concepts and to conduct the experiment plan. As a result, data is generated that includes all the relevant information and features for the subsequent data analytics task to reach the defined business goals.
3. Technical Implementation. The goal of this task is to enable the evaluated model to be provided with run-time data during production. Therefore, the data acquisition method from the "Technical Realization" phase of DMME is transformed into a run-time capable infrastructure.

2.3.2 Methodology for Deep Reinforcement Learning

Hillebrand [27] proposed a methodology for deep reinforcement learning (DRL) in autonomous systems. This methodology was conceived for the necessity of compromise a design method and consecutive design decisions for the development of systems using DRL. The methodology is based upon the basic principles of the V-Model [27]. The process contains four consecutive interactions and is shown in Fig. 6.

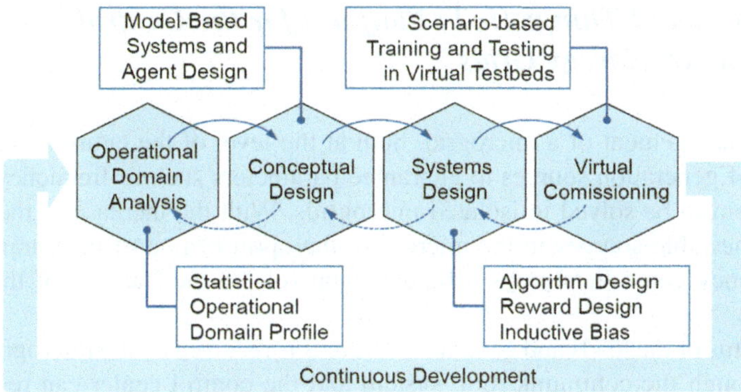

Fig. 6 Design process for DRL [27]

3 Results

The National Program for Automation, Robotics and AI has a total of 30 projects headed by 16 national entities between Universities, Research Institutes and Industries. The Interdisciplinarity between knowledge branches of the program is shown in Fig. 7.

In this, it´s showed that 37% of the total of projects are doing in an interdisciplinary way. The principal source of interdisciplinarity is the junction of AI and Automation branches. Is necessary to remark the tendency of the junction of AI, Automation and Robotics that also is a growing tendency in the scientific world. These numbers can indicate that the program in their first year has a good relation of interdisciplinarity.

Below is describe three examples of projects developed in the program where can see how is implemented the interdisciplinarity and the necessary methodologies for the obtention of planned results.

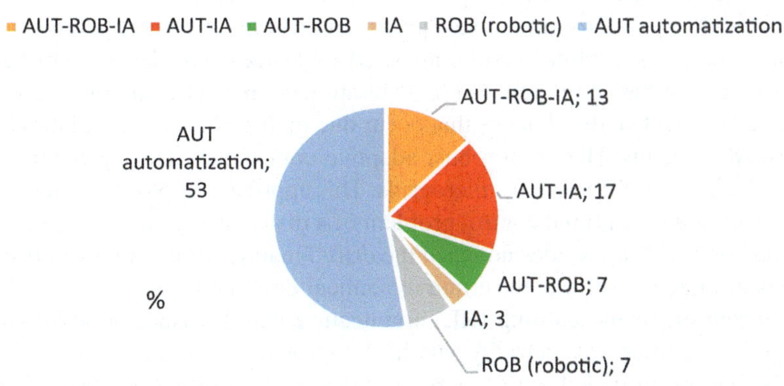

Fig. 7 The interdisciplinarity of the ARIA program

3.1 Internet of Things (IoT) Platform for the Control of Smart Energy Grids

Efficient management of a microgrid, both at the level of the entire system and at the level of generation sources to guarantee parameters such as frequency stability is a problem to be solved in isolated microgrids. With the increase in the penetration of renewable sources in the microgrid, the optimization of the parameters by the frequency controllers plays a fundamental role in the stability of the general frequency.

The status of the loads and generation sources is received by the microgrid control center through the communication system and the control center can regulate the output power of renewable energy sources and/or the power of storage devices. The isolated microgrid existing on Isla de la Juventud, in the Republic of Cuba, can be modeled considering the sources of photovoltaic energy, wind power, diesel generators, batteries and loads. Hardware devices controlling primary sources, both renewable and non-renewable are SCADAs.

An electricity microgrid (MG) is an energy system consisting of local electricity generation, local loads (or energy consumption) and storage capacities. MGs often face difficulties in supplying demand due to the lack of sufficient energy generation sources.

This problem is caused by the uncertain nature of renewable energy sources (such as wind and photovoltaic (PV) generations), market prices as well as loads with new options (such as heat pumps, e-vehicles, storage systems) that lead to difficulties in ensuring power quality and in balancing generation and consumption. To tackle these problems, MGs should be managed by an energy management system (EMS) that facilitates the minimization of operational costs, emissions and peak loads while satisfying the MG technical constraints.

The different operation modes of the microgrids can result in severe frequency and output voltage degradations. Electrical switches enable microgrids to alternate between the different modes of operation (islanded and grid-connected). During islanded mode, the microgrid delivers power and its output voltage is regulated to a specific nominal level, while in grid-connected mode, the controller reacts to the intermittent nature of the acting renewable generation source. The microgrid controller is required to regulate the changes that occur due to disturbances or variations in the active load demands. Herein, a robust adaptive control scheme is generalized and proposed to control autonomous microgrids. This approach employs a value iteration process to solve the optimal control problem of a distributed generation source. The resulting control strategy does not need the drift dynamics of the micro-grid system, which is ultimately useful in uncertain dynamical environments.

Deep reinforcement learning (RL) architecture may be used for addressing the problem of operating an electricity MG in a stochastic environment.

MGs that are provided with flexible options such as different types of storage devices to be able to address both short- and long-term fluctuations of electricity

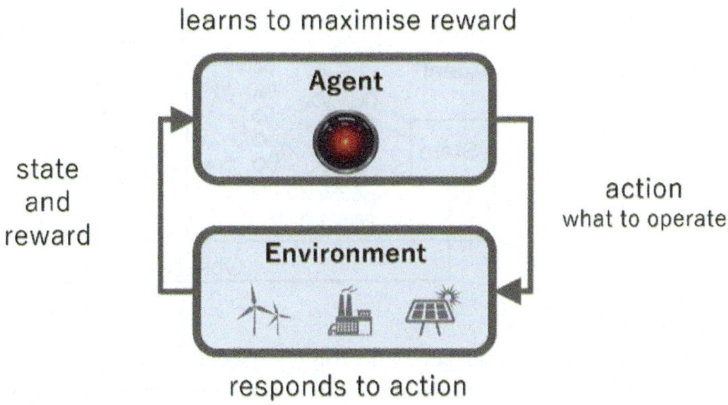

Fig. 8 Agent-environment interaction in RL [28]

production using PV panels (typically, batteries for short-term fluctuations and diesel generator for long-term fluctuations).

The control strategy of a hybrid renewable energy system (HRES) is necessary for improving the productivity and reliable operation of a power system working under the uncertainties of the RE resources and the dynamic loads. Under this system, maximum power point tracking (MPPT) control is used to improve the conversion efficiency of solar and wind energy systems while an energy management system (EMS) is developed for controlling the power flows among system components and handling reliable operation. The recent development and achievement in the fields of machine learning (ML) and reinforcement learning (RL) [28] leads to new challenges and opportunities for energy management and MPPT control. RL-based control systems can learn and act just following experiences when interacting with the environment. In contrast, traditional methods need particular mathematical models of the system and environment, which requires highly control knowledge, data and domain expertise. The RL controller can be considered as an agent and the agent can learn how to act based on the reward and current states received from the environment (Fig. 8) [28].

3.2 Deep RL Solutions for Sequential Decision Making

Optimally operating an MG can be formalized as a partially observable Markov Decision Process (MDP), where the MG is considered as an agent that interacts with its environment [28] (Fig. 9).

The development of the web application for the analysis of the frequency deviations of the Isla de la Juventud electrical system enables specialists to have a tool capable of centralizing all the information on the behavior of energy sources at the time of the frequency deviation. This brings the benefit of taking advantage of a

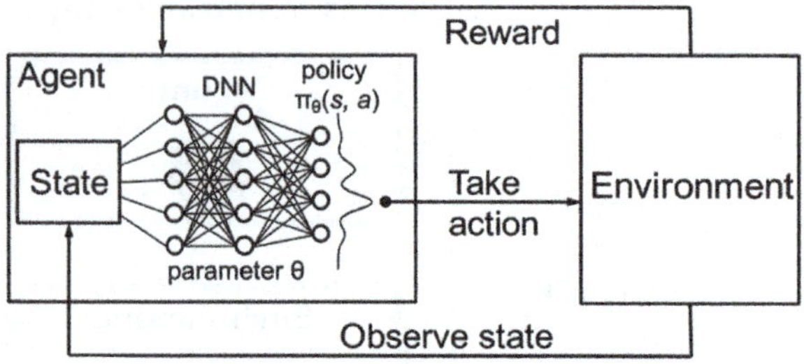

Fig. 9 Deep RL design [28]

lot of data that is kept daily and yet is no longer used in a fairly high percentage, through the processing of this data valuable information is extracted to provide a tool that supports the decision-making of specialists. Another benefit is obtaining faster and more accurate analyzes of several parameters at the same time, these analyzes improve the control and operation of the electrical system.

In Fig. 10, the frequency is displayed every one second through a line graph, this reflects its behavior for one day. This allows the specialist to carry out various analyzes on how this behavior influenced the electrical system, what was the meteorological situation at that moment that a frequency deviation occurred, or what could have caused this deviation to occur [29].

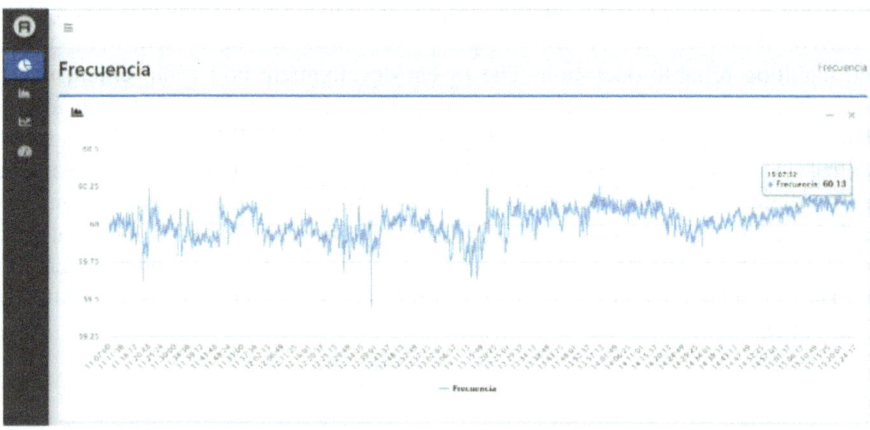

Fig. 10 Frequency of electric system [29]

3.3 Conception of an IoT Reference Model for Flight Operational Safety Using an Analytical Platform of Data Mining Flight Data Recorder

The safety of commercial airlines has grown considerably since its insertion over one hundred years ago. However, without more complete information, it is very difficult to achieve a substantial improvement of these indicators. The safety challenges of the coming years require the development and understanding of new types of data to improve the safety of other segments of commercial aviation, moving from the traditional reactive methods of incident-accident analysis to more proactive and predictive methods based on systems.

It remains to talk about the use of new technologies in the search for new security solutions, within these technological advances includes safety management under an Internet of Things (IoT) concept. With this, a system of systems is proposed that allows to collect, store, process and analyze all the necessary information, either in real-time or historical, to make better decisions about flight safety [30].

Anomaly detection is an edge of data analysis for civil aviation. This aspect consists of finding patterns that are not part of the expected behavior. The formulation of the problem depends on the nature of the data, availability of labels and the type of anomaly to be found. The anomalous data is also known as outliers, discordant observations, exceptions, aberrations, surprises, peculiarities, or contaminants.

The application areas of machine learning have extended to the area of flight safety. The digitization of the parameter collectors has generated a large amount of data, which today is the fundamental raw material used to raise the quality of the risk management process. Conducting the knowledge extraction processes in the database is the objective of the developed analytical platform.

Based on the conceptualization of the grouping task and the study of the operation of the K-means, SOM and DBSCAN algorithms, a platform was created by using Knime for pre-processing, obtaining the models and diagnosing the anomalies. The three algorithms applied allowed the flights to be grouped according to the similarity in the behavior of a small sample of five parameters in the take-off and landing stages.

The system offers the possibility of using four services aimed at research. They are:

1. Anomaly detection with data mining techniques from flight parameter collectors.
2. Implementation of a Markov model for risk assessment in flight operations with self-organized maps.
3. Discovery of flight safety anomalies in unstructured databases (texts) with data mining techniques.
4. Data mining patterns for weather predictions.

Table 1 Results of the analysis with K-means for the VRTG parameter

Phase	Group	Objects	Label
Take-off	Cluster_0	492	Normal
	Cluster_1	21	Outlier
	Cluster_2	45	Outlier
Landing	Cluster_0	43	Outlier
	Cluster_1	446	Eventual
	Cluster_2	60	Outlier

3.4 *Diagnosis of Data Anomaly*

An example of the work carried out in diagnosing the anomaly is the analysis of the aircraft's rotation speed data during take-off. In this validation, three different techniques have been used to group the flights according to their behavior in the take-off and landing stages. The use of the three variants allows the results obtained to be contrasted, therefore, actions have been carried out on the output data to standardize them and facilitate the pertinent comparisons.

The analysis is divided by parameter, phase and algorithm. Therefore, a flight on take-off, according to the VRTG values, can be diagnosed anomalous by K-means, while using the same technique for the corresponding landing parameter, it can be diagnosed if the parameter is normal (Table 1).

In the diagnosis of anomalous data, emphasis is placed on searching for thresholds, rankings, or any measure that contributes to the effectiveness of the analysis. In this case, the percentile calculation is applied. The percentile is a parametric measure that does not require the distribution of the data to be normal, it is not very sensitive to outliers and its results are robust in data extracted from real contexts [31].

3.5 *Parkinson's Disease Classification and Medication Adherence Monitoring Using Smartphone-Based Gait Assessment and Deep Reinforcement Learning Algorithm*

For the diagnosis and classification of Parkinson's patients, the unified scale of Parkinson's patients (Unified Parkinson's Disease Rating Scale—UPDRS) is used, which requires the patient to perform a series of tests among which the biomarkers of speech, the facial expression, the hand movement and walking analysis are considered, after which the doctor diagnoses the patient whether or not he has Parkinson's disease according to the score obtained.

The project proposes a system for monitoring patients with the use of cell phones and their automatic classification according to the data collected by them. The system starts from the budget that Parkinson's patients have different abnormalities when walking if they do not follow the required medication. The cell phone collects the data passively while the patient has his cell phone in his pocket. After that, the data preprocessor helps to extract the walking cycles that this Parkinson-related biomarker contains. The algorithm proposed for the classification and Medication Adherence Monitoring is Deep Reinforcement Learning, it demonstrates the feasibility of using cell phones to monitor the biomarker walking in Parkinson's patients and the possibility of Passive Medication Adherence Monitoring and Dynamic Treatment Regimes [32].

4 Conclusions

Automation, Robotics and artificial intelligence are disciplines that by themselves constitute the basis for the modern development of a country. The integration of these disciplines in a National Program created and directed by CITMA allows generating a synergy that enables the promotion of these disciplines thanks to interdisciplinary nature, the use of common development methodologies and addressing interrelated theoretical elements such as Iterative Learning control, reinforcement learning, deep reinforcement learning.

In the results, was exposed the analysis of real interdisciplinarity obtained in the Program and presenting some projects applications like the control of intelligent energy networks, the detection of anomalies in the operational safety of flights and patient monitoring of Parkinson's for the determination of the state of the disease and the ingestion of the corresponding medicines that showed how the methodologies were used and the real implementation and beneficiaries of interdisciplinarity, in fields that before were developed by other methods. This is the principal novelty of this paper.

References

1. Cosca, V.: Informe voluntario de Cuba. Informe Nacional sobre la implementación de la AGENDA 2030 (2019)
2. PCC.: Documentos del 7mo. Congreso del Partido aprobados por el III Pleno del Comité Central del PCC (2017)
3. GOE.: Constitución de la República de Cuba, Gaceta Oficial No 5 Extraordinaria, Ministerio de Justicia, 10 abril 2019, 2019-GOC-2019-406-EX5 (2019)
4. Wang, W., Siau, K.: Artificial intelligence, machine learning, automation, robotics, future of work and future of humanity: A review and research agenda. J. Database Manage. **30**(1). https://doi.org/10.4018/JDM.2019010104 (2019)
5. Bristow, D., Tharayil, M., Alleyne, A.: A survey of iterative learning control: A learning-based method for high-performance tracking control. IEEE Control Syst. Mag. **26**(3), 96–114 (2006)
6. Freeman, C.T., Tong, D., Meadmore, K., Cai, Z., Rogers, E., Hughes, A.M., Burridge, J.H.: Phase-lead iterative learning control algorithms for functional electrical stimulation-based stroke rehabilitation. Proc. Inst. Mech. Eng. Part I-J. Syst. Control Eng. **225**(I6), 850–859 (2011)

7. Chien, C., Tayebi, A.: Further results on adaptive iterative learning control of robot manipulators. Automatica **44**(3), 830–837 (2008)
8. Bolder, J., Kleinendorst, S., Oomen, T.: Datadriven multivariable ILC: enhanced performance by eliminating L and Q filters. Int. J. Robust Nonlinear Control **28**(12), 3728–3751 (2018)
9. Longman, R.W., Peng, Y., Kwon, T., Lus, H., Betti, R., Juang, J.: Adaptive inverse iterative learning control. J. Chin. Soc. Mech. Eng. **32**(6), 493–506 (2011)
10. Kinoshita, K., Sogo, T., Adachi, N.: A relationship between iterative learning control using the gradient method and stable inversion. Trans. Soc. Instrum. Control Eng. **36**(12), 1117–1124 (2000)
11. Amann, N., Owens, D., Rogers, E.: Iterative learning control using optimal feedback and feedforward actions. Int. J. Control **65**(2), 277–293 (1996)
12. Sutton, R.S., Barto, A.G.: Reinforcement Learning: An introduction. MIT press (2018)
13. Silver, D., Hubert, T., Schrittwieser, J., Antonoglou, I., Lai, M., Guez, A., Lanctot, M., Sifre, L., Kumaran, D., Graepel, T., et al.: A general reinforcement learning algorithm that masters chess, shogi and Go through self-play. Science **362**(6419), 1140–1144 (2018)
14. Der, A., Ditlevsen, O.: Aleatory or epistemic? Does it matter? Struct. Saf. **31**(2), 105–112 (2009)
15. Kurmeoto, T., et al.: A Reinforcement Learning System with Multi-Layered Fuzzy Neural Network. The Institute of Industrial Applications Engineers, Japan. https://doi.org/10.12792/icisip2017.081 (2017)
16. Lage, A.: La Osadía de la Ciencia. Editorial Academia, Cuba, ISBN 978-959-270-398-8 (2018)
17. Núñez, J., Montalvo, L.F., Figaredo, F.: Pensar Ciencia, Tecnología y Sociedad. Compilación de artículos. Editorial Felix Varela, La Habana, Cuba. ISBN 978-959-07-1008-7 (2008)
18. Apostel, L.: Interdisciplinarity: Problems of Teaching and Research in Universities. OECD, Paris (1972)
19. Wilthagen, T., et al.: A Time for Interdisciplinarity. Tilburg University (2018)
20. LERU.: Interdisciplinarity and the 21st Century Research-Intensive University (2016). [Online] Available on: https://www.leru.org/files/Interdisciplinarity-and-the-21st-Century-Research-Intensive-University-Full-paper.pdf [Accessed: 21 May 2021]
21. Fiore, S.M.: Interdisciplinarity as teamwork. How the science of teams can inform team science. Small Group Res. **39**(3), 251–277 (2008). [Online] Available on: https://doi.org/10.1177/1046496408317797 [Accessed: 12 May 2021]
22. Klein, J.T: Interdisciplinarity. History, theory & practice. Wayne State University Press, Detroit (1990)
23. Plasencia, A., et al.: Consideraciones para la elaboración de una metodología de proyecto de minería de datos para el control, la supervisión y el diagnóstico. Reporte de Investigación ICIMAF (2010)
24. Segovia, J.: Definition and instantiation of an integrated Data Mining Process Jornadas de Seguimiento de Proyectos. Universidad Politécnica de Madrid, Programa Nacional de Tecnologías Informáticas (2007)
25. Gal, A., et al.: A new vision over Agile project management in the internet of things era. Procedia. Soc. Behav. Sci. **238**, 277–285 (2018)
26. Huber, S., et al.: Data mining methodology for engineering. Procedia CIRP **79** (2019).
27. Hillebrand, M., Lakhani, M., Dumitrescu, R.: A design methodology for deep reinforcement learning in autonomous systems. Procedia Manufacturing **52**, 266–271 (2020)
28. Tomin, N., et al.: Deep reinforcement learning for energy microgrids management considering flexible energy sources. EPJ Web Conf. **217**, 01–16 (2019). [Online] Available on: https://doi.org/10.1051/epjconf/201921701016 [Accessed: 21 May 2021]
29. Rivero, Y., et al.: Web application for the analysis of the frequency deviations of the electrical system on the Isle of Youth. X Taller Internacional De Cibernética Aplicada, La Habana, Cuba (2020)
30. Suárez, B., et al.: Conception of an IoT Reference Model for Flight Safety. IX Taller Internacional De Cibernética Aplicada, La Habana, Cuba (2019)

31. Plasencia, A., et al.: Methods and Algorithms for anomalies detection in the flight safety from the Flight Data Recoreder. VII Taller Internacional De Cibernética Aplicada, La Habana, Cuba (2017)
32. Plasencia, A., Shichkina, Y., et al.: Parkinson's disease classification and medication adherence monitoring using smartphone-based gait assessment and deep reinforcement learning algorithm, in *14th International Symposium «Intelligent Systems», INTELS'20*, Moscow, Russia (2020)

Chapter 19
Tendencies in Augmented Reality

Nayma Martín Amaro, Pedro Y. Piñero Pérez, Dagoberto Félix Pérez Montesinos, and Iliana Pérez Pupo

Abstract The expansion of the internet and new information and communication technologies have changed people's lives in many ways. Augmented Reality (AR) is a promising alternative in the midst of the great technological revolution. These technologies are projected in the field of decision-making as a trend that could transform current practices. The objective of this research is to create a theoretical framework for the subject that allows researchers to analyze trends in the methods for using augmented reality, the use strategies used in research and the main areas of application. As conclusions, is identified the need to improve methods to use augmented reality, as well as the possibility of using augmented reality in decision-making and in project management.

Keywords Augmented reality · Decision-making · Project management

1 Introduction

Augmented Reality exists as a new technology since the eighties and since then it has remained in constant evolution, although always related to the world of video games and entertainment, the rebirth of gadgets and their popularization has allowed their migration to other applications such as education [1, 2].

N. M. Amaro (✉)
Centro de Capacitación, Empresa de Telecomunicaciones ETECSA, Sucursal Sancti Spiritus, La Habana, Cuba

P. Y. Piñero Pérez · I. P. Pupo
Grupo de Investigaciones en Gestión de Proyectos, Universidad de las Ciencias Informáticas, 54830 La Habana, CP, Cuba
e-mail: ppp@uci.cu

I. P. Pupo
e-mail: iperez@uci.cu

D. F. P. Montesinos
Departamento de Informática, Grupo Empresarial CIMEX, Sucursal Sancti Spiritus, Cuba
e-mail: dagofpm@cimex.com.cu

P. Y. Piñero Pérez et al. (eds.), *Artificial Intelligence in Project Management and Making Decisions*, Studies in Computational Intelligence 1035,
https://doi.org/10.1007/978-3-030-97269-1_19

Although its use began in the early 1990s [3], has become popular today due to its cheaper and simplified equipment and devices necessary for its use [4]. Thus, this technology is being investigated in various sectors [5].

As it is a relatively recent concept, an approximation to its definition is necessary for its correct understanding. Thus, since the late twentieth century, authors such as Azuma [6] define Augmented Reality as a technology that improves the user's sensory perception by superimposing virtual objects to the real world, thus it seems that these virtual elements belong and coexist in the same space as the real objects.

The concept of Augmented Reality can be confused with another term with which it is related, Virtual Reality [7]. Both terms refer to different levels of immersion of users in virtual environments where the real and the virtual coexist [8]. The fundamental difference that distinguishes both technologies is that while Augmented Reality combines virtual elements in real environments or contexts, Virtual Reality completely replaces a real environment with a virtual one. Therefore, when researching the subject and implementing it in different contexts, it is important to distinguish between the two concepts.

Alternative realities are enjoyed through Virtual Reality and Augmented Reality, which are created from scratch, either as completely fictional worlds or as an adaptation of real environments. This implies that, in essence, infinite worlds can be created: alternate realities offer infinite possibilities [9, 10]. Augmented Reality is a technology that mixes the real world with the virtual world. In it, different layers of information are superimposed on the physical world, offering a new experience of reality. This experience offers a multitude of application possibilities, since information can not only be accessed but also interacted with [11, 12]. It is also a valid alternative that allows connecting the real world with digital content. This feature allows the user to reinforce their knowledge by associating it with the real world [13, 14].

In several consulted sources they speak that Augmented Reality fits perfectly as tools to support the training of professionals, since they make it possible to improve teaching and learning processes achieved in conjunction with traditional educational approaches [15, 16].

Increasingly affordable technology and resources can be an obvious path to educational equity and quality, mitigating many of today's shortcomings [17].

Augmented Reality is currently a technology that has been introduced in most sectors of society.

In the consulted bibliography, few review works associated with this topic were identified, although several works on its application were identified, for example in fields such as higher education and medicine.

Based on this problematic situation, the objective of this work is: to build a theoretical referential framework through a tertiary review that allows to analyze the evolution and trends in the development of augmented reality and its applications. The work is organized as follows: the second Methodology section explains the experimental design used, the Development section shows the results of the analysis and finally the Conclusions are presented.

2 Methodology

The study that is presented constitutes an exploratory investigation [18] in it, he develops the following design of experiments:

Step 1 Design of a systematic review protocol that allows the construction of a theoretical referential framework associated with the subject.
Step 2 In-depth analysis of the bibliography consulted by each of the established categories.
Step 3 Application of descriptive statistics and meta-analysis techniques for the presentation of research results.
Step 4 Identify open lines of research.

In line with the silver, a systematic review protocol is designed [19] and the historical evolution of augmented reality is analyzed in different stages: publications before 2000, publications between 2000–2017 and from 2017 to 2021.

Systematic review protocol applied:

1. Definition of the object of study, field of action and objective of the research.
   a. Research object: augmented reality.
   b. Course objective: Build a theoretical referential framework associated with augmented reality, trends and applications in project management.
   c. Research field: structures, methods and applications of augmented reality.
2. Definition of a bibliographic manager.
3. Define sources of academic information for the development of the review: "Semantic Scholar", "Google Scholar" and "Scopus" and other meta-search engines based on open science.
4. Definition of the following set of key phrases to carry out the searches: "Augmented Reality" AND ("decision-making" OR Usability), BIM + Augmented Reality, Augmented Reality + decision-making.
5. Define the goals of the bibliometric analysis in the form of research questions and inclusion—exclusion criteria:
   a. How has been the trend of publications per year?
   b. Who are the main authors?
   c. What are the affiliations and countries of the main authors?
   d. How are the publications distributed considering the types of documents in: articles, books, theses and conference or congress proceedings?
   e. Exclusion of works published in spaces with a low level of arbitration.
6. Sort and filter posts in the following set of categories:
   - Classics: refers to pioneering publications in augmented reality, where the fundamental principles that set guidelines in the theory are exposed.

- Extensions to theories: refers to publications that extend the theory proposed in the publications understood as classic, do not set guidelines that significantly change the previously proposed methods, although they do develop contributions to knowledge.
- Application results: publications that focus on the use of existing theory in concrete practical scenarios.
- Tertiary reviews: refers to review articles on trends and developments in the subject in question.

7. Synthesize the main trends.
   As a second step in the methodology used, a detailed analysis of each of the works is carried out and they are characterized with respect to the following elements: Areas of application of the proposal.

Then, descriptive statistics techniques and meta-analysis are applied for the presentation of the research results [19]. Finally, open lines are analyzed and proposed in the object of study analyzed.

3 Development

From the study carried out, the majority of publications in Augmented Reality were identified in 2019 [20]. On the other hand, most of the articles are written by varied authors. By consulting the journals that publish scientific papers, it can be seen that the papers have been published in different countries (Table 1) and therefore, in different languages.

The authors with more publications in this topic are Lai, A., Chen, C., Gutiérrez, J. M., Fernández, M. D., Toledo-Morales, P., Sánchez-García, J., Liu, Y., Lu, S., Kao, C., Chung, L., Tan, K., Henssen, J. A., van den Heuvel, L., De Jong, G., Vorstenbosch, A., van Cappellen and V., Van den Hurk. We can see in the review that Augmented Reality is independent of any discipline, since there are studies of its treatment in different disciplines, such as the development of projects associated with medicine, computer science, engineering and education.

As for the *h*-index, there are varied coefficients, especially the one from the Journal of Computer Assisted Learning and the British Journal of Educational Technology.

A better characterization of the bibliography consulted for this work is shown below:

- Most of the researches consulted are articles in journals, it is identified that in this thematic area there are few published doctoral or other theses, see Fig. 1.
- Most of the publications consulted are indexed in *Scopus* and *Web Of Science* see Fig. 2.
- The main sources of information and the main publishers for the researches consulted were IEEE, Elsevier and Springer respectively.

Table 1 Magazines and countries with more papers related to augmented reality

Journals	Countries
British Journal of Educational Technology	UK
Journal of Computer Assisted Learning	
Latin American Journal of Educational Technology—Relatec	Spain
Spaces	
Journal of Media Education and TIC	
NORIA Educational Research	
Journal of Sport and Health Research	
Pixel-Bit, Media and Education Magazine	
RIED. Ibero-American Journal of Distance Education	
International Journal of Engineering Business Management	Croacia
International Journal of Engineering Education	Ireland
Anatomical sciences education	U.S
Advances in Intelligent Systems and Computing	Germany
Revista Alteridad	Ecuador
Polytechnic	Mexico
Journal Between Science and Engineering	Colombia
Research Technology. TIA Academy	
Journal of Computer Engineering Teaching and Learning	Argentina

In the bibliography consulted, a set of elements that characterize or form part of the subject area is identified.

Augmented reality has two functions: to increase the perception of reality (it shows reality, but chooses what can be seen and what not) and to create an artificial environment (it shows what is not real, allowing us to see the imaginary), with an increased perception of useful information that helps decision-making and actions. Augmented Reality systems are characterized by having real and virtual elements in a real environment with a high degree of interactivity, with the option of registering and positioning the virtual information taking into account the three-dimensionality of the real world.

There are four technical aspects that are needed to integrate a system with Augmented Reality, these are: capture of the real scene, identification, mixture of reality and virtual information and visualization. Regarding the hardware requirements, there are: computer or mobile device, monitor or display device, camera, tracking or sensing system, GPS, accelerometer, connection infrastructure and markers, among others [21].

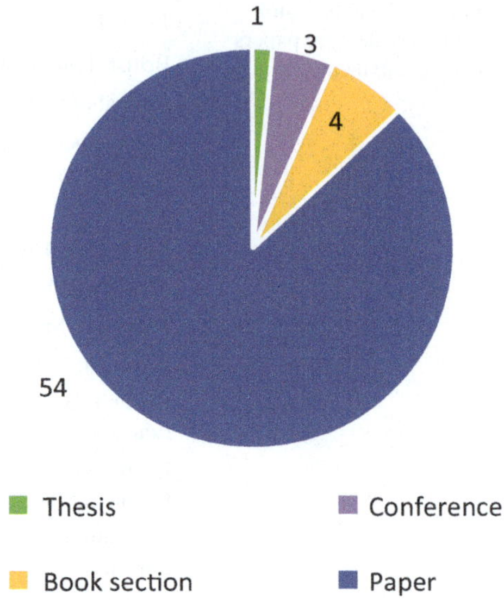

Fig. 1 Bibliography by type of publication

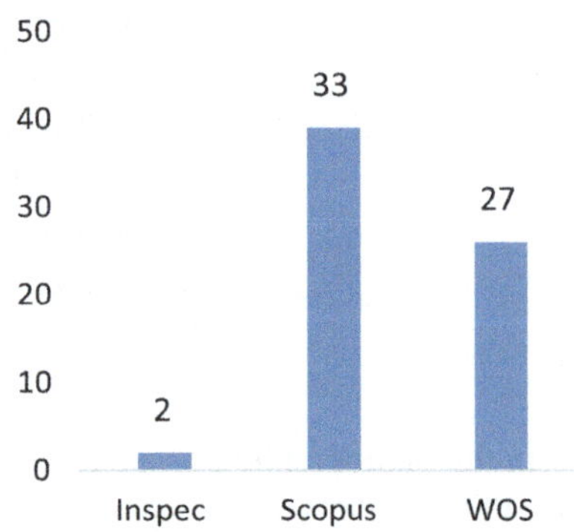

Fig. 2 Indexing of the publications in magazines and conferences analyzed

3.1 Example of Results of Application of Augmented Reality in the Development of Projects Associated with Medicine

In comparative studies between augmented reality simulation boxes such as the ProMIS and virtual reality boxes such as the LapSim, the results are slightly better in the augmented reality simulation station, having been tested with groups of novice and expert doctors. This situation occurs mainly due to the realism in haptic feedback, which turns out to be a challenge in virtual reality devices, while in augmented reality systems the doctor operates real instruments, just as he would have them in a real laparoscopic operation [22].

Advances in Augmented Reality are not limited to the visual field but also to the tactile field, recreating virtual tactile sensations, which opens the way to augmented reality applied to haptic or haptic Augmented Reality, envisioning new research panoramas aimed at the mixture of sensations of the real and virtual world and expanding sensations such as strength, stiffness, flexibility, roughness, temperature and others [22, 23].

Some systematic reviews have concluded that technical aspects such as tracking and recording are the most important issues for Augmented Reality applications. In turn, various studies have shed light on the augmented reality monitoring process, but the challenges and limitations of these processes have not yet been clearly reported [24]. For this reason, one of the areas of greatest development in the future will be the monitoring and spatial localization of the position of surgical instruments in real time. Also, many works propose the use of magnetic sensors and optical sensors, also including ultrasound and preoperative computed tomography, or magnetic resonance imaging for such monitoring [25], while other studies mention relevant issues such as accuracy, portability and cost.

The combination of augmented reality in real time and wearable computing devices such as Google glass is a trend in the field of surgery [25]. This will be one of the future development vectors, since the deployment in augmented reality systems continues to be a limiting factor [26].

Currently, a lot of work is being done on displays that provide enough brightness, resolution, field of view and contrast to appreciate the real and imaginary world. The proposed technologies still have many limitations and are not small enough, light enough or low cost [22].

Regarding augmentation applications in Augmented Reality in minimally invasive surgery, the studies consulted show that augmented reality has not only been limited to the visual area, as currently some developments in tactile and auditory augmentation are covered, to recreate the sensations of open surgery. One of the most interesting augmentation applications is to project images onto the patient, or onto the organs, either veins or grids, which will guide the surgeon with needles, catheters and cuts [22].

Within the research areas in Augmented Reality, the monitoring and spatial location of the position of surgical instruments turns out to be a subject of research trend, as well as the monitoring or tracking in intra-abdominal operations, due to the effect of breathing that requires a dynamic tracking system of the movement of organs in specific anatomies [22].

3.2 Example of Results of Augmented Reality in the Development of Projects Associated with Education

In the bibliography consulted [27–34], we find Augmented Reality studies to support the teaching of concepts related to the subject of technical drawing.

The applications were built from the practical exercises taught in the Representation Systems subject [35].

In this study we found an Augmented Reality (AR) system developed in a mobile application made for the Android operating system, which can be downloaded from the Google Play store [36] The Unity tool was used for its development [37] with the addition of the Vuforia library [38] for RA. For its part, the 3D models used were made in Blender 3D and the markers were built with the computer-aided design tool LibreCad [39].

Augmented reality fits very well as a training support tool. Quoting Hirokazu Kato, original developer of the ARToolKit project, a library that is used to create augmented reality applications: "I believe that augmented reality is the best way to connect the real world and digital content. This feature allows the user to reinforce learning educational content through its association with the real world" [40].

With reality, you can leave the classroom and learn from what you observe, it is a different concept of discovery-based learning. In the same way, it is a very useful tool in the classroom, it helps to reinforce and consolidate knowledge through the visualization of 3D models [40, 41].

Magic Book, from the New Zealand HIT group. The student reads a real book through a handheld viewer and sees virtual content on the real pages; in this way, when the student sees an augmented reality scene that they like, they can enter the scene and experience it in an immersive virtual environment [40].

The Massachusetts Institute of Technology (MIT) and Harvard are developing applications of augmented reality in games format in their education programs and groups; these games seek to engage high school students in situations that combine real-world experiences with additional information presented to them on their mobile devices [40].

Construct3D is a 3D geometric construction tool designed specifically for math and geometric education. It is based on the augmented reality mobile collaboration system. Anecdotal evidence supports the claim that using Construct3D is easy to learn and encourages experimentation with geometric constructions [40].

Ninus is an interactive product for early childhood education created by Btripple. This educational tool, intended for children, allows interactive group work in the classroom. With this product, children can learn to act autonomously, to think, to communicate, to have initiative and to discover the environment, also, it encourages participation and the capture of children's attention [40].

Ar-Books.com was born with the aim of reinventing the way of reading books, complementing the illustrations and photographs that are traditionally interspersed with textual content and enabling a three-dimensional interactive experience. The AR-Books.com collection will be made up of a series of books with a variety of topics, both didactic and narrative, where the common thread of the collection will be the use of augmented reality to enrich its reading and comprehension [40].

Talentum Schools is an educational initiative of Telefónica that aims to promote the creation of technology at an early age through free workshops given in Movistar stores [20]. In them, children from four years of age and young people up to eighteen learn programming concepts with tools such as augmented reality or the MIT Scratch programming language, robotics with the Atti robot developed by SK Telecom, or the creation of Android mobile applications with App Inventor [40].

In some of the bibliographies consulted we can find applications based on Augmented Reality, which in the educational field have had a high impact, among them we mention:

- Arloon: Application package for both Primary and Secondary Education, focused on various areas of science. It has augmented reality, practical content, etc.
- Chromville: This set of applications, aimed at Early Childhood Education and Primary Education courses, allows students to color each of the cards proposed and thanks to the application, we will see how the drawings come to life. Its interactivity increases, thanks to cards that allow modifying and playing with them through the application.
- Elements 4D: The study of chemical compounds as well as stoichiometric reactions are more accessible thanks to this application. Through its cubes and the union between them we can access all the information of the element or compound.
- Aurasma: An important element is the possibility of creating content based on augmented reality without prior programming knowledge. Aurasma offers this possibility, obtaining truly impressive results. Adapted to both the online version and the app, Aurasma has great potential both for any educational level and for any subject.
- SpaceCraft 3D: The exploration of our Solar System and the Universe in general, is carried out thanks to satellites and probes. This application developed by the National Aeronautics and Space Administration (NASA), allows the user to know in detail some of the missions sent for this purpose. Its graphics and interactivity stand out.
- Anatomy 4D: The detailed study of the human body through the use of augmented reality is already possible thanks to this application. Its elaborate graphics allow the user to easily explore the human body.
- Zookazam: The study of the animal world, knowing their customs or habitat is the highlight of this great application. It has numerous resources, among which the audio explanations of the different sections stand out.
- Quiver: The plastic expression takes center stage with this application. Thanks to its interactive cards, the drawings come to life, allowing you to interact with them.
- ARFlashcards: Learning the alphabet in English for young learners is made more accessible with this app. Thanks to its audio, the student will be able to make a correct pronunciation.
- Plickers: Carrying out group activities based on augmented reality is possible with this application.

The following list presents different examples of applications of Augmented Reality techniques [42, 43]:

- Microsoft Hololens glasses that allow you to perceive the perfect combination between the physical world and virtual information.
- Google Glass from the Google rand.

- Ikea with its furniture catalog that allows you to locate the furniture in your own home and perceive how it would look through augmented reality.
- Volkswagen with its Augmented Reality Apps, used to compare the energy use of an electrical appliance with the energy use of its cars.
- Strabucks Electric Café, where people can send Valentine messages over their cup of coffee.
- Mercedes Benz is its augmented reality app, where its users can check what a new car would look like.
- Carrefour with its new augmented reality catalogs.
- Audi with the use of augmented reality in the user manuals for those who buy their cars.

3.3 In the Bibliography Consulted Some of the Augmented Reality Applications in Business and Advertising are Exposed

Calvin Klein with its *Get it Uncesored* campaign that, through the use of smartphones, allows users to access through a QR code located on giant billboards in the cities of New York and Los Angeles.

The public can scan the code, allowing them access to video of the campaign, starring Lara Stone, A. J., Sid Ellisdon, Grayson Vaughan and Eric Anderson; after watching the video, users can share the QR code through Facebook and Twitter [40].

Durex, with its baby campaign for iPhone, uses QR codes for advertising in which it tries to explain how hard it can be to have a baby, especially when it comes to an unwanted pregnancy and how the use of condoms can relieve from such a heavy "burden" [44].

Layar from Firefox is an "augmented reality browser available for iPhone and Android operating system smartphones from all countries of the world" [45]. It is an application that identifies the real image of the user's position (which is detected by a GPS receiver) and combines it with the compass incorporated in the devices to know where the user is looking or heading. In the process, the smartphone camera captures and reproduces the image on the screen and, at the same time, the software superimposes on that image information related to what is seen on the screen; that is, if the user points his camera at a historic building, extra information related to what is being seen automatically appears on the screen [45].

3.4 Augmented Reality Tools and BIM (Building Information Modeling) Methodology

Technology evolves rapidly and constantly, playing a very important role in our society, as it optimizes resources, time and money. In civil engineering, the most current example is the use of Building Information Modeling (BIM) [46], a collaborative methodology to design, build and manage more efficiently. The model created with BIM can be linked to Augmented Reality (AR), which allows user interaction with it. The AR superimposes that 3D model to reality to increase the information.

BIM is a collaborative work methodology in which projects are created and managed through a multidisciplinary digital 3D model. This model forms a large database with information, such as the geometry of the object, physical data, cost, execution time, data on maintenance.

Augmented Reality (AR) is a technology that improves the visualization of elements and their understanding and even allows interaction with them.

When working with BIM, a 3D model is created that can be used by the AR and it facilitates the visualization and compression of any model, mainly, the design, construction, operation and maintenance processes.

In the literature we found different Augmented Reality Software, such as:

- ARki; AUGmentecture Aumentaty; UniteAR; WakingApp; Zappar
- Unity (allow interaction with BIM information)
- Unreal Engine (allow interacting with BIM information).

An Augmented Reality system basically consists of five hardware components: a central unit equipped with image mixing and rendering software; a screen-based display system or HMD device; a system of location, follow-up and tracing (tracking); the camera's recording sensors; and, finally, additional input devices. The main problem of AR applications, in general, lies in the exact superposition of the digitally generated image on the image of the real visual world of the observer, that is, the system must recognize the exact position of the person and the direction of the gaze (visual ray) before the synthetic images can be juxtaposed [47].

In the field of Augmented Reality, there are basically two different visualization methods: the so-called optical see-through (OST) and the so-called video see-through (VST). The difference between them is that in the first case the screen of the device is transparent and the digital overlay is projected directly onto the direct view of the user, while in VST devices the view of the device consists of a real-time video source. Captured by a camera located on the HMD device and then projected onto the screens installed on the display device. In this case and although the image corresponds in scale, position and perspective with the vision of the real world [47].

In addition to the above, other methods can be mentioned: conventional screen-based systems (Monitor Based AR) that we can consider a variant of the VST, but without the need for HMD devices, which is why it is of great interest in applications related to teaching or the so-called Spatial Augmented Reality (Spatial Augmented Reality) also called Projective (Projective AR) based on the projection of digital images on a real environment [47].

3.5 Classification of Augmented Reality (AR)

According to various authors [48, 49], different RA levels are classified depending on the type with different RA levels depending on the type of interactivity type:

- Level 0: QR codes. They are hyperlinks that take us to Web spaces or provide us with information in the form of text, sound, etc.
- Level 1: Augmented reality with markers. It is the most used and uses images as a link element to obtain the augmented element.
- Level 2: At this level is geolocated augmented reality. The development of geolocation devices allows the creation of an augmented reality in a specific situation.
- Level 3: Level in which the use of augmented reality is found thanks to the use of HDM devices such as Hololens.

Another level that we find in the study of augmented reality [50]. It consists of the creation of new models of human–computer interaction. This line of research can be applied to people with communication problems, disabilities or degenerative diseases such as Alzheimer's. Identifying a series of patterns and acting through the indications provided by the device.

The bibliography consulted [42], allowed to describe the augmented reality process, for it to be possible it is necessary:

- A device that takes the actual images that users observe.
- A device on which the combination of real and virtual images is projected.
- A processing device to interpret the information from the real world and generate the virtual information to combine it.
- An augmented reality activator, for example, GPS, QR codes, among others. The latter currently integrated into any Smartphone device, ablet, laptop, some cameras and glasses such as the Hololens.

The concatenation of these processes results in a system with the following characteristics, which define augmented reality:

- Combine real and virtual objects in new integrated environments.
- The signals and their reconstruction are executed in real time.
- The applications are interactive.
- Real and virtual objects are registered and geometrically aligned with each other and within space, to give them spatial coherence (UNAM).

3.6 Types of Augmented Reality

Augmented reality based on the recognition of patterns or marks: uses markers, which can be symbols or images that are superimposed when a specific software recognizes them. "The Augmented Reality software tracks the pattern or marker, allowing the position of the 3D model that appears on the screen to be adjusted when we move or rotate it" [51].

Augmented reality based on the recognition of markeless images: use images of the environment as triggers to place virtual content on them. "Instead of patterns or markers, the activating element is the image itself and therefore there are no intrusive elements in the scenes" [51].

Augmented reality based on geolocation: in this type of AR, a point of interest is located and displayed on the device screen.

According to the bibliography consulted [52], the following Classification of Augmented Reality supports is presented.

There are three ways to present Augmented Reality technology, with the traditional computer, with miniaturized portable devices and with specific augmented reality equipment.

Management of Augmented Reality in traditional computer. The camera connected to the computer equipment digitizes the captured image, incorporating layers with other data, still or moving images, texts and/or sounds. This information is added from software that selects specific documents from its own database or from access to data on the Internet. A screen presents the synthesis, shows in a synchronized way both the real images captured, along with the rest of the overlapping data synchronized in size, position and in real time. The virtual elements are intended to be precisely coordinated with the real objects and their position, since a small orientation error causes a perceptible mismatch.

Augmented Reality Management on laptop. Advanced mobile devices are small mini computers that incorporate mobile imaging cameras. With this, the augmented reality effect can be reproduced and the result of Augmented Reality displayed on the device screen. In this case, information can be added from the GPS geographic positioning systems, necessary to be able to precisely locate the user's situation.

Augmented Reality Management with specific equipment. Sometimes there are specific capture and vision devices that are integrated into special glasses and allow the user to see reality through the lens and to superimpose and display other graphic information.

3.7 Classification of Augmented Reality Technologies Developed

Augmented Reality works based on the superposition of information on reality from three basic technological resources that sometimes complement each other: the software trigger patterns, geo-location and interaction with the Internet [52, 53].

Software RA shooting patterns. Certain images can be the digital spring that launches the operation of an application on that same image. That is, a brand, a specific drawing or image, an iconic bar or digital code can be triggers for the image, video, text, sound and/or Internet link, incorporated into the captured image.

Geo-location. The possibility of Global Positioning System (GPS) detection in portable devices places the user anywhere on Earth. Thus, thanks to the calculation of the distance relative to geostationary satellites, a captured image can be located with a precision of a few meters at a certain height, direction, longitude and latitude. Once the place is known, a spatial reference is superimposed on the image: north, south, east and west, which can be related to a stored map and indicate nearby places.

Interaction with the Internet. There are applications that link the captured images with similar ones in Internet databases. The team looks for the essential elements of the real image and overlays the hyperlinks with complementary information from the Internet.

3.8 Hardware and Software Components Necessary for the Implementation of an Augmented Reality System

Augmented reality systems require hardware components on which the desired actions are to be carried out [54].

The basic elements necessary to implement an augmented reality system are similar to those of any traditional computer system. Thus, components for processing, storage, communications, etc. are necessary. In fact, the difference in the basic equipment is in the purpose of the system.

Consequently, the hardware elements necessary for any augmented reality system are: a camera, a display screen or holographic equipment in 2D or 3D, a processing and calculation equipment, storage components and, finally, transmission and reception devices of data in case they are necessary. It should not be forgotten that the benefits of each of them depend fundamentally on the type of system to be implemented [54]. Table 2 contains a summary of the common hardware components in an augmented reality system.

Table 2 Summary of the hardware components of an augmented reality system

Technique	Hardware component
Capture video-through	All kinds of video cameras
See-through capture	Integrated video cameras, HMD (Head Mounted Display)
Geo-positioning recognition	GPS antenna, motion sensors
Hybrid recognition	GPS antenna, motion sensors, image capture devices
Image treatment	Processors with processing speed of at least 2 GHz
Storage	Conventional hard drive
Local communications	Network card, RJ45 connectors, wireless antenna, access point
Mobile communications	GSM equipment
Video-through display	Video screens, monitors, projectors
See-through display	HMD, mobile phones

Table 3 Summary of the software requirements of an augmented reality system

Technique	Software component
Capture scene	Video camera drivers
Visual recognition	Image recognition libraries
Geo-positioning recognition	Digital compass, GPS software
Hybrid recognition	Digital compass, image recognition libraries, GPS software
Image treatment	Image processing libraries
Storage	Database
Local communications	Network Driver
Mobile communications	GSM controller
Content display	Multimedia content playback software, image processing libraries

Once the hardware components have been described, the necessary software to implement an augmented reality system will be exposed. As happens when choosing the right hardware, the decision to use one software or another is determined by the performance and characteristics of the system.

In general terms, an augmented reality system needs recognition software, either by positioning or by images (with or without markers), image processing libraries (if needed), at least one database, libraries of increase of digital content and, if necessary, communications software. Table 3 shows a summary of the generic software required for each stage [54].

It should be noted that most image processing, like ARToolkit, Java3D, libreries implement functions for the recognition, processing and manipulation of visual content.

3.9 Application Areas of Augmented Reality

Augmented Reality has been applied in numerous domains and in different types of application, whether it is a descriptive application, such as for prediction or classification. The Fig. 3 summarizes the analysis carried out in this regard.

From the analysis carried out, it is identified that the area with the greatest number of applications in the period 2000–2017 was associated with the environment, medicine and energy sectors and projects. However, the trend in the last five years (2017–2021) is concentrated in the development of projects associated with medicine, engineering, science and social sciences.

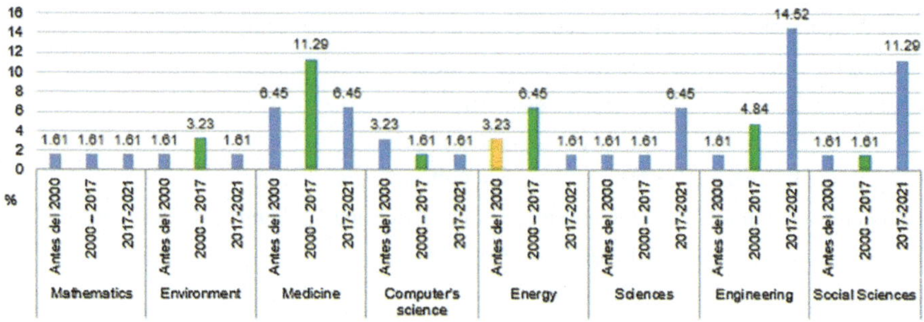

Fig. 3 Analysis of publications regarding application environments

4 Conclusions

In this work, a review of the state of the art of augmented reality has been carried out from the following perspectives: application areas and their evolution, classification of Augmented Reality (AR) technologies developed, classification of AR supports, types of AR, AR classification, AR-based applications.

Augmented reality has become an innovative trend applicable to countless areas. In recent years there has been a growth in the application of augmented reality, in many areas, such as commercial, scientific, educational, tourism, automotive and social. The foregoing allows us to speak of a technological revolution that provides an expansion in the fields of application of augmented reality, as well as its use by a large number of users.

It can be determined that there are still limitations for the use of augmented reality, such as its high prices and necessary technological developments that are tied to the use of infrastructure and networks for optimal use.

Augmented Reality has generally been used in a descriptive way, the most exploited application areas being medicine, engineering, energy, science and social sciences. However, as of 2017 there is a significant increase in applications in engineering and social sciences. The use of Augmented Reality in decision-making and project management is identified as an open line for research.

References

1. Kirkley, J.K.: Creating next generation blended learning environments using mixed reality, video games and simulations **49**, 42–53 (2004)
2. Altomari, A.P.: Realidad virtual y realidad aumentada en la educacion, una instantanea nacional e internacional. Econ. Creat. **7**, 34–65 (2017)
3. Mizell, D.: Augmented reality: an application of head-sup display technology to manual manufacturing processes. En Proc. Twenty-Fifth Hawaii Int. Conf. Syst. Sci. (1992)
4. Akçayır, G.M.: Advantages and challenges associated with augmented reality for education: a systematic review of the literature. Educ. Res. Rev. **20**, 1–11 (2017)

5. Pallavicini, F., Pedroli, E., Serino, S., Riva, G.: Augmented reality: a brand new challenge for the assessment and treatment of psychological disorders. Comput. Math. Methods Med. 1–12 (2015)
6. Azuma, R.: A Survey of Augmented Reality.Presence: Teleoperators and Virtual Environments. **6**(4), (1997)
7. Rodríguez-Jiménez, C., Ramos, M.: Virtual reality in physical education area. J. Sport Health Res. **11**, 177–186 (2019)
8. Ibáñez, M.B., Delgado, C.: Impact of an augmented reality system on students' motivation for a visual art course. Comput. Educ. **68**, 586–596 (2013)
9. Pacheco, A.: The development of augmented reality systems in informatics higher education. Procedia Comput. Sci. **25**, 179–188 (2013)
10. Dieker, L.: Mixed reality environments in teacher education: development and future applications. Online Real Time Using Web 20 for Distance Educ. Rural Spec. Educ. 116 (2016)
11. Marquez, I.: La simulación como aprendizaje: educación y mundos virtuales. II Congr. Int. Comun. **3** (2010)
12. Ye, Y.: Design and implementation of digital art teaching system based on interactive virtual technology. Int. J. Emerg. Technol. Learn. IJET **11**, 49–54 (2016)
13. Wu, H.K.: Current status, opportunities and challenges of augmented reality in education. Comput. Educ. **62**, 41–49 (2013)
14. Howe, C., McCredie, N., Robinson, A., Grover, D.: Augmented reality in education—cases, places and potentials, in *IEEE 63rd Annual Conference International Council for Educational Media (ICEM)*, pp. 1–11 (2013)
15. De La Torre, J.L.: Aplicación de tecnologías gráficas avanzadas como elemento de apoyo en los procesos de enseñanza-aprendizaje del dibujo, diseño y artes plásticas (2013)
16. Alvarez-Marin, A.: Realidad Aumentada como Apoyo a la Formación˜de Ingenieros Industriales. Form.˜universitaria **10**, 31–42 (2017)
17. Durall, E.: Perspectivas tecnológicas: educación superior en Iberoamérica (2012)
18. Hernández-Sampieri, R., Torres, C.P.: Metodología de la investigación. McGraw-Hill Interamericana México DF, vol. 4 (2018)
19. Littell, J.H., Corcoran, J., Pillai, V.: Systematic Reviews and Meta-Analysis. Oxford University Press (2008)
20. Kacprzyk, J., Zadrożny, S.: Linguistic database summaries using fuzzy logic, towards a human consistent data mining tool (20), 10 (2009)
21. Xiao, D., Jakimowicz, J.J., Albayrak, A., Buzink, S.N., Botden, S.M., Goossens, R.H.: Face, content and construct validity of a novel portable ergonomic simulator for basic laparoscopic skills **71**(1), 65–72 (2014)
22. Aguilar, E., Vivas, A., Sabater, J.: Una Aproximación a la Realidad Aumentada y sus Aplicaciones Quirúrgicas. Entre Cienc. E Ing. **12**(24), 15–24 (2018). https://doi.org/10.31908/19098367.3811
23. Ruiz, S.: Enseñanza de la anatomía y la fisiología a través de las realidades aumentada y virtual. Innov. Educ. México **19**, 57–76 (2019)
24. Rabbi, I., Ullah, S., Khan, S.U.: Augmented reality tracking techniques—a systematic literature. IOSR J. Comput. Eng. IOSRJCE **2**(2), 23–29 (2012)
25. Harris, A., Bello, F., Kneebone, R.: Simulation and training in minimal access surgery. Train. Minimal Access Surg. 35–47 (2015) (Springer)
26. Jeon, S., Choi, S., Harders, M.: 10 haptic augmented reality. Fundam. Wearable Comput. Augment. Real 227 (2015)
27. Caber, J.: Aplicaciones de la Realidad Aumentada en educación. Rev. Educ. Mediática TIC **6**(1), 4–8 (2017)
28. Rodríguez, A.M.: Diseño e implementación de una experiencia para trabajar la interculturalidad en Educación Infantil a través de realidad aumentada y códigos QR. Educar **55**(1), 59–77 (2019)
29. Vincenzi, M.A.: La Realidad Aumentada en la Educación. Catalogación de Aplicaciones Educativas (2019)

30. Blázquez, A.: Realidad aumentada en educación. Rey Juan Carlos (2019)
31. Ortiz, E.: La realidad aumentada a través de una secuencia didáctica digital. Prax. Educ. ReDIE **21**, 104–120 (2020)
32. Villalustre, L., del Moral, M.: Percepción docente sobre la realidad aumentada en la enseñanza de Ciencias en Primaria. Análisis DAFO. Rev. Eureka Sobre Enseñ. Divulg. Las Cienc., **16**(3). https://doi.org/10.25267/Rev_Eureka_ensen_divulg_cienc.2019.v16.i3.3301 (2019)
33. Roig, R., Lorenzo-Lledó, A., Mengual, S.: Utilidad percibida de la realidad aumentada como recurso didáctico en Educación Infantil. Campus Virtuales **8**(1), 19–35 (2019)
34. Angarita, J.: Apropiación de la realidad aumentada como apoyo a la enseñanza de las Ciencias Naturales en educación básica primaria. Lenguaje, formación y cultura. Rev. Bol. Redipe 7(12) (2019)
35. Alvarado, Y., Jofré, N., Rosas, M., Guerrero, R.: Aplicaciones de Realidad Virtual y Realidad Aumentada como soporte a la enseñanza del Dibujo Técnico. Enseñ. Aprendiz. Ing. Comput. **9** (2019)
36. https://play.google.com/store/apps/details?id=com.UNSL.SR
37. https://unity3d.com/es
38. https://www.vuforia.com
39. http://librecad.org
40. Melo, I.M.: Realidad aumentada y aplicaciones. Real. Aumentada Apl. **6**(1), 28–35 (2018)
41. Cárdenas, H.A.: Realidad aumentada (RA): aplicaciones y desafíos para su uso en el aula de clase. Educ. Ciudad **35**, 137–148 (2018)
42. Rigueros, C.: La realidad aumentada: lo que debemos conocer **5**(2), 257–261 (2017)
43. Ierache, J., Igarza, S., Nahuel, A., Mangiarua, E., et al.: Herramienta de Realidad Aumentada para Facilitar la Enseñanza en Contextos Educativos Mediante el Uso de las TICs. Rev. Latinoam. Ing. Softw. **2**(6), 365–368 (2014)
44. Ruiz, S., Polo, D.: La realidad aumentada como nuevo concepto de la publicidad a través de los Smartphone. Razón Palabra 80 (2012)
45. Layar: Augmented reality: layar reality browser. Recuperado Httpwww-layarcom (2011)
46. de Diego, F.D., Hindi, R., Lozano, J.A.: Aplicación de herramientas de Realidad Aumentada a modelos de puentes elaborados con BIM. Hormig. Acero Av. (2020). https://doi.org/10.33586/hya.2020.2171
47. Fernández, A.J.: De Las Arquitecturas Virtuales A La Realidad Aumentada: Un Nuevo Paradigma De Visualización Arquitectónica. APEGA X Congreso Internacional Expresión Gráfica aplicada a la Edificación. Graphic Expression applied to Building International Conference, pp. 111–120 (2010)
48. Lens-Fitzgerald, M: Augmented Reality Hype Cycle (2009)
49. Fombona, J., Pascual, M.J., Madeira, M.F.: Realidad aumentada, una evolución de las aplicaciones de los dispositivos móviles. Rev. Medios Educ. **41**, 197–210 (2012)
50. Schmorrow, D., Stanney, K.M., Reeves, L.M.: Foundations of augmented cognition: Augmented cognition—past present and future. Analsysis Inc., Strateg (2006)
51. Muñoz, A., Ortega, A., Rodríguez, A., Carrillo, A., Ortiz, C., López, C.: Recursos educativos aumentados una oportunidad para la inclusión. Cartagena Indias Sello Editor. Tecnológico Comfenalco (2016)
52. Cadavieco, F., Sevillano, M., Ferreira, M.F.: Realidad Aumentada, Una Evolución De Las Aplicaciones De Los Dispositivos Móviles. Pixel-Bit Rev. Medios Educ. **41**, 197–210 (2012)
53. Contreras, M.S., Barrios, E.M., Rodriguez, J.C.: Reconocimiento y rastreo de imágenes en aplicaciones de Realidad Aumentada. **40**(34), 7 (2019)
54. Lópe, H.: Análisis Y Desarrollo De Sistemas De Realidad Aumentada. Máster en Investigación en Informática Facultad de Informática Universidad Complutense de Madrid (2010)

Chapter 20
Integration of Augmented Reality and BIM in Early Stages to Improve Design Management in Engineering and Construction Projects

Alexis Ibarra-Araya, Boris Heredia-Rojas, Luis Alvarado-Acuña, and Ingrid Jamett-Aranda

Abstract Thanks to the significant growth of information technologies, several tools allow improving skills and coordination between different project actors, among which augmented reality (AR) and building information modeling (BIM) stand out. This research aims to verify current technological advances that allow integrating these tools within a design process. Firstly, from initial bibliometric analysis, a conceptual model is obtained that incorporates AR and BIM technologies within the design management process to be applied in engineering and construction projects. Secondly, to validate the relevance and applicability of the conceptual model, the Delphi methodology is applied with experts from different industries and experience in early-stage project design. This study determines the state of the art in using BIM and AR in Architecture, Engineering and Construction (AEC) industry through the methodologies used. In addition, it is possible to obtain a preliminary diagnosis of how these technologies are contributing to the industry. Furthermore, the results obtained confirm that AR greatly enhances the visualization of the BIM confederate models. The interviewed experts agree that adding virtual elements in the real world provides a better vision of the project, communicating effectively between the different project actors, phases and processes.

Keywords Augmented reality · Building information modeling (BIM) · Design management · Integration

---

A. Ibarra-Araya · B. Heredia-Rojas (✉) · L. Alvarado-Acuña
Universidad Católica del Norte, Av. Angamos 0610, Antofagasta, Chile
e-mail: bheredia@ucn.cl

I. Jamett-Aranda
Universidad de Antofagasta, Av. Universidad de Antofagasta, 02800 Antofagasta, Chile

P. Y. Piñero Pérez et al. (eds.), *Artificial Intelligence in Project Management and Making Decisions*, Studies in Computational Intelligence 1035,
https://doi.org/10.1007/978-3-030-97269-1_20

1 Introduction

Design management is a critical stage in which the ideas and thoughts of the client are conceptualized, which will later become the final objectives of a project [1]. The project's success will be evaluated according to the guidelines marked in this stage, becoming a fundamental process to define the absolute requirements that will allow a project to be completed successfully, both for the client and for all interested parties.

On the other hand, the continuous advancement of technologies provides tools that improve processes and automate tasks. Thus, companies must adapt their methods to these technological changes to stay competitive in their markets.

Augmented Reality (AR) is a new technology that allows the merging of virtual elements in the real world [2]. AR has a potential advantage when complementing with Building Information Modeling (BIM) for the development and management of projects, generating a significant impact by visualizing projects in early stages, reducing uncertainties and possible failures and interferences between areas or disciplines.

Currently, several factors affect the coordination between different areas and disciplines of a project. The lack of communication and well-defined processes at the beginning of a project are recurrent issues [3]. The use of BIM supports the collaboration and traceability of the project in all areas [4]. However, the visualization of the parts of a project is done using 3D models on flat screens; although it contributes enormously to visualization, it fails to deliver the dimensionality factor to the project. In addition, there is no integrated vision of the project that allows all participants to integrate their image in a single environment in a simple way that allows better coordination between the different areas. Suppose coordination between areas/disciplines is not improved in the early stages and we do not implement management processes defined at the beginning of the project. In that case, problems will likely continue to arise, which will negatively impact delivery times and the cost of the project, which will continue to increase. Also, the work environment could deteriorate and likely cause the loss of key personnel and widespread project failure.

This research proposes to investigate how design management is related to BIM and AR integration in engineering and construction projects. Verifying its advantages and disadvantages and measuring its integration in the early stages of projects will be valuable for companies interested in incorporating this new model to manage their projects effectively. Although, at first glance, the advantages are varied, a possible combination of BIM and AR can improve communication between areas/disciplines of a project, encourage collaboration between areas and improve interfaces through the visualization of the finished project. Continuous improvement is the basis for increasing the competitiveness of companies and technology has undoubtedly come to revolutionize the industry in general. With the incorporation of AR and BIM, there is an excellent potential for improvement in the management of project design process that can be exploited. In addition, it is proposed to use the ISO 9001 standard in item 8.3, which has already established processes that allow better plan and control of projects at the beginning and improve the design process [3].

This research contributes to centralizing information and knowledge of the potential of these tools, seeking to identify methods that could improve coordination between areas and disciplines in engineering and construction projects. Therefore, this research has a practical justification; a specific conceptual framework aims to solve a problem in a particular area for effective project management.

2 Literature Review

In recent years, there has been an increase in engineering and construction projects developed by multidisciplinary teams with specific technical knowledge, where proper coordination of these is vital to generate successful designs in the early stages that allow meeting the customer's expectations, reducing future design changes.

According to the results obtained from the study carried out by Yap et al. [4], 39 reasons for design changes were identified during the execution of the projects. Among the leading ten reasons are the lack of coordination between professionals from different disciplines, changes in the requirements and specifications, omissions and increase in the project's scope, discrepancies and errors in the design documents, omissions, incomplete drawings, etc. Due to the abovementioned, projects become increasingly complex as they have different interconnected disciplines, presenting problems mainly related to the lack of interaction between them, causing wrong estimates and rework that translate into higher costs at the end of the project [5].

To solve these problems, the need to implement technological tools that allow integrating the different areas of a project is born, one of these in BIM, which is a platform that integrates architecture, engineering and construction. This tool is an information base that allows, together with the three-dimensional modeling of models, to associate the parametric information to each of the elements and components that make up the project. As a result, it allows integrating the relevant information from different specialties in a single source through the construction of a centralized model that improves coordination and communication between all the actors, reducing failures and providing better details of the work.

BIM is a concept born in the late 70 s by Professor Charles Eastman [6]. Since its development, it has expanded to various perspectives such as design, construction process estimation, life cycle, performance and technology. One of the significant changes occurred in 2000 when the ArchiCAD company developed Revit. This software creates a parametric engine that allows adding time dimension [7].

According to BIM Forum Chile [8], intercommunication between the different areas of the team is one of the essential characteristics, so it is critical to creating this environment through formal channels that allow a fluid exchange of information and collaboration. However, its use brings many challenges, mainly a cultural change in the organization, thus creating a culture of cooperative work.

One of the characteristics of the BIM methodology is its fourth dimension (4D), which allows analyzing and controlling the progress times in the construction of a project. BIM gives team members the ability to control the temporary progress of project activities from the stage of the design and this allows to acquire significant advantages, such as optimization of the times, detection of errors and planning. Although this tool has great potential in its use, the study by Boje et al. [9] shows that the adoption rate in developed countries remains low, mainly due to a low diffusion of the tool. This point agrees with Romigh et al. [10], who carried out a qualitative study on the use of BIM 4D during the construction phase concerning the visualization and communication of planned information at the site of the operation. As a result, 67% of the interviewees were familiar with the tool and only 33% used 4D in their projects. These interviews also suggest that some participants highlighted that 4D is used mainly as a marketing tool for BIM. In general, they are not interested in upgrading to 4D models during the execution of projects that are constantly changing.

Although BIM 4D has many advantages, its use is not very well spread in the industry, so it is interesting to promote its use.

In recent times, another tool echoing in the AEC industry is AR. AR is a visualization technology that allows virtual elements to be superimposed on the real world [2]. Its use allows physically interacting with virtual design concepts on an accurate scale, like interacting with the real physical element—the results in a theoretical opportunity where it can be used to enhance the understanding of different requirements. With the help of the superposition of components, a better understanding of all the parts that make up some aspects of the project is achieved [11].

Its use has many advantages, specifically in the AEC industry. According to the study by Noghabaei et al. [12], it is predicted that within the next 5–10 years, it will be used in most projects. Due to this, it is interesting to see how the integration of both tools, BIM and AR, could result in improving coordination in the early stages between different teams from different areas.

This study initiates with a systemic review of the current literature, which defines the application and integration of BIM and AR technologies. As goals, this literature review finds out about: 1. Verify the advances of BIM and AR in the AEC industry; 2. Identify tools that facilities BIM and AR integration.

According to these points, information sources are defined from Web of Science (WoS). The defined search period is between 2010 and 2020. 'BIM AND BIM' is defined as the first search criteria. With this criteria, 9185 results related to BIM are subsequently obtained from the set of articles found. Then, new search criteria, 'AR' is entered, resulting in 110 entries.

To further narrow the search, new filters are applied, specifically in the WoS categories, specifically in the following fields: civil engineering; construction building technology; industrial engineering; management; multidisciplinary engineering; computer science engineering and information systems. As a result of the search, 90 investigations were identified, of which most of the results are in the areas of civil engineering, construction building technology and multidisciplinary engineering.

Before of a detailed analysis, the scientific articles found were cleaned. The objective was to find articles relevant to the investigation that provides information regarding BIM and AR integration. For this, 57 articles were grouped in a pivot table and classified according to their main topic: AR + BIM; and AR and BIM in engineering and construction projects (see Table 1).

Table 1 Relevant studies from the literature review

Year	Title	Author(s)
2020	1.BIM-based augmented reality inspection and maintenance of fire safety equipment	Chen et al
	2.From BIM to extended reality in AEC industry	Alizadehsalehi et al. [4]
	3.Conceptual framework for integrating BIM and augmented reality in construction management	Machado and Vilela
	4.State-of-the-art review on mixed reality applications in AEC industry	Cheng et al
	5.Technology acceptance model for augmented reality and building information modeling integration in the construction industry	Elshafey et al
	6.Augmented and virtual reality in construction: drivers and limitations for industry adoption	Delgado et al. [18]
	7.A research agenda for augmented and virtual reality in architecture, engineering and construction	Delgado et al. [18]
2019	1.(BARMS) as an intelligent instruction platform for complex plumbing facilities	Diao and Shih
	2.Indoor localization for augmented reality devices using BIM, point clouds and template matching	Herbers and Konig
	3.Collaborative welding system using BIM for robotic reprogramming and spatial augmented reality	Tavares et al
	4.Augmented reality system for facility management using image-based indoor localization	Baek et al
	5.Impact of augmented reality and spatial cognition on assembly in construction	Kwiatek et al
	6.Remote interactive collaboration in facilities management using BIM-based mixed reality	El Ammari and Hammad
	7.BIM-based and AR application combined with location-based management system for the improvement of the construction performance	Ratajczak et al
	8.Dynamic BIM-augmented UAV safety inspection for water diversion project	Liu et al
	9.BIM integration in augmented reality model	Saar et al
	10.A BIM-based location aware AR collaborative framework for facility maintenance management	Chen et al
	11.Development of BIM-AR-based facility risk assessment and maintenance system	Wang and Piao
	12.Augmented reality for identifying maintainability concerns during design	Khalek et al
	13.Leveraging mobile augmented reality devices for enabling specific human behaviors in design and constructability review	Alsafouri and Ayer

(continued)

Table 1 (continued)

Year	Title	Author(s)
	14.Quantitative evaluation of overlaying discrepancies in mobile augmented reality applications for AEC/FM	Gomez-Jauregui et al
	15.A web-based BIM-AR quality management for structural elements	Mirshokraei et al
	16.Trends and research issues of augmented reality studies in architectural and civil engineering education-a review of academic journal publications	Diao and Shih
	17.Mobile augmented reality to influence design and constructability review sessions	Alsafouri and Ayer
	18.Design assessment in virtual and mixed reality environments: comparison of novices and experts	Wu et al
	19.Effect of varying task attributes on AR aided point layout	Chalhoub and Ayer
	20.Use of augmented reality technology to enhance comprehension of construction assemblies	Bademosi et al
	21.Exploring the performance of an augmented reality application for construction layout tasks	Chalhoub and Ayer
	22.A review on using opportunities of augmented reality and virtual reality in construction project management	Ahmed
2018	1.Integrating mobile building information modelling and augmented reality systems: an experimental study	Chu et al
	2.An interactive augmented reality tool for constructing free-form modular surfaces	Fazel and Izadi
	3.Review of ICT implementations for facilitating information flow between virtual models and construction project sites	Alsafouri and Ayer
	4.Using mixed reality for electrical construction design communication	Chalhoub and Ayer
	5.Improvement of realism of 4D objects using augmented reality objects and actual images of a construction site	Kim et al
	6.A systematic review of augmented reality content-related techniques for knowledge transfer in maintenance applications	del Amo et al
	7.Mobile augmented reality applications for construction projects	Zaher et al
	8.Characteristics of models that impact transformation of BIMs to virtual environments to support facility management operations	Zou et al
2017	1.Information modeling for virtual and augmented reality	Patti et al
2016	1.Investigating human and technological requirements for successful implementation of a BIM-based mobile augmented reality environment in facility management practices	Gheisari et al
	2.Integrated collaborative tools for precast supply chain management	Abedi et al

(continued)

Table 1 (continued)

Year	Title	Author(s)
2015	1.Bridging BIM and building: from a literature review to an integrated conceptual framework	Chen et al
	2.BIM2MAR: An efficient BIM translation to mobile augmented reality applications	Williams et al
	3.A critical analysis of building information modelling systems used in construction projects	Abanda et al
	4.Measuring the potential of augmented reality in civil engineering	Meza et al
	5.Mobile augmented reality for building maintenance	Woodward et al
	6.Using augmented reality in a multiscreen environment for construction discussion	Lin et al
	7.Using augmented reality to facilitate piping assembly: an experiment-based evaluation	Hou et al
2014	1.A defect management system for reinforced concrete work utilizing BIM, image-matching and augmented reality	Kwon et al
	2.Integrating augmented reality with building information modeling: onsite construction process controlling for liquefied natural gas industry	Wang et al
	3.Building Information Modeling (BIM) application framework: The process of expanding from 3D to computable nD	Ding et al
	4.Component based engineering of a mobile BIM-based augmented reality system	Meza et al
2013	1.A conceptual framework for integrating building information modeling with augmented reality	Wang et al
	2.Towards cloud Augmented Reality for construction application by BIM and SNS integration	Jiao et al
	3.A framework for proactive construction defect management using BIM, augmented reality and ontology-based data collection template	Park et al
	4.A framework for construction safety management and visualization system	Park and Kim
	5.Augmented reality supporting user-centric building information management	Olbrich et al
2012	1.On-site building information retrieval by using projection-based augmented reality	Yeh et al

Then, an analysis of the results obtained with the VOSviewer tool is carried out. This tool is a social network analysis software that allows the creation, visualization and exploration of maps based on bibliometric data networks, displaying clusters of information that support the classification of results [13].

The keyword co-occurrence analysis was carried out with all the selected investigations. For the study, a minimum of five words was configured to consider a link between the terms. The analysis resulted in a total of 24 keywords out of a total of 421. It resulted in a total of three different clusters.

The co-occurrence analysis shows a strong relationship between these two technologies, where it turns out that, of the 90 investigations processed, 37 articles relate these terms. In other words, there has been a growth in research that addresses the possibility of integrating these tools in the last five years. When reviewing the publications, the trend of works published to integrate this tool is growing. The decreased number of publications found is because it is an emerging technology that is in the process of development that has been increasing in recent five years. It is interesting to note that publications of this nature have been somewhat reduced during this year, which could be the effect of the global pandemic that affected the planet.

To answer one of the research questions about the tools that allow this integration, of the articles reviewed, 75% use Autodesk Revit, one of the oldest and most reliable software on the market [14].

For BIM Integration with AR, Unity 3D shows a significant boom in recent times in the research reviewed. The works of Ratajczak et al. [14] and Getuli et al. [15] obtained favorable results with the use of this tool. At the end of 2019, Unity launched an integration plugin with Revit called Unity Reflex, making importing models and displaying them in a virtual environment much more manageable. It can be processed by Unity Pro and generate an AR environment. The latter must be accompanied by a tracking method that inserts the model into the real world. This tool helps connect design and construction, connecting multiple BIM models with metadata to one view of real-time visualization information.

Unity Reflex keeps all the workflow in the same line, improving the lines of communication and iterations of change. Also, it leads to enhanced decision-making in the project development stages, creating a collaborative environment in real-time, integrating the best tools of BIM and consolidating its visualization in a single work environment.

It is important to note that this tool is relatively new. To the date of this research, there are no studies on its use. However, solving the integration of AR and BIM quickly and reasonably promises positive results.

3 Development of Conceptual Model

This study proposes to use an integrated BIM and AR tools to improve decision-making in early stages for the management of design in engineering and construction projects, however it is interesting and promising to integrate these concepts with some design methodology that ensures a reduction in errors. forced and set a trend and innovation in obtaining requirements, their review, their validation and approval by all stakeholders involved in the project.

The ISO Standards present different aspects of quality management providing guidance and tools for companies and organizations that want to ensure that their products and services meet customer requirements where quality is constantly improved. The ISO 9001:2015 establishes the requirements for a quality system as a set of interrelated elements of an organization that interact to define policies, objectives and processes to achieve those objectives [3]. As a framework for the integration of BIM and RA, this study uses item 8.3 of design and development management to establish, implement and maintain an adequate design and development process to ensure the subsequent provision of required products and services.

As a result of the systematic literature review, the creation of a model is obtained to be used as input for the design and development of engineering projects, this process will have to go through a validation and verification cycle in which it proposes to improve decision-making and refine the client's requirements for a correct management of elements that will be part of the project, allowing a transversality of all the areas involved. The conceptual model is presented in Fig. 1.

The proposed model begins with the obtaining of the requirements by the client, which will then be used as inputs for the initial design and development process. This survey carries out using Autodesk Revit software where the different stakeholders associated with the project areas involved in its development. This stage does not require high polygon modeling (LOD1, LOD2) because at the beginning of this a scheme will be used general for the approval of the requirements. Then, as a second stage, we proceed to the creation of the virtual AR environment using the integration with Unity Reflex in Unity Pro.

This process generates an AR viewer as an output that can be used as an initial tool for verifying the client's requirements in conjunction with the different areas. The verification process allows everyone involved to verify the changes in real time on the construction site and the final visualization of the different stages of the project.

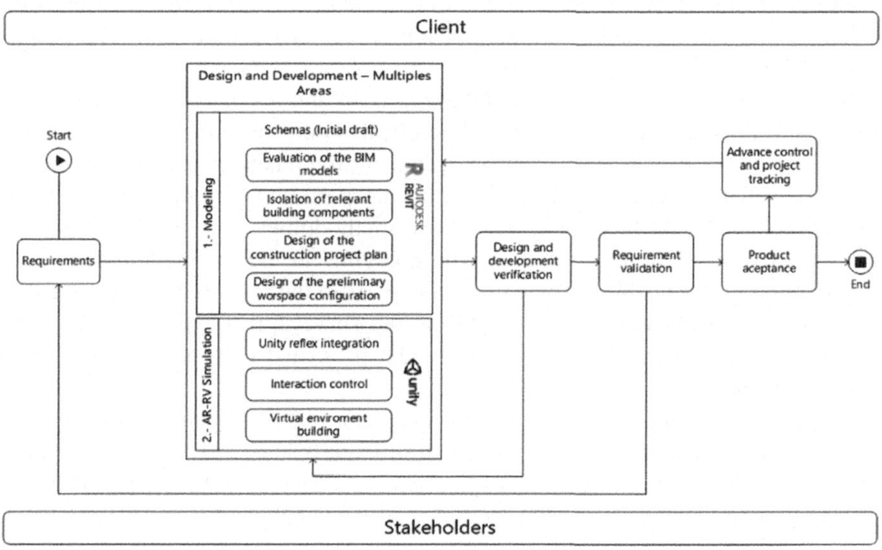

Fig. 1 BIM and AR integration model in the design process

The interaction of the parties and possible inconsistencies that may be found during the validation can be captured in this stage.

Additionally, the generated viewer has the option of incorporating comments by users at the time of being immersed in it, which remains in the AR environment and will also be stored in a list of comments as exchange controls.

As it is an integrated system, changes can be viewed in real time, which provides flexibility and little rework in the review stage and feedback from customers and stakeholders. After incorporating the changes, the validation process is proceeded which, upon approval, with the lifting of the models with a higher polygonate which will be part of the final project's requirements.

The processes generated in the previous stages serve to project control in the construction stage, so the initial work is a guide for the progress control in the project.

This model is not intended to replace the defined internal processes of an organization, but to be an addition that will allow to enhance the making of requirements and decisions in early stages, reduce communication errors, improve customer participation in the validation of requirements and integrate the different project's areas.

4 Research Design

Two types of methodologies are addressed in this study. Initially, the result is developed under a bibliographic review to determine the subject's relevance for constructing the BIM-AR integrated model. Then, it empirically measures the advances over time, obtaining a global vision of the current state of the issue under research [16].

Subsequently, the conceptual model about BIM and AR integration model proposed is validated, according to the theoretical nature on which the research is based. To coincide with what is investigated by hypothetical reality, it is necessary to consider the suggestions and thoughts of experts and professionals who cover this research field. At this point, the second methodology is inserted to carry out the respective validation, proceeding with applying the Delphi methodology [17]. This method motivates the participation of experts in the investigated topic through different associated iterations. When filling out surveys, this expert's panel steers validates and, if required, modifies the model presented to obtain a more robust model validated.

One of the most important decisions to consider is identifying and selecting potential members who will be part of the panel of experts [17]. For this study, experts with at least five years of project's experience are considered, of an age range that ensures generational diversity and with clear concepts about AR and BIM in AEC industry.

On the other hand, the questionnaire designed in the validation process contains 32 questions to which the study experts are asked to answer individually. To analyze and measure the experts' level of agreement or disagreement, two types of questions are used. The first is based on the 5-point Likert scale (strongly agree, somewhat agree, neither agree nor disagree, somewhat disagree, strongly disagree). The second group consists of open questions, where the expert is invited to express his opinion according to a particular point. This action obtains specific feedback that allows improving the proposed model and envisioning some additional strategy that has not been identified in the study or the proposal for improving the model. Based on the feedback received, the second round of consultation is carried out, where the comments from the previous stage are incorporated. Finally, this process is repeated until when the consensus of all the experts is obtained.

5 Multicriterial Experiment

At the end of the first round, the 32 responses are compiled and grouped by analyzing the experts' answers. Then, the results of the consensus analysis are processed to determine the level of consistency in the responses. To measure the degree of experts' consensus, the method used is the proposed by Hurtado-Delgado et al. [18], which is displayed in the following Eq. (1):

$$Cc = [1 - Vn/Vt] * 100 \tag{1}$$

where:

Cc: the consistency coefficient expressed as a percentage.

Vn: the number of experts in disagreement.

Vt: the total number of experts.

According to Hurtado-Delgado et al. [18], the consistency coefficient is considered acceptable if the value obtained exceeds 70%. The questions below this limit are eliminated due to the experts' low consistency or little consensus. To interpret the first survey results, each question is listed from Q1 to Q32, corresponding to the order they appear in the questionnaire.

All the questions evaluated obtain a coefficient over 70% acceptance. Therefore, it is sufficient to accept the model in general [18]. However, a second round of the study is considered, focusing mainly on the five questions that averaged Cc under 80%, i.e., Q10, Q11, Q12, Q13, Q22 and Q29. The general idea was to rephrase some questions and confirm the score obtained in the first round. The coding and scores of questions under 80% can be seen in Table 2. Subsequently, Cronbach's alpha coefficient is applied to evaluate the consistency of the results, which through a formula allows estimating the reliability of a questionnaire [19]. Reliability is one of the leading technical qualities of measuring instruments linked to the validity of their content. This calculation expresses the internal consistency based on the covariance between the items.

Table 2 Results of delphi round 1

Exp	Questions (Q)																		
	10	11	12	13	14	16	17	18	20	21	22	23	24	25	26	27	28	29	30
1	1	5	5	1	5	5	5	5	3	5	1	1	5	5	3	3	5	3	5
2	4	5	3	5	5	5	5	5	5	5	3	5	5	5	5	5	5	5	5
3	5	5	5	4	4	4	4	5	5	5	5	5	5	5	5	5	5	5	5
4	5	5	5	5	5	5	5	5	5	5	5	5	5	5	5	5	5	5	5
5	5	5	5	5	5	5	5	5	5	5	5	5	5	5	5	5	5	5	5
6	1	3	1	2	2	4	4	5	5	5	5	5	5	3	4	4	5	3	5
7	3	3	5	4	5	3	5	5	5	5	5	5	5	5	5	5	5	5	5
%Cc	71	71	71	71	86	86	100	100	86	100	71	86	100	86	86	86	100	71	100

There are various formulas to obtain the calculation of the coefficient. This study uses the following Eq. (2) [19]:

$$\alpha = k((\sum s - i^2/s - t^2))/(k - 1) \tag{2}$$

where:

k: the number of items in the instrument,

$s - i^2$: the variance of the score on item i.

$s - t^2$: the variance of the total scores of the questionnaire or test.

To facilitate this calculation, statistical software called SPSS is used. The result is a Cronbach's Alpha of 0.885 applied to the 32-questions. Nunnally [20] indicates that the minimum accepted score requires values equal to or greater than 0.80 or 0.90.

Then, the second round is carried out in the same way as the first round. An email is sent with the questions selected from the previous round. These questions were reformulated to confirm the results, estimating 2 weeks to answer the survey. However, this stage is completed in one week. Table 3 shows the results of round 2.

As a result of the second round, all the questions about 70% acceptance are obtained. According to Hurtado-Delgado et al. [18], the model is accepted and the experts conclude the review stage. The experts' comments collected during this validation stage are included and thus the final validated model is reached and presented in Fig. 2. Finally, unlike the model proposed in the first Delphi round, the validated model incorporates a previous step to design and development, which consists of creating an architecture draft that considers carrying out an initial survey before modelling and integrating the AR viewer integrated with BIM.

Table 3 Results of delphi round 2

Exp	Questions (Q)					
	10	11	12	13	22	29
1	5	4	4	5	6	5
2	5	3	5	5	3	5
3	5	5	5	5	4	4
4	4	5	5	4	4	4
5	5	5	5	5	5	4
6	5	3	4	5	5	5
7	5	4	5	4	4	5
%Cc	100	71	100	100	86	100

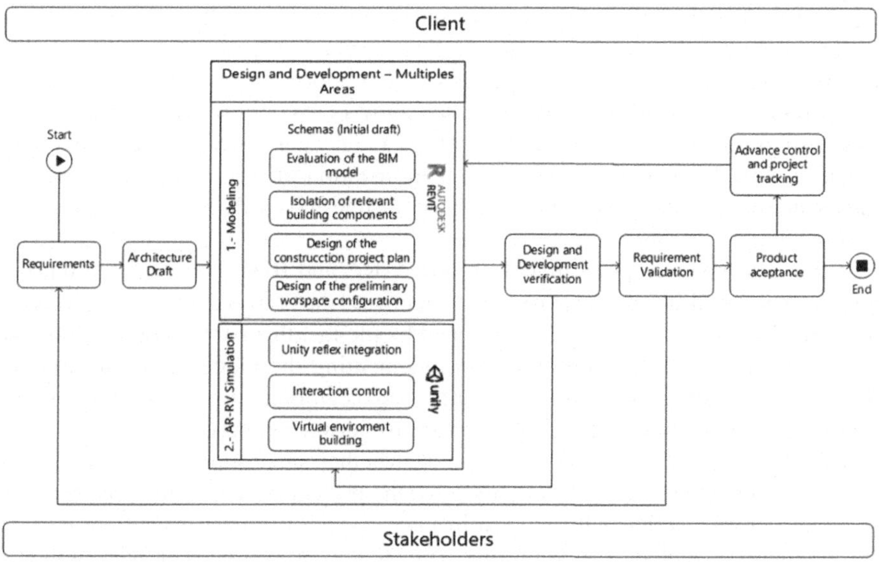

Fig. 2 Validated conceptual model

6 Results and Analysis

Initially, the bibliographic analysis results show a genuine interest in the scientific area of integrating AR and BIM as a process of validation of procedures related to the field of AEC (Architecture, Engineering, and Construction). However, although studies associated with AR and BIM in projects have been increasing, it is not until recent years that the market has become interested in generating tools that allow automating this process. This situation would somehow facilitate the importation and validation of elements incorporated in projects, such as the case of Unity Reflect [21].

Unity Reflect [21] provides a solution to the problems presented in studies. For example, Getuli et al. [15] invested considerable time integrating the BIM models with increased reality. In general, due to the technological advances of recent times, following the study proposed by Volk et al. [22], which envisioned the remarkable growth of technologies such as cloud computing and AR, evidenced in this study. Everything points that the integration of AR and BIM is possible.

Regarding whether integrating these tools improves the process of design management in the early stages, the current literature does not contemplate methodological proposals that indicate using these tools in an integrated approach in a relatively simple way and effortlessly on engineering projects.

Based on the Delphi feedback analysis, the experts unanimously agree that incorporating AR and BIM into the design management process improves the project visualization because visualizing virtual elements in the real world helps better perceive the design proposal generated in the previous stages. Furthermore, it is essential to note that Freire and Alarcón [1] highlights the importance of conceptualizing the design process as a flow of information that allows the coordination of independent flows and integration of design with suppliers and the construction stage. This issue is achieved in the model, thanks to the centralization of the changes proposed by clients and stakeholders made directly, in a real-time and easily accessible integrated system.

Regarding the conceptual model proposed, RA and BIM are integrated under a consensus reached among the experts who approved the proposal; however, it was not absolute. This deviation could be explained by the fact that some experts expressed doubts about the cost and duration of implementing the model in a real process. Hence, incorporating the architecture draft step as an initial input was evaluated in the second round, obtaining the total consensus between experts.

The experts are unanimous in analyzing whether superimposing 3D models in the real world delivers more information for the project managers to avoid interferences and problems that will be seen in later stages of project execution. Among the comments collected, one stands out "The client is always present, who 'will see' what he is thinking and allows us to continue advancing in the project. That way, make sure you meet his expectations." Additionally, in this way, the incorporation of AR with BIM in the design management process in the early stages provides the project key actors with a more global vision of all the areas that compose it.

6.1 Theoretical Implications

By incorporating the BIM and AR integration within the verification and validation process suggested by ISO:2015, a workflow more organized is obtained. The ideas conceived in the draft architecture allow the incorporation of basic pre-engineering, generating the initial models visualized using RA and validated by stakeholders.

This loop can be repeated as many times as necessary until a final approved version, which will provide the basis for the work to execute.

Some comments delivered by the experts expose the importance of identifying the problems that need a simulation in real-time to evaluate movements and interferences, where the model would come to solve these points. In addition, the continuous and early client involvement is vital in the design management process.

Although the client needs to be involved in the design process, this does not ensure that their presence in the entire project decreases making mistakes. This evidence agrees with other studies [23]; [8], stating that effective communication between the interested actors should be treated as a crucial project factor. Therefore, it is critical to creating these environments through formal channels that allow a fluid exchange of information and collaboration between the different disciplines. The proposed model addresses the improvement of the communication of all project stakeholders, based on the possibility of leaving physical comments that can be viewed by all users within the BIM and AR environment. Also, it provides a greater understanding of the client's vision and other stakeholders on a specific point under evaluation.

According to the opinion of the experts on the major causes of change in project design, we can observe that within 80% of the most significant reasons we have: changes in requirements and specifications, incorporation of new requirements, errors and discrepancies in design documents, omission of design, incomplete drawings, and lack of coordination between disciplines. The results coincide with Yap et al. [4], where they conclude that changes in requirements and lack of coordination of the areas are among the most common causes found in the design process. They emphasize that proper coordination between the different consultants is critical to obtaining good results in the initial design. Furthermore, design discrepancies often lead to mistakes in the construction field, resulting in rework and increased project costs.

The study proposed by Getuli et al. [15] implements an AR viewer from BIM models intending to improve the activity and safety planning process. However, the results obtained in its demo required considerable effort regarding importing the virtual elements to the AR viewer. Applying the model proposed in this study that considers Unity Reflect, which integrates natively with Revit, would considerably reduce the work time in creating the AR environment, thus optimizing the times associated with this task. Likewise, applying the model proposed in the Getuli et al.'s study would give continuity to the conceptual model, seeing the changes in real-time activities. At the same time, the operators immerse themselves in the virtual environment.

6.2 Practical Implications

According to what was stated by one of the experts: "During construction, there are problems that need a simulation to evaluate movements, interferences, continuity and AR can contribute to the visualization of the problem and provide good solutions."

The results presented in this study show a genuine interest in implementing new technologies to solve everyday engineering problems. The incorporation of proposals such as creating simulators to analyze different factors would enhance the design and incorporate new elements to evaluate.

In general, the ability to generate value, based on the use of technological tools, is essential to solving problems in the design stages, where the commitment and support of the project manager, clients, and stakeholders are crucial to promote this type of initiative and carry them out [24].

7 Conclusions

The integration of information from the different areas of a project directly produces benefits and improvements in the design phase, leading to a better vision in the project's pre-construction. In addition, having a more dynamic exchange of information allows accurate information to be shared with the level of detail that each specialty deems appropriate, thus achieving cross-sectional management and development of the project design.

Implementing a new methodology requires an adaptation process that should have a variable duration, depending on different internal factors, such as the number of people who are part of the project. In addition, it is important to consider training plans for the client and stakeholders to understand how the new process works.

Applying these new technologies contributes to the growth of innovation within the project management processes in the early stages, thus presenting significant advantages in contrast to traditional methods.

The benefits of a design change in early stages, based on BIM and AR model, are immediately visible, as this virtual world of information exchange and global vision of project components is generated. One of the outstanding contributions of AR and BIM integration is the collision detection, where virtual inspections could be carried out to manually identify possible interferences that could put the project at risk.

While these technologies have many potential payoffs, its implementation can be a bit tricky. In turn, the complement proposed in this study allows relatively easy integration and creation of virtual environments with BIM and AR. However, it is limited to the functionalities provided by the tool.

If we want to generate something much more specific to a project, new resources will need to be applied to create something more tailored. Furthermore, using this model in different projects requires an injection of resources that perhaps minor organizations would not be willing to disburse because software licenses tend to be expensive and there is a lack of competent personnel.

The proposed method of this study resulted in the construction of a conceptual model, which integrates the use of AR and BIM under the validation and verification process proposed by ISO 9001, item 8.3. Although experts empirically verified the theoretical foundations of its use and potential improvements in team coordination, it would be interesting to apply this study in an organization belonging to the

AEC industry. Furthermore, this implementation will provide a more practical understanding associated with your real opportunities in the early stages of the project, measuring the improvements in team coordination more tacitly.

Finally, the use of artificial intelligence and Machine Learning is revolutionizing the field of AR. Incorporating these tools to identify and predict possible errors or clashes between different areas and processes would greatly help prevent and avoid making errors in the early stages.

References

1. Freire, J., Alarcón, L.: Mejoramiento del proceso de diseño en proyectos. Revista Ingeniería en Construcción **16**(1), 61–71 (2001)
2. Milgram, P., Kishino, F.: A taxonomy of mixed reality visual displays. IEICE Trans. Inf. Syst. **77**(12), 1321–1329 (1994)
3. Ahmed, S.: A Review on using opportunities of augmented reality and virtual reality in construction project management. Organ. Technol. Manage. Construct. Int. J. **10**(1), 1839–1852 (2018)
4. Alizadehsalehi, S., Hadavi, A., Huang, J.: From BIM to extended reality in AEC industry. Automat. Construct. **116**(1) (2020)
5. ISO.: ISO 9001—Quality Management Systems: Requirements (2015)
6. Yap, J., Hamzah, A.R., Wang, C., Skitmore, M.: Exploring the underlying factors inducing design changes during building production. Prod. Plann. Control **29**(7), 586–601 (2018)
7. Jackson, S.: Project cost overruns and risk management. de Proceedings of Association of Researchers in Construction Management 18th Annual ARCOM Conference. Northumber University, Newcastle, UK (2002)
8. Reddy, P.: BIM for building owners and developers: making a business case for using BIM on projects. John Wiley & Sons (2011)
9. Latiffi, J., Brahim, Syazli, M.: The development of building information modeling (BIM) definition. Appl. Mech. Mater. **567**, 625–630 (2014)
10. BIM.: Forum Chile: La Propuesta de BIM y la Colaboración (2017). [Online] Available: http://www.bimforum.cl/wp-content/uploads/2017/07/BIM-Forum-Chile-_.pdf. Accessed 22 May 2021
11. Boje, V., Bolshakova, A., Guerriero, S., Kubicki, Halin, G.: Semantics for linking data from 4D BIM to digital collaborative support. Frontiers of Engineering Management, pp. 1–13 (2020)
12. Romigh, J., Kim, Sattineni, A.: 4D Scheduling: A Visualization Tool for Construction Field Operations de 53rd ASC Annual International Conference Proceedings (2017)
13. Khalek, J., Chalhoub, S.: Ayer: augmented reality for identifying maintainability concerns during design. Adv. Civil Eng. **2019**, 1–12 (2019)
14. Noghabaei, M., Heydarian, A., Balali, V., Han, K.: Trend analysis on adoption of virtual and augmented reality in the architecture. Eng. Construct. Ind. Data **5**(26), 1–18 (2020)
15. Buonocore, E., Picone, F., Russo, G., Franzese, P.: The scientific research on natural capital: a bibliometric network analysis. Environ. Account. Manage. **6**(4), 381–391 (2018)
16. Ratajczak, J., Riedl, M., Matt, D.: BIM-based and AR application combined with location-based management system for the improvement of the construction performance. Buildings **9**(5) (2019)
17. Getuli, V., Capone, P., Bruttini, A., Isaac, S.: BIM-based immersive virtual reality for construction workspace planning: a safety-oriented approach. Autom. Constr. **114**, 1–20 (2020)
18. Gomez-Jauregui, V., Manchado, C., Del Castillo-Igareda, J., Otero, C.: Quantitative evaluation of overlaying discrepancies in mobile augmented reality applications for AEC/FM. Adv. Eng. Softw. **127**, 124–140 (2019)

19. Hsu, C.C., Sandford, B.: The delphi technique: making sense of consensus. Pract. Assess. Res. Eval. **12**(1), 1–7 (2007)
20. Hurtado, Z., Correa, Conde, Y.: The role of a public university in a global environment: networks and externalities of the R&D of the Cauca University. Estudios Gerenciales **29**(129), 396–405 (2013)
21. Rodríguez, J., Reguant, M.: Calcular la fiabilidad de un cuestionario o escala mediante el SPSS: el coeficiente alfa de Cronbach. REIRE Revista d'Innovació i Recerca en Educació **13**(2), 1–13 (2020)
22. Numally, J.: Psychometric theory 3E. Tata McGraw-hill education (1994)
23. Unity: Unity Reflex (2020). [Online] Available: https://unity.com/products/unity-reflect. Accessed 21 May 2021
24. Volk, R., Stengel, J., Schultmann, F.: Building information modeling (BIM) for existing buildings—literature review and future needs. Autom. Constr. **38**, 109–127 (2014)
25. Omran, A., Omran, Pakir, A.: Critical success factors that influencing safety program performance in Malaysian construction projects: case studies. J. Acad. Res. Econ. **2**(1) (2010)
26. Chan, J.: Project commitment: how to get it and keep it! PM Netw. **14**(6), 77–84 (2000)

Chapter 21
Design of a Proposal for the Implementation of Scope, Schedule and Cost Management in Fast-Track Projects Using BIM Methodology in 3D, 4D and 5D

John Cardoch-Álvarez, Luis Alvarado-Acuña, Boris Heredia-Rojas, and Stephanny Acevedo-Acevedo

Abstract Competitiveness in engineering companies is supported by the use of ICTs, especially in complex projects where the proper use of information systems, techniques and organizational tools that collaborate with the objectives associated with both the product and the project is essential, among others. Those that stand out the Fast-track projects and the BIM methodology. This Research seeks to consolidate a referential framework that describes the theories of project management, BIM methodology and project acceleration through Fast-track, to then develop an analysis of the state of the art of project management and its correlation with respect to Fast projects, -track and BIM. With the result of this analysis, a first conceptual model is generated called: "Handbook of good practices for the implementation of scope, schedule and cost management in Fast-track projects using BIM methodology in 3D, 4D and 5D". Subsequently, the model is subjected to a series of iterations using the Delphi method for its modification and subsequent validation by consulting experts, obtaining a theoretically supported and empirically validated good practices manual.

Keywords 3D-4D-5D BIM · BIM methodology · Fast-track · Knowledge management · Project management

1 Introduction

In the development of highly competitive Fast-Track construction projects in the Antofagasta region, Chile, weaknesses are observed in scope, schedules and costs. Fast-Track projects are characterized by having limited and dynamic deadlines, with multidisciplinary tasks and activities in parallel, overlapping and overlapping. In addition, we must consider that there are many factors that lead to a project of this

J. Cardoch-Álvarez · L. Alvarado-Acuña (✉) · B. Heredia-Rojas · S. Acevedo-Acevedo
Universidad Católica del Norte, Av. Angamos, 0610 Antofagasta, Chile
e-mail: lualvar@ucn.cl

P. Y. Piñero Pérez et al. (eds.), *Artificial Intelligence in Project Management and Making Decisions*, Studies in Computational Intelligence 1035,
https://doi.org/10.1007/978-3-030-97269-1_21

execution modality that does not comply with the scope and exceed schedules and costs, so it is necessary to manage them from the technical direction of projects, implementing a methodology appropriate to dynamism and speed of a project of this modality throughout the entire life cycle [1].

The design of the guide for the implementation of schedule and cost management using BIM methodology must be sustainable, include continuous improvement and the result, in addition to implementing 3D, 4D and 5D modeling, it focuses on maintaining contractual deadlines and not exceeding costs [2]. Based on the above, the general research question arises: How can delays in contractual deadlines and costs of a fast-track project be avoided?

Where, BIM is a set of methodologies, technologies and standards that allow the design, construction and operation of a building or infrastructure collaboratively in a virtual space. Since BIM 3D is a virtual model that contains data information for each of its components and construction elements, the 4D model links the project timeline and the 5D model links costs [3].

Therefore, the challenge of this research is to make a guide for the implementation of scope, schedule and cost management in Fast-Track projects using BIM methodology in 3D, 4D and 5D.

The following research hypotheses are proposed:

- The management of a project contributes to improvement by incorporating BIM modeling in 3D, 4D and 5D, which allows to achieve the objectives associated with the product and the project.
- Managing the scope, deadlines and costs using the BIM methodology leads to better project management, avoiding or reducing cost overruns and delays.

Figure 1 shows the step by step of the research methodology, it will be a mix between exploratory and descriptive, since it is a first approach to Fast-Track project management with the use of BIM methodology. To later, identify characteristics and define behaviors and verify the association of the variables. In addition, the result of the investigation can be used for future investigations. Along the same lines, the

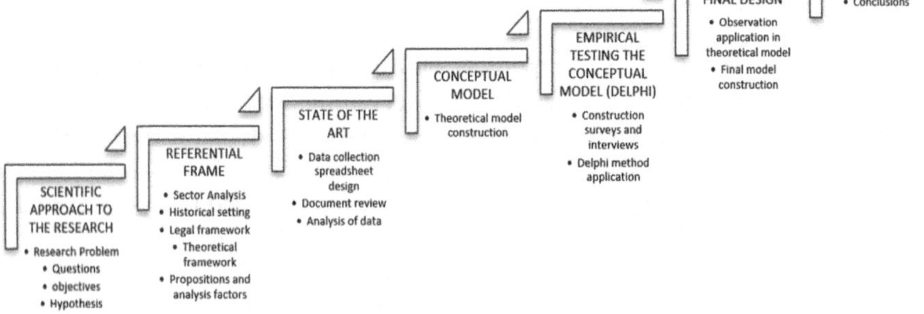

Fig. 1 Step by step research methodology

research is inductive and deductive, because the theoretical framework and state of the art are used in the application of it to a concrete reality, in order to design an implementation manual. Then there is a compilation of the state of the art and preparation of a conceptual model for the development of the guide. Finally, using the Delphi Methodology [4] the guide is validated through expert validation.

2 Literature Review

The research will be built based on four theories: Life cycle; BIM 3D, 4D and 5D methodology; Management of scope, costs, schedules; and, Knowledge management.

According to PMI [5], the life cycle of the project is a set of phases of the project, generally sequential and sometimes overlapping, whose name and number are determined by the management and control needs of the organization or organizations that participate in the project. Project, the nature of the project and its area of application. Scope management includes the processes required to ensure that the project includes all required work. Project Cost Management includes the processes involved in estimating, budgeting and controlling costs so that the project is completed within the approved budget Project Deadline Management includes the processes required to manage project completion within contractual times. Project Schedule Management includes the processes required to manage the completion of the project in the contractual times. BIM according to Autodesk [6], is a process based on an intelligent model that generates a vision that helps to plan, design, construct and manage constructions or buildings. The BIM Methodology consists of a set of methodologies, technologies and standards that allow the design, construction and operation of a building or infrastructure collaboratively in a virtual space. Where, 3D BIM is a virtual model that contains data information on each of its components and construction elements [7]. The 4D model links the project timeline [8] and the 5D model links the project costs [9]. Figure 2 shows the main milestones that mark the development of the BIM methodology in Chile [10].

Knowledge management is the process of using existing knowledge to generate new knowledge to achieve project objectives and contribute to organizational learning PMI [5] and [11].

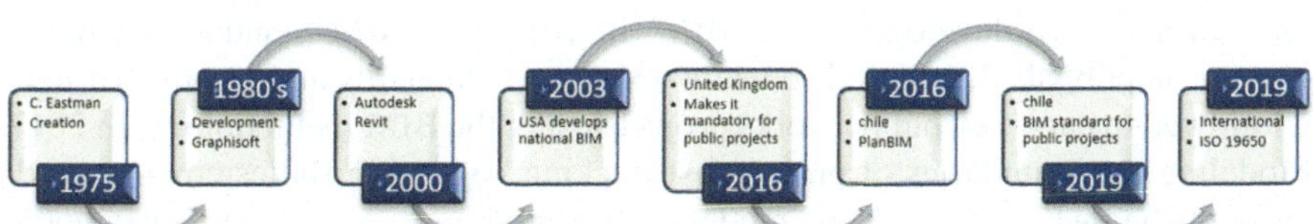

Fig. 2 Historical framework BIM methodology

Table 1 Theoretical propositions and their associated analysis factors

PT	Theoretical propositions	FA	Analysis factors
PT1	The project life cycle helps us define the phases and deliverables of the project	FA.1.1	Gestation
		FA.1.2	Development engineering
		FA.1.3	Design
		FA.1.4	Construction
		FA.1.5	Closing
PT2	Project management helps us to obtain success in projects by reducing extra costs and delays	FA.2.1	Scope
		FA.2.2	Cost
		FA.2.3	Schedule
		FA.2.4	Knowledge
PT3	The acceleration of projects are techniques that allow increasing competitiveness and reducing times	FA.3.1	Fast-track
PT4	BIM methodologies help us plan and manage costs and time graphically	FA.4.1	3D model
		FA.4.2	4D model
		FA.4.3	5D model

According to Khoueiry in 2013 [12], he says that a project of the Fast-Track type compresses the total duration, speeding up its execution, overlapping activities that are normally sequential such as, for example: design with construction, construction with pre-operation and pre-operation tests with commissioning and operation. If good coordination and management can be achieved, the project completion can be achieved efficiently. Therefore, Fast-Tracking is one of the most common project acceleration techniques [1].

Based on these theories, the theoretical propositions and their associated analysis factors are developed, which are shown in Table 1.

3 State of the Art and Development of the Conceptual Model

Since 2016, the BIM Plan began in Chile [10], a plan that seeks that by 2025 all public works projects are developed using BIM. In 2019 the "BIM Standard for public projects" is published, which deals with a series of standards and norms that must be followed to carry out public works projects using the BIM methodology. In 2017, "Modeling the boundaries of project fast-tracking" by Pablo Ballesteros-Pérez [1] was published where a stochastic model of Fast-tracking-type projects is developed where they quantify the non-linear relationship between cost and superposition, to finally be they compare manual calculations with the results of a generic algorithm through the methodology of the case.

Table 2 Selection of publications for bibliometric analysis, before 2015

Authors	Tittle	Source
Nawari O. Nawari	BIM standard in off-site construction	Journal of architectural engineering© ASCE, 2012
Kristen Barlish, Kenneth Sullivan	How to measure the benefits of BIM —a case study app	Automation in construction 2012
David Bryde et al	The project benefits of building information modelling (BIM)	International journal of project management 2012
Salman Azhar et all	Building information modeling (BIM): now and beyond	Australasian journal of construction economics and building 2012
Rebekka Volk, Julian Stengel et al	Building information modeling (BIM) for existing buildings—literature review and future needs	Automation in construction 2013
McGraw-Hill construction	National BIM standard—United States	APEC Indonecia 2013
Mauricio Loyola V	Encuesta Nacional BIM 2013: Informe de resultados	Departamento de Arquitectura, Universidad de Chile, 2013
Darius Migilinskasa et al	The benefits, obstacles and problems of practical bim implementation	Procedia engineering 2013
Julio César Acosta Prado et al	Ba: espacios de conocimiento. Contexto para el desarrollo de capacidad de innovación. Un análisis desde la gestión del conocimiento	Revista EAN 2014

The following is a list of the publications selected for the construction of the theoretical model based on the publications of the following authors: [1–5, 7–9, 11, 13–36], see Tables 2, 3 and 4.

Nazia Rasul, Muhammad Sohail Anwar Malik, Beenish Bakhtawar and Muhammad Jamaluddin Thaheem publish the 2019 "Risk assessment of fast-track projects: a systems based approach" quantifying the risks and through iterations they manage to find the best mitigations reducing the impacts in time, cost, scope, quality and safety [13].

In 2018, Min–Yuan Cheng and Nai—Wen Chang developed the research work "Dynamic construction material layout planning optimization model by integrating 4D BIM" where they develop a model for the optimization of the storage area of materials for construction projects using the methodology BIM in 4D [8].

From the analysis of the documents listed in the bibliography using the Mentimeter® word cloud tools and the VOSViewer® relationship map, see Fig. 3. In

Table 3 Selection of publications for bibliometric analysis. Period 2015–2017

Authors	Tittle	Source
Qinghua He, Ge Wang et al	Mapping the managerial areas of building information modeling (BIM) using scientometric analysis	International journal of project management 2015
Eyzaguirre Vela, Raúl Ralph	Potenciando la capacidad de análisis y comunicación de los proyectos de construcción, mediante herramientas virtuales BIM 4D durante la etapa de planificación	Pontificia Universidad Católica del Perú, 2015
Juan Ocampo	La gerencia bim como sistema de gestión para proyectos de construcción	National University of Colombia, 2015
Reza Dehghan, Kamran Hazini et al	Optimization of overlapping activities in the design phase of construction projects	Automation in construction, 2015
Yan Liu, Sander et al	Understanding effects of BIM on collaborative design and construction: An empirical study in China	International journal of project management 2016
Mercedes Reguant-Álvarez et al	El método Delphi	REIRE 2016
Alcántara Rojas, Paúl Vladimir	Metodología para minimizar las deficiencias de diseño basada en la construcción virtual usando tecnologías BIM	Universidad Nacional de Ingeniería 2016
Mostafa Ali et al	A method for clustering unlabeled BIM objects using entropy and TF-IDF with RDF encoding	Advanced engineering informatics 2017
Dermot Kehily et al	Embedding life cycle costing in 5D BIM	Journal of information technology in construction, 2017
Ignacio Morales et all	Guía inicial para implementar BIM en las organizaciones	BIM forum Chile, 2017
Pablo Ballesteros-Pérez	Modelling the boundaries of project fast-tracking	Automation in construction 2017
Carolina Briones	Metodologia BIM y su adaptación a nivel nacional e internacional	Plan BIM 2017
Carolina Briones	La Revolución del BIM en la coordinación de proyectos	Plan BIM 2017
PMI	PMBOK 6ta Edición	PMI 2017

Table 4 Selection of publications for bibliometric analysis, last 3 years

Authors	Tittle	Source
Timothy O. Olawumi et al.	A scientometric review of global research on sustainability and sustainable development	Advanced engineering informatics 2018
Min-Yuan Cheng et al.	Dynamic construction material layout planning optimization model by integrating 4D BIM	Engineering with computers 2018
Eloi Coloma Picó	Introducción a la Tecnología BIM	Universitat Politècnica de Catalunya 2018
Peter Lovea et al.	Project controls for electrical, instrumentation and control systems: enabling role of digital system information modelling	Automation in construction 2019
Nazia Rasul et al.	Risk assessment of fast-track projects: a systems based approach	International journal of construction management 2019
Amjed N. Hasan et al.	The benefits of and challenges to implement 5D BIM in construction industry	Civil engineering journal 2019
Jesús M. de la Garza et al.	Workflow process model for flash track projects	American society of civil engineers 2018
Bryan Franz, John Messner	Evaluating the impact of building information modeling on project performance	Journal of computing in civil engineering 2019

Fig. 3 Keywords cloud

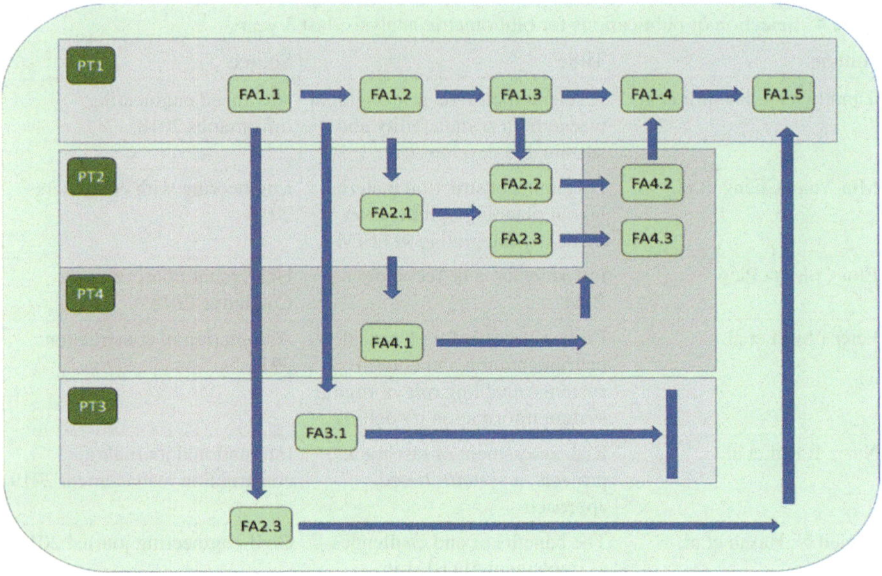

Fig. 4 Conceptual model

this figure you can see the cloud created using all the keywords of the research documents used. From this image you can see that the central theme is Building Information Model or BIM and that the words or concepts that surround it are Construction, Concurrent Engineering, Collaboration and Fast-Tracking. To determine the concurrency and relationships, the VOSViewer® tool is used, as mentioned in the previous paragraph. The relationship map shown above reveals the high use and linkage that BIM has in Architectural Design and that is the first concept used for the creation of the conceptual model.

The conceptual model is generated from the propositions and analysis factors. Based on the aforementioned table together with the bibliographic analysis matrix, a conceptual map shown in Fig. 4 is created of good industry practices to finally form the deliverable of the research work "proposal for implementation of scope management, schedules and costs in Fast-track projects using BIM methodology in 3D, 4D and 5D".

4 Empirical Testing of the Conceptual Model

Fast-Tracking is one of the most common project acceleration techniques, where activities that are normally sequential overlap, such as: design with construction, construction with pre-operation tests and pre-operation tests with commissioning. Up and running [1].

This study consists of answering an "online" survey to validate a conceptual model developed based on the analysis of the state of the art. The answers to the questions are at five levels: Strongly agree, Agree, Moderately agree, Disagree and strongly disagree.

4.1 Bibliometric Analysis

The following points describe the detailed procedure of the empirical study for the validation of the model and its results. It should be noted that the theoretical propositions (PT 1) are excluded from the survey, since it uses the life cycle of a project as a basis.

In the investigation authors begins with the theoretical study, for which, the frame of reference is developed defining the theoretical propositions and analysis factors, to support the state of the art, from which the bases that support these are obtained, ending with a conceptual model. Then we continue with the empirical study, where a survey is carried out to the different units of analysis to evaluate the analysis factors and with it the theoretical propositions.

Table 5 Units of analysis shows the experts that will constitute the study grouped into 5 units of analysis.

Table 5 Units of analysis

UA.1	BIM expert	Cofundador academia BIM: Ricardo Zepeda
UA.2	MeGIP Professor	Juan Huidobro Arabia
UA.3	F-T Project manager	César Torres
	F-T Project manager	Omar Valencia
	F-T Project manager	Cristian Abarca
UA.4	F-T Project planner	Nixon Avila
	F-T Project planner	Mariana Coubernas
UA.5	F-T Electrical engineer	Daniel Cruz
	F-T Electrical engineer	Ronald Pizarro
	F -T Mechanical engineer	Marco Hernandez
	F-T Civil engineer	Guillermo Flores
	F-T Civil engineer	Dante Sepulveda
	F-T electronic engineer	Rodrigo Erazo

4.2 Expert Evaluation Report by Analysis Factors

Next, the results of the study are shown by analysis factor and then grouped by theoretical propositions in the next section. For the validation of the Analysis Factors, the following formula is considered:

$$\%\ \text{validation} = \%\ \text{strongly agree} + \%\ \text{agree} + (\%\ \text{moderately agree})/2 \quad (1)$$

For an analysis factor to be considered valid, the validation percentage must be greater than 50%.

The analysis factor FA.2.1 is part of the technical preposition PT2 that refers to the fact that Project Management helps us to obtain success in projects by reducing extra costs and delays focused on the Scope of the project. FA.2.1 is validated by having 90% acceptance by the analysis units.

Analysis Factor FA.2.2 is part of the technical preposition PT2 that refers to the fact that Project Management helps us to obtain success in projects by reducing extra costs and delays focused on the Cost of the project. FA.2.2 is validated by having 80% acceptance by the analysis units.

Analysis Factor FA.2.3 is part of the technical preposition PT2 that refers to the fact that Project Management helps us to obtain success in projects by reducing extra costs and delays focused on the Project Term. FA.2.3 is validated by having 79% acceptance by the analysis units.

Analysis Factor FA.2.4 is part of the technical preposition PT2 that refers to the fact that Project Management helps us to obtain success in projects by reducing extra costs and delays focused on Project Knowledge. FA.2.4 is validated by having 90% acceptance by the analysis units.

The analysis factor FA.3.1 is part of the technical preposition PT3 that refers to the fact that the acceleration of projects (Fast-track) are techniques that allow increasing competitiveness and reducing times. FA.3.1 is validated by having 93% acceptance by the analysis units.

The Analysis Factor FA.4.1 is part of the technical preposition PT4 that refers to the fact that BIM Methodologies help us to plan and manage costs in a better graphical way. It is shown that FA.4.1 is validated by having 94% of acceptance by the units of analysis.

The analysis factor FA.4.2 is part of the technical preposition PT4 that refers to the fact that BIM Methodologies help us to plan and manage costs in a better graphic way. FA.4.2 is validated by having 94% acceptance by the units of analysis.

The Analysis Factor FA.4.3 is part of the technical preposition PT4 that refers to the fact that BIM Methodologies help us to plan and manage costs in a better graphic way. FA.4.3 is validated by having 98% acceptance by the analysis units.

Table 6 Results of technical preposition

PT.2			PT.3		PT.4		
Analysis factor	Results (%)		Analysis factor	Results (%)	Analysis factor	Resultssss (%)	
FA2.1	90		FA3.1	93	FA4.1	94	
FA2.2	80	85			FA4.2	94	95
FA2.3	79				FA4.3	98	
FA2.4	90						

4.3 Expert Evaluation Report by Theoretical Proposition

As can be seen from the previous section, all the analysis factors were validated in the first iteration of the survey. Therefore, prepositions are validated in the same way.

The technical preposition PT.2 that refers to the fact that Project Management helps us to obtain success in projects by reducing extra costs and delays. Table 6 shows that PT.2 is validated by having 85% acceptance by the units of analysis.

The technical preposition PT.3 that refers to the fact that the acceleration of projects are techniques that allow increasing competitiveness and reducing times. In analysis PT.3 is validated by having a 93% acceptance by the analysis units Table 6.

The technical preposition PT.4 refers to the fact that BIM Methodologies help us to plan and manage costs in a better graphic way. Table 6 shows that PT.4 is validated by having 95% acceptance by the units of analysis.

In general, an analysis of the results of the survey is also carried out without considering the units of analysis. In general, Fig. 5 shows a graph with the global results of the conceptual model at the 5 levels of evaluation.

The following list and Table 7 shows the results of this analysis, of which, like the previous exercises, the model is fully validated, with an acceptance percentage of 90%.

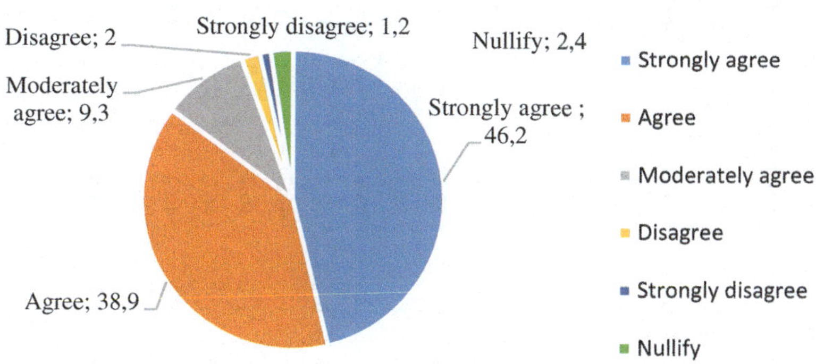

Fig. 5 Global summary

Table 7 Percent of results' model is fully validated of technical preposition

Analysis factors		Strongly agree %	Agree %	Moderately agree %	In disagreement %	Strongly disagree %	Emp Ty %	Validated %		
FA.2.1	a	61.5	23.1	15.4	0.0	0.0	0.0	Yes	92.3	90
FA.2.1	b	46.2	46.2	0.0	7.7	0.0	0.0	Yes	92.3	
FA.2.1	c	61.5	15.4	15.4	0.0	7.7	0.0	Yes	84.6	
FA.2.1	d	53.8	30.8	7.7	0.0	7.7	0.0	Yes	88.5	
FA.2.1	e	7.7	76.9	15.4	0.0	0.0	0.0	Yes	92.3	
FA.2.2		30.8	38.5	15.4	7.7	7.7	0.0	Yes	76.9	
FA.2.3		38.5	30.8	15.4	15.4	0.0	0.0	Yes	76.9	
FA.2.4		38.5	46.2	7.7	7.7	0.0	0.0	Yes	88.5	
FA.3.1	f	38.5	53.8	0.0	0.0	0.0	7.7	Yes	92.3	
FA.3.1	g	53.8	38.5	7.7	0.0	0.0	0.0	Yes	96.2	
FA.3.1	h	38.5	53.8	0.0	0.0	0.0	7.7	Yes	92.3	
FA.4.1	i	46.2	38.5	7.7	0.0	0.0	7.7	Yes	88.5	
FA.4.1	j	46.2	38.5	7.7	0.0	0.0	7.7	Yes	88.5	
FA.4.2	k	46.2	38.5	7.7	0.0	0.0	7.7	Yes	88.5	
FA.4.2	l	46.2	30.8	15.4	0.0	0.0	7.7	Yes	84.6	
FA.4.3	m	76.9	23.1	0.0	0.0	0.0	0.0	Yes	100	
FA.4.3	n	53.8	30.8	15.4	0.0	0.0	0.0	Yes	92.3	
PT2	o	53.8	30.8	15.4	0.0	0.0	0.0	Yes	92.3	
PT2	p	38.5	53.8	7.7	0.0	0.0	0.0	Yes	96.2	

- FA.2.1

  a. For scope management it is advisable to develop a WBS (work breakdown structure) as detailed as possible.
  b. For the control of construction progress, it is considered that graphic reports using the 3D model: global, by systems and specialties, by subsystems and specialties, and production lines will deliver valuable information to all interested parties.
  c. A WBS (work breakdown structure) should be organized by defining production lines based on systems and subsystems.
  d. The WBS (work breakdown structure) should be divided into activities segregated for each discipline (electrical, instrumentation, mechanical, hydraulic, civil)
  e. Each activity must be assigned to the subsystems, systems and production lines in order to conform to the WBS (work breakdown structure) of the project.

- FA.2.2 The management of the budget and its development must be based on the accelerated program.
- FA.2.3 To manage the schedule, it is recommended to develop the work plan on the WBS (work breakdown structure) using classical methodology.
- FA.2.4 Any new type of implementation must be adapted to the environmental and organizational factors of each organization and capitalize on the knowledge using the Nonaka method.
- FA.3.1

  f. For Fast-Track-type construction projects, KPIs must be generated for progress, time and costs: global, by systems and specialties, by subsystems and specialties, and production lines.
  g. The acceleration of the program is carried out by superimposing activities, so that the process of construction, pre-commissioning, and commissioning are directed at the end of subsystems.
  h. The accelerated program must maintain the same structure as the WBS (work breakdown structure).

- FA.4.1

  i. For the control and management of the project, the following views must be created: global views, subsystem views, system views, views by disciplines in each subsystem view and views by disciplines in each systems view.
  j. The development of the 3D BIM model must be done following the structure of the WBS (work breakdown structure), where each element of the model must contain or be associated with an activity defined in the WBS.

- FA.4.2

  k. For the 4D Model: The accelerated program must be integrated on the 3D model.

l. For the control of the schedule, it is considered that graphic reports using the 4D model: global, by systems and specialties, by subsystems and specialties, and production lines will deliver valuable information to all interested parties.

- FA.4.3

m. 12. For the 5D Model: The budget must be integrated into the 3D model, valuing and quantifying each of the elements of the model.
n. 15. For budget control, it is considered that graphic reports using the 5D model: global, by systems and specialties, by subsystems and specialties, and production lines will deliver valuable information to all interested parties.

- PT2

o. KPI's are a good tool for decision making.
p. The best strategy to visualize different performance indicators is to set up a remote access platform with information in the form of KPI's Dashboards that integrates different systems such as Assemble, PI vision, corporate website, among others.

5 Results: Final Validated Model

Written guide for the implementation of scope, schedule and cost management in Fast-track projects using BIM methodology in 3D, 4D and 5D.

1. For the scope, a WBS should be developed as detailed as possible, for this it is suggested:
   - Define the project in production lines based on systems and subsystems.
   - Divide by disciplines: electrical, instrumentation, mechanical, hydraulic, civil, etc.
   - For each discipline create activities
   - For each subsystems and, systems the activities created in the previous point are assigned.
2. The program or schedule must be developed on the WBS using classical methodology from start to finish, it is suggested to use Primavera P6.
3. The Fast-track acceleration process must be carried out on the schedule, overlapping the stages that allow it during the development of the project.
   - The construction and assembly process, here the subsystems must be completed with all the disciplines involved completing the production lines.
   - The constructive tests and quality plan must be applied by subsystems grouping them in systems, later in production lines.
   - The pre-commissioning tests are developed in parallel to the construction in the constructively completed subsystems, then grouped by systems to finally continue with production lines.

- Commissioning tests must be carried out on finished production lines in parallel with construction and pre-commissioning.

4. New accelerated fast-track program in spring P6, must maintain the same structure of the WBS.
5. The budget must be developed on the accelerated schedule.
6. 3D model: The development of the BIM model must be done following the structure of the WBS.
   - Each element of the model must be associated with an activity.
   - Create global view.
   - Create subsystem views.
   - Create system views.
   - Create views by disciplines in each subsystem view.
   - Create views by disciplines in each systems view.
7. Model 4D: The accelerated program must be integrated on the model.
8. Model 5D: The budget must be integrated into the model, valuing and quantifying each element of the model.
9. Control construction progress graphically using the 3D model, creating graphical reports.
10. Control schedule graphically using the 4D model.
11. Control budget graphically using the 5D model.
12. Generate KPI's based on progress, times and costs.
13. Assemble platform with remote access showing KPI Dashboards in Assemble, PI Vision, corporate website, etc.
14. Each organization that applies this manual must adapt it to its environmental factors and capitalize on the knowledge using the SECI method.

6 Conclusions

Research begins with posing a problem and finding its solution. For which, the research question is created: How can delays in contractual deadlines and costs of a fast-track project be avoided? Giving way to three specific questions How to perform 3D, 4D and 5D modeling in a fast-track project? How to implement the BIM methodology in 3D, 4D and 5D in a fast-track project? How to use the BIM methodology to meet contractual deadlines and costs in a fast-track project? Being their answers the objectives of this work, where the main objective is the Design of a guide for the implementation of the BIM methodology in 3D, 4D and 5D for the management of scope, schedules and costs of fast-track projects.

To achieve the objective of the research and solve the problem, first degree hypotheses are created. The management of a project contributes to improvement by incorporating BIM modeling in 3D, 4D and 5D, which allows achieving the objectives associated with the product and to the project and second grade The management of the scope, schedules and costs using the BIM Methodology leads to a better management of a project, avoiding or reducing cost overruns and delays.

The hypotheses are substantiated through the study of the referential framework, to later use it as the basis of the state-of-the-art shield, from which the conceptual model and its validation using the Delphi method are born.

The answer to this question is the manual of good practices "Written guide for the implementation of scope management, schedules and costs in Fast-track projects using BIM methodology in 3D, 4D and 5D".

Fulfilling and achieving the objective and hypotheses raised in the research, since, it was demonstrated in a theoretical way and through the validation of experts that the integration of the BIM methodology to project management will contribute to improve project management.

For the state of the art, use is made of a matrix of scientific documents, reports, publications, articles, etc. of the last years grouping them in the analysis factors of this research, to then perform a relationship analysis between the keywords of the documents based on the analysis factors. From which the following points emerge, which are the basis for the development of the conceptual model.

- BIM is strongly related to architecture projects.
- In recent years, BIM has been joining construction projects and increasing its use.
- The use of BIM for the design and management of projects can be identified.
- One of the most used project acceleration methods is Fast-Tracking.

With the use of the Delphi methodology for consulting experts, it was possible to validate the theoretical propositions and analysis factors of the research with which the conceptual model was built, through the responses of an online survey. With this, the model is considered validated in the first query iteration with an acceptance percentage of 91%.

For the contrast and its validation, a tabulation by analysis units and a global tabulation were used, where for the first case the questions were grouped into analysis factors and for the analysis only the responses of the analysis units with expertise in them were considered, discarding the answers of the experts in other matters. Then, the results of the analysis factors were grouped into the theoretical propositions, making a simple average of the results for each proposition and thus validating them. Finally, with the validated prepositions, consider the conceptual model validated. In addition, a global tabulation was performed where the responses of all the experts were considered without grouping them into units of analysis, resulting in an acceptance of 90%.

The methodology used in this research is the exploratory and descriptive study with inductive and deductive methods, ending with the opinion of experts using the Delphi method.

In detail, a theoretical framework was made based on the exploration and description of all the theories used: Life cycle of a project, Management of scope, time, cost and knowledge and 3D, 4D and 5D BIM Methodology. From which, both the theories propositions and the analysis factors were deduced and induced. The latter being the engine of the study of the state of the art, which was carried out through a bibliometric study analyzed a matrix of documents. Shaping the conceptual model that is finally validated with the use of expert opinion.

The implementation of the good practice's manual "Written guide for the implementation of scope, schedule and cost management in Fast-track projects using BIM methodology in 3D, 4D and 5D" will be of great help for the development of fast-tracking projects. In the city of Antofagasta and in Chile, since, being multidisciplinary projects with multiple jobs and parallel stages, it is necessary to correctly manage the schedule in order to avoid delays and accidents, for this the guide proposes to create performance indicators and reports graphics that will allow you to take corrective or mitigating actions on time.

Along the same lines, one of the points of failure of this type of projects are the extra-costs, with 5D modeling it will be possible to manage visually which will be of great help as well as for the management of the schedule the guide will give evaluation tools for correct decision-making.

To achieve all this, the project scope must be clear and managed in such a way as to be able to comply with what is required by the project. That is why the first steps of the guide are focused on the definition, development and measurement of the scope.

Finally, with the use of the SECI Methodology, it is possible to capitalize on the knowledge and gradually improve the performance of the good practice guide.

The projection of the research is varied and offers us several lines of research. For example:

- The implementation of the guide in fast-tracking construction projects, measuring the results of both the project and each step of the guide.
- Optimize the good practice guide.
- Carry out an investigation using the methodology of the case with the application of the guide presented in this work.
- Study if it can be applied to other types of construction projects.
- Give it a focus on agile methodology, both for construction and other projects.

References

1. Ballesteros, P.: Modelling the boundaries of project fast-tracking. Autom. Constr. **84**, 231–241 (2017)
2. Volk, R., Stengel, J., Schultmann, F.: Building information modeling (BIM) for existing buildings—literature review and future needs. Autom. Constr. **38**, 109–127 (2014)
3. Nawari, N.O.: BIM standard in off-site construction. J. Archit. Eng. **18**(2), 107–113 (2012)
4. Reguant, M., Torrado, M.: El método Delphi (2016). [Online] Available on: http://diposit.ub.edu/dspace/handle/2445/110707. Accessed 12 May 2021
5. PMI.: The standard for project management and a guide to the project management body of knowledge seventh edition. Ed.: Project management institute. ISBN: 978-1-62825-664-2, pp. 370 (2021).

6. BIM.: Autodesk: Autodesk BIM 360 (2016)
7. He, Q., Wang, G., Luo, L., Shi, Q., Xie, J., Meng, X.: Mapping the managerial areas of building information modeling (BIM) using scientometric analysis. Int. J. Project Manage. **35**(4), 670–685 (2017)
8. Cheng, M.Y., Chang, N.W.: Dynamic construction material layout planning optimization model by integrating 4D BIM. Eng. Comput. **35**(2), 703–720 (2019)
9. Kehily, D., Underwood, J.: Embedding life cycle costing in 5D BIM. J. Inform. Technol. Construct. **22**, 145–167 (2017)
10. Corporación de Desarrollo Tecnológico—Cámara Chilena de la Construcción: Plan BIM Chile: Guía inicial para implementar BIM en las organizaciones, (2020) [Online] Available on: https://planbim.cl/biblioteca/documentos/. Accessed 21 May 2021
11. Prado, J.C., Torres, R.A., Fischer, A.L.: Espacios de conocimiento, contexto para el desarrollo de capacidad de innovación. Un análisis desde la gestión del conocimiento. Revista Escuela de Administración de Negocios, pp. 44–63 (2014).
12. Khoueiry, Y., Srour, I., Yassine, A.: An optimization-based model for maximizing the benefits of fast-track construction activities. J. Oper. Res. Soc. **64**(8), 1137–1146 (2013)
13. Rasul, N., Malik, M.S., Bakhtawar, B., Thaheem, M.J.: Risk assessment of fast-track projects: a systems-based approach. Int. J. Construct. Manage., pp. 1–16 (2019)
14. Ali, M., Mohamed, Y.: A method for clustering unlabeled BIM objects using entropy and TF-IDF with RDF encoding. Adv. Eng. Inform. **33**, 154–163 (2017)
15. Olawumi, T.O., Chan, D.W.: A scientometric review of global research on sustainability and sustainable development. J. Clean. Prod. **183**, 231–250 (2018)
16. Guia Inicial para Implementar BIM en Las Organizaciones | PDF | Modelado de información de construcción | Software. Scribd (2021). [Online] Available on: https://es.scribd.com/document/347302644/Guia-Inicial-Para-Implementar-BIM-en-Las-Organizaciones. Accessed 2 May 2021
17. Barlish, K., Sullivan, K.: How to measure the benefits of BIM—a case study approach. Autom. Constr. **24**, 149–159 (2012)
18. Coloma, E.: Introducción a la tecnología BIM. Universitat Politècnica de Catalunya. Departament d'Expressió Arquitectònica I (2008). Accedido: 08 Nov 2021. [Online] Available on: https://upcommons.upc.edu/handle/2117/12226. Accessed 21 May 2021
19. McGraw-Hill: Construction, National BIM Standard—United States. APEC Indonesia (2013)
20. Dehghan, R., Hazini, K., Ruwanpura, J.: Optimization of overlapping activities in the design phase of construction projects. Autom. Constr. **59**, 81–95 (2015)
21. Loyola, M.: Encuesta Nacional BIM 2013: Informe de Resultados. Universidad de Chile, Departamento de Arquitectura (2013)
22. Loyola, M.: Encuesta Nacional BIM 2016: Reflexiones y visiones de expertos (2016)
23. Loyola, M.: Encuesta Nacional BIM (2019). Accedido: 08 Nov 2021. [Online] Available on: https://www.academia.edu/43099372/Encuesta_Nacional_BIM_2019. Accessed 12 May 2021
24. Briones, C.: Metodología BIM y su adaptación a nivel nacional e internacional. Plan BIM Chile (2017)
25. Briones, C.: La Revolución del BIM en la coordinación de proyectos. Plan BIM Chile (2017)
26. Love, P.E., Zhou, J., Matthews, J.: Project controls for electrical, instrumentation and control systems: enabling role of digital system information modelling. Autom. Constr. **103**, 202–212 (2019)
27. Hasan, A.N., Rasheed, S.M.: The benefits of and challenges to implement 5D BIM in construction industry. Civil Eng. J. **5**(2), 412 (2019)
28. Bryde, D., Broquetas, M., Volm, J.M.: The project benefits of building information modelling (BIM). Int. J. Project Manage. **31**(7), 971–980 (2013). https://doi.org/10.1016/j.ijproman.2012.12.001
29. Liu, Y., Van, S., Hertogh, M.: Understanding effects of BIM on collaborative design and construction: an empirical study in China. Int. J. Project Manage. **35**(4), 686–698 (2017)
30. Garza, J.M., Pishdad-Bozorgi, P.: Workflow process model for flash track projects. J. Constr. Eng. Manage. **144**(6), 06018001 (2018). https://doi.org/10.1061/(ASCE)CO.1943-7862.0001501

31. Azhar, S., Khalfan, M., Maqsood, T.: Building information modeling (BIM): now and beyond. Australasian J. Construct. Econ. Build. **12**(4), 15–28 (2012)
32. Franz, B., Messner, J.: Evaluating the impact of building information modeling on project performance. J. Comput. Civ. Eng. **33**(3), 04019015 (2019). https://doi.org/10.1061/(ASCE)CP.1943-5487.0000832
33. Migilinskas, D., Popov, V., Juocevicius, V., Ustinovichius, L.: The benefits, obstacles and problems of practical bim implementation. Procedia Eng. **57**, 767–774 (2013). https://doi.org/10.1016/j.proeng.2013.04.097
34. Rojas, A., Vladimir, P.: Metodología para minimizar las deficiencias de diseño basada en la construcción virtual usando tecnologías BIM. Universidad Nacional de Ingeniería (2013). Accedido: 08 Nov 2021. [Online] Available on: http://oatd.org/oatd/record?record=oai%5C%3Acybertesis.uni.edu.pe%5C%3Auni%5C%2F3760. Accessed 21 May 2021
35. Eyzaguirre, R.R.: Potenciando la capacidad de análisis y comunicación de los proyectos de construcción, mediante herramientas virtuales BIM 4D durante la etapa de planificación (2015)
36. Hurtado, J.G.: La gerencia bim como sistema de gestión para proyectos de construcción. Gerencia Tecnológica Informática **14**(38), 2 (2015)

Chapter 22
Hybrid System (M-SALD) of Multicriterial Analysis as a Decision Support Tool for the Selection of Areas for the Construction of Hydraulic Structures

Solangel Rodríguez Vázquez and Natalia V. Mokrova

Abstract The identification of feasible areas for the location and possible construction of dams allows to guide and optimize the conservation efforts of the natural ecosystems, as well as the social and economic means that are around the area of analysis. The objective of this article is to present the proposal of a hybrid system that, using multicriteria and georeferenced analysis, allows the selection of possible potential areas to locate/build dams in them. In addition, a definition of the proposed hybrid system is proposed and a general methodological outline is established that guides the conduct of future studies. During the development, qualitative techniques based on simple analysis and the experience of experts were analyzed; and quantitative techniques based on the use of statistical and optimization models; in addition to spatial techniques used as tools for the realization of simpler quantitative processes. Of these, the latter have advantages, since they consider the execution of more detailed analyses, both of criteria and of restrictions and threats. The minimum phases that must be carried out are the definition of the objective to be prioritized, the selection and processing of criteria; the identification of restrictions and threats; as well as the choice of areas through optimization methods. Knowledge of the different methodologies makes it easier for decision makers to choose or combine the most suitable ones for the main objective.

Keywords Dam construction · Decision making · Georeferenced analysis · Geospatial techniques · Multi-criterial analysis · Project management

S. R. Vázquez (✉) · N. V. Mokrova
Universidad de las Ciencias Informáticas, CP 54830 La Habana, Cuba

S. R. Vázquez
Moscow State University of Civil Engineering, Moscow, Russia

P. Y. Piñero Pérez et al. (eds.), *Artificial Intelligence in Project Management and Making Decisions*, Studies in Computational Intelligence 1035,
https://doi.org/10.1007/978-3-030-97269-1_22

1 Introduction

Many countries today invest human and material resources to study the phenomena associated with floods and develop methods, techniques and instruments that help prevent them, with the aim of preserving human lives and the natural, social and economic resources that are usually affected by this type of event. The construction of dams is one of the means used to prevent climatological phenomena (hurricanes, cyclones, tropical storms, heavy rains) and to ensure a reliable and uninterrupted supply of water to the sectors of the economy and the population of the country.

The process of 'exhaustively' examining an area for dam construction is still done manually, say, based on visual analysis of digital elevation models (DEM), aerial photographs or satellite imagery. Then, for the most promising sites, hydrological models and/or data are developed to evaluate the performance of a dam at selected locations. An inadequate selection of the location would lead to non-compliance with the main objective of the dam, poor efficiency in terms of its exploitation and problems in the land planning processes, causing economic and social impacts. A correct search and planning process ensure that the benefits obtained justify the future environmental and social impacts to which the area of influence would be subjected. Many are the contributions that specialists in this subject have made through the analysis of cartographic, topographic information. Authors such as: Chezgi et al. [1], Noori et al. [2], Dai [3], Jozaghi et al. [4], Muhammad et al. [5], Minatour et al. [6, 7], Iftikhar [8], Peng [9, 10], Şen [11], Barkhordari [12], Ibrahim et al. [13], Forzieri et al. [14], Kharazi [15], Yazdani [15], Khazealpour [15], Petheram [16], Gallant [16], Read [16], Abushand [17], Alataw [17], Elhag [18], Eljack [18], Baban [19–22], Singh [23–25], Singh [23–25], Litoria [24], among others, incorporate in their research studies of the areas and location of dams, as well as channels and areas located for water supply. Research is also carried out on communal areas and crop areas to assess the need to create or bring dams to these places.

Through the traditional approach, hydrological information is processed and analyzed independently, causing a large volume of variables to be analyzed at the time of decision making, this makes it a complex and inefficient process. In addition, the analysis is complicated by a number of factors: a high level of uncertainty about initial information, future conditions, decision-maker preferences (DM); a weak structuring. Taking into account the salient features of the problem of selecting areas for the location of dams, there is a need to further improve the methodology and software to support the decision-making process.

2 Methodology and Methods

To carry out this research, an analysis of the state, problems and development prospects of the hydraulic industry in Cuba was carried out: concluding that it is necessary to build new dams and reservoirs, corresponding to the points analyzed in the Final Report "Integrating management of watersheds and coastal area in Caribbean small island developing states" [26]. On the watershed of the analysis of normative documents, scientific and methodical works the outline of the generalized process for decision-making in the location of dams is elaborated. To a lesser extent, issues of choice of areas for the location of dams have been resolved to date. However, the greatest difficulty lies in the multicriterial analysis of each of the alternatives (A_i) to be evaluated. To select an element (A_i) it is necessary to pre-select several area options (analysis of alternatives (Watershed)) and delimit through a distance analysis the protected areas of the points of the rivers (analysis of alternatives (Rivers)) that will be evaluated, Fig. 1.

The quality of choice of river alternatives depends on the outcome of comparison of river watershed alternatives. The analysis of each of the alternatives is based on the use of multiple criteria that allow to establish the problems in the area of influence under analysis, recognizing the existence and potential of the area's resources. There are many studies related to the analysis of the watersheds specifically and that use interchangeably parameters in relation to the type of research they carry out. Despite this, they all match in a series of parameters that are of importance for this type of analysis: (the *Evaluation variables* are the expert, weighted according to their opinion, the *Variables provided by experts* are variables that remain constant values for all of the alternatives and the *variables Other* variables that are not of importance, but nevertheless are necessary for the calculation of the variables of evaluation) see Fig. 2.

The task of selecting the destination for the construction of dams is shown as a task of rational choice or as a decision-making problem defined in an environment of uncertainty, multicriteria, in which dynamic situations usually occur and in which the following elements are considered:

- A group of experts $E = \{e_k | k \in (1, \ldots, q)\}$ who assess each criterion (C_i) against the defined objective and each of the sub-criteria $\left(P_{ij}\right)$ against the criteria (C_i).
- A set of criteria $C = \{c_i | i \in (1, \ldots, m)\}$ that characterize the defined objective.

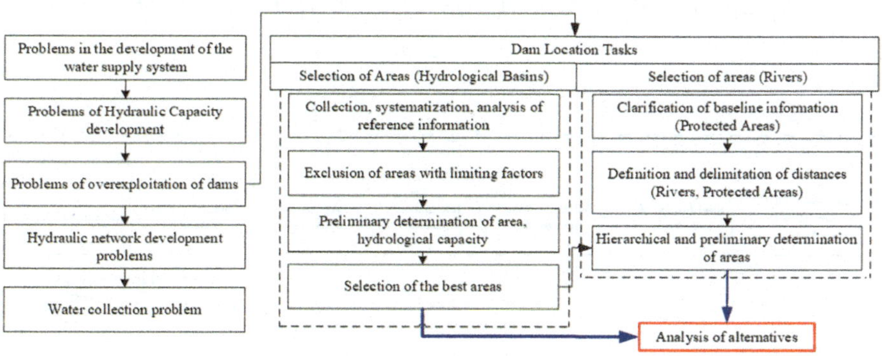

Fig. 1 Summary outline of the decision-making process on the location of dams

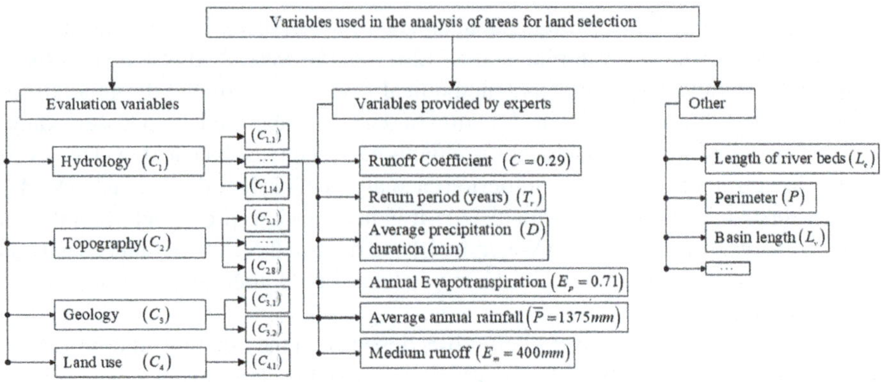

Fig. 2 General scheme of variables (criteria and sub-criteria) of analysis

- A set of sub-criteria $P = \{p_{ij} | i \in (1, \ldots, m); j \in (1, \ldots, n)\}$ associated to each of the criteria (C_i) established, so it is set to $P \subset C$ and $p_{ij} = C_{ij}$.
- A set of alternatives $A = \{a_u | u \in (1, \ldots, r)\}$ that characterize the areas of analysis.

The generalized decision-making process is based on the realization of 8 phases (Fig. 3) of which 7 are part of the decision Analysis process and the latter is not necessarily based on rational factors, because they can influence among other factors, those of a subjective and/or emotional type. The problem for the selection of areas for the location of dams (SALD) is mainly reflected in the phases of 3–5. This is because the collection of information and the assessment of alternatives are the two fundamental aspects that it is proposed to improve as part of the hybrid system proposal that is proposed in this research.

To improve the decision-making process and solve SALD problems, the task of choosing alternatives is formulated. $A = \{A_1, A_2, \ldots, A_u\}$ it is a set of alternatives that are evaluated from a set of sub-criteria associated with a set of specific criteria $F = \{(P_{1j} \subset C_1) \cup (P_{2j} \subset C_2) \cup (\ldots) \cup (P_{ij} \subset C_i)\}$. Each element of the set P

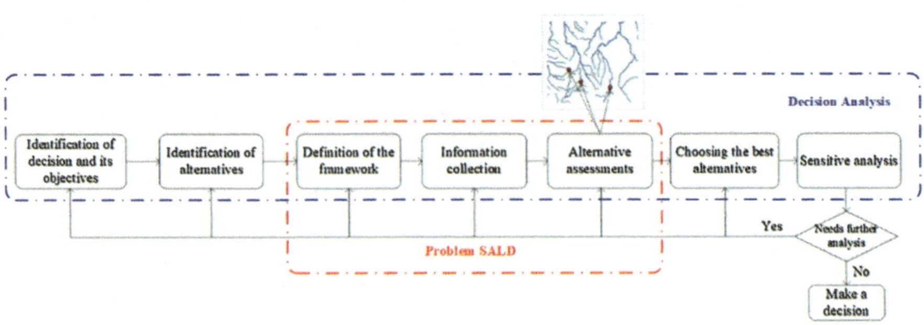

Fig. 3 Outline of the decision-making process

and C is evaluated by an expert E and sorted by preference. It is necessary that these alternatives be ordered in a hierarchical manner, taking into account the multicriterial estimates obtained from the analysis carried out.

The present research proposes to develop a hybrid system composed of the AHP-TOPSIS methods of multicriterial analysis to solve the SALD problems. The steps of its implementation are shown in Fig. 4. The hybrid method used in this study is based on the integration of hierarchical ordering methods AHP and TOPSIS modified mainly to be used in solving SALD problems. First, the AHP method is applied in order to determine which criteria and sub-criteria should be taken into account and the weight they should have when evaluating possible alternatives (Steps 1–5). Subsequently, the TOPSIS method (Steps 6–11) is applied in order to select the solution closest to the ideal before all the possible alternatives that exist, making use of a similarity index that is constructed by combining proximity to the positive ideal and distance from the negative ideal.

In step 1, the objective is formulated, the criteria of the set C and the sub-criteria of the set P are defined to evaluate each of the alternatives A_i. The standard way to perform step 2 is for the expert at the time of weighing to make a paired comparison between each of the c_i, p_{ij} and a_u respectively. For example, each of the criteria must be evaluated with respect to the other criteria on the watershed of a defined scale, the problem increases when the number of criteria, sub-criteria and alternatives to be assessed is large and therefore would increase the inconsistency of the evaluation $(C_1 \geq C_2, C_2 \geq C_3, C_3 \geq C_1)$. In the case of SALD problems, the number of variables that influence decision-making is quite large, as well as the number of possible alternatives that are analyzed. Because of them they proposed to make modifications in step 2. The modifications made are based on the expert only evaluating the criteria and sub-criteria based on his opinion regarding the influence that the criterion or sub-criterion has on the defined objective, thus eliminating the need to compare them with each other. This assessment is made using the scale proposed in Table 1.

In the case of the evaluation of alternatives in correspondence with the sub-criteria, the present research made modifications as shown in Fig. 5 regarding the

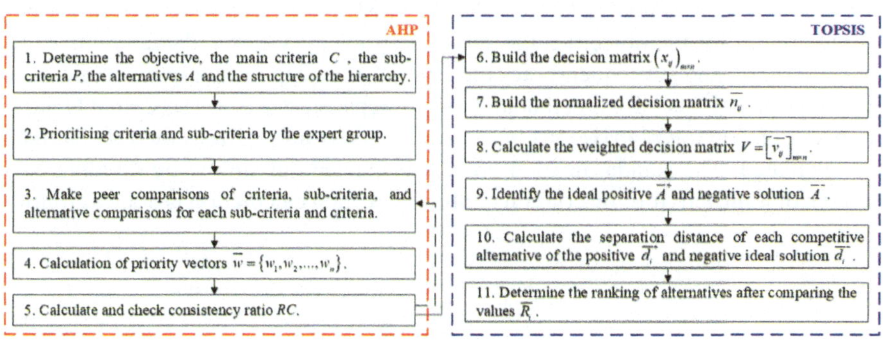

Fig. 4 Hybrid method of multicriterial analysis

Table 1 Proposed scale for direct assignment by the expert

Qualitative	Very low importance	Low importance	Moderately Important	Strong importance	Maximum importance
Quantitative	1	3	5	7	9

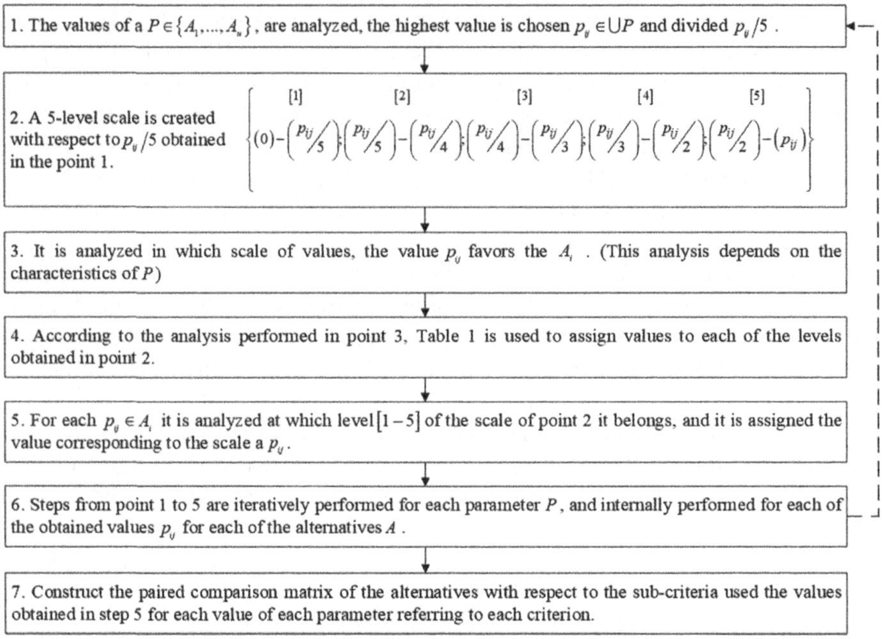

Fig. 5 Measurement methodology used for the evaluation of $P_{ij} \in A$, with a view to the subsequent comparison between the alternatives with respect to each P

standard methodology. Once the variables have been evaluated, the paired comparison matrices are formulated (step 3, Fig. 4) by M-SALD (proposed hybrid system) for the criteria for the objective, the sub-criteria for the criteria, the alternatives for the sub-criteria and subsequently for the criteria. By using the comparison matrices, the vector of weights ($\overrightarrow{w}$) is calculated in two steps (step 4, Fig. 4) see Eq. (1). First, the peer comparison matrix, $A \times w = \lambda_{\max} \times W$ is normalized, then the weights are calculated. The standardization process is carried out simply by dividing each element of joint activities by column totals. The weight calculation is done as follows [13]:

$$w_i = \sum_{j=1}^{n} \frac{x_{ij}}{n} \tag{1}$$

$$CI = \frac{\lambda_{\max} - n}{n - 1} \tag{2}$$

Table 2 Random consistency index (RI) based on matrix dimension (n)

n	1	2	3	4	5	6	7	8
RI	0	0	0.525	0.882	1.115	1.252	1.341	1.404
n	9	10	11	12	13	14	15	16 ≤
RI	1.452	1.484	1.513	1.535	1.555	1.570	1.583	1.595

In the AHP, peer comparisons in a judgment matrix are considered adequately consistent (step 5, Fig. 4), if the corresponding RC is less than 10%. The coefficient of RC is calculated after the consistency index (CI). CI is defined and the numerical calculation is performed as follows [27], Eq. (2):

Next, the RC is obtained by dividing the value of CI by the random consistency index (RI). The values of RI are shown in Table 2. Then the value of RC is calculated using the Eq. (3):

$$RC = \frac{CI}{RI} \tag{3}$$

The consistency test is completed when RC is computed numerically. If $RC <$ 10%, the data obtained are consistent and therefore you can continue with the realization of step 6. If $RC \geq$ 10%, the obtained data is inconsistent, the original values in the peer comparison matrix must be reconsidered and revised and therefore step 2 will be returned.

To carry out step 6, a matrix $\left[x_{ij}\right]_{m \times n}$ consisting of m alternatives and n criteria is established with the intersection of each alternative and criteria given as x_{ij}. Subsequently normalization (Step 7) is performed to obtain comparable scales. Vector normalization is defined below [28].

$$\overline{n_{ij}} = \frac{x_{ij}}{\sqrt{\sum_{j=1}^{m} (x_{ij})^2}}; \quad i = 1, \ldots, m; \quad j = 1, \ldots, n \tag{4}$$

Next, the weighted normalized value $\overline{v_{ij}}$ of the weighted normalized decision matrix (Step 8) $V = \left[\overline{v_{ij}}\right]_{m \times n}$ is calculated as:

$$\overline{v_{ij}} = w_j \otimes \overline{n_{ij}}; \quad i = 1, \ldots, m; \quad j = 1, \ldots, n \tag{5}$$

The positive ideal value set $\overline{A}^{+}$ is the one that maximizes the benefit criteria (or attributes) and minimizes the cost criteria, while the negative ideal value set $\overline{A}^{-}$ (also called anti-ideal solution) maximizes the cost criteria / attributes and minimizes the benefit criteria / attributes. The negative ideal solution consists of the worst performance values, while the best alternative is the one closest to the ideal solution (Step 9) [29].

$$\overline{A}^{+} = \left\{\overline{v_1}^{+}, \ldots, \overline{v_n}^{+}\right\} = \left\{\left(\max_i \overline{v_{ij}}, j \in J\right)\left(\min_i \overline{v_{ij}}, j \in J'\right)\right\}; \quad i = 1, \ldots, m \tag{6}$$

$$\overline{A}^{-} = \left\{\overline{v_1}^{-}, \ldots, \overline{v_n}^{-}\right\} = \left\{\left(\min_i \overline{v_{ij}}, j \in J\right)\left(\max_i \overline{v_{ij}}, j \in J'\right)\right\}; \quad i = 1, \ldots, m \tag{7}$$

For the realization of step 10, the separation distance of each competitive alternative of the ideal solution and the negative ideal solution is calculated.

$$\overline{d_i}^{+} = \left\{\sum_{j=1}^{n} \left(\overline{v_{ij}} - \overline{v_j}^{+}\right)^2\right\}^{\frac{1}{2}}; \quad i = 1, \ldots, m \tag{8}$$

$$\overline{d_i}^{-} = \left\{\sum_{j=1}^{n} \left(\overline{v_{ij}} - \overline{v_j}^{-}\right)^2\right\}^{\frac{1}{2}}; \quad i = 1, \ldots, m \tag{9}$$

To determine the ranking among the alternatives as a final step (step 11) it is necessary to compare the values $\overline{R_i}$. The relative proximity $\left(\overline{R_i}\right)$ to the ideal solution can be expressed as stated in Eq. 10, the closer the value $\overline{R_i}$ is to 1, the higher priority A_i will have.

$$\overline{R_i} = \frac{\overline{d_i}^{-}}{\overline{d_i}^{+} + \overline{d_i}^{-}}; \quad i = 1, \ldots, m \tag{10}$$

Using the AHP-TOPSIS hybrid method with the proposed modifications in the schematics (Figs. 4 and 5), a correct multicriterial analysis associated with SALD problems can be performed. The uncertainty of DM in terms of criteria and sub-criteria can be taken into account by applying the appropriate methodology (Fig. 5). Translation of estimates will reduce the number of evaluations by the DM and thus avoid errors in the analysis. The M-SALD hybrid system is not intended to replace the work of specialists, but to be part of the support tools that are usually used in the planning of hydraulic projects. The model developed is not limited to the analysis of a specific type of dam, but allows any type of dam project to be evaluated through it. M-SALD as shown in the following diagram (Fig. 6), is composed of three main phases: The Inputs, the Components that are based on the analysis and evaluation of the criteria and parameters (AHP-TOPSIS) and finally the Output.

For the realization of the *Phase "Inputs"* it is necessary to carry out a series of steps. First, it is necessary to make the decision of which is the candidate project (purpose with which you want to build the dam). Layers are then added to the model. *shp* necessary for the analysis, from which a set of evaluation variables $C_{i.j}$ =

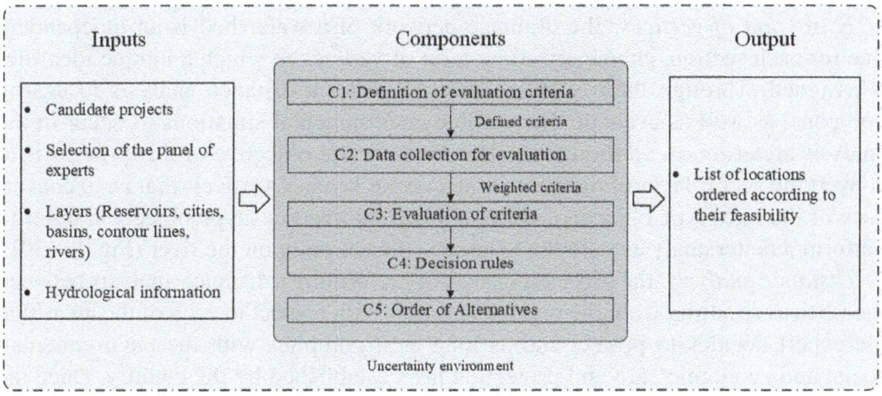

Fig. 6 General scheme of the M-SALD model

$\{C_{1.1}, \ldots, C_{i.j}\}$, will be extracted, in addition to a set of variables provided by the expert $P = \{p_1, p_2, \ldots, p_n\}$.

Each of these variables will be used in the evaluation of the set $A = \{A_1, A_2, \ldots, A_u\}$. At this phase, the group of experts $E = \{e_1, \ldots, e_k\}$ will be selected and the criteria and sub-criteria that will be taken into account for the analysis of alternatives will be defined according to the defined objective.

The *Phase "Components"* consists of 5 internal components that structure the steps resulting from Fig. 4. K_1 includes step 1 and takes care of structuring the problem as a hierarchy. In K_2 is carried out the process of collecting the information necessary for the analysis and extraction of the parameters of the layers. K_3 on the other hand, it is the component that includes steps 2–8, K_4 it includes steps 9–10 and finally K_5 it is in charge of the entire ordering process (step 11) of the alternatives that will finally be proposed to the expert by the model. Once the input variables to M-SALD were established, the multicriterial evaluation was carried out using the hybrid multicriterial analysis method AHP-TOSIS with the modifications provided by the author (Fig. 9). The use of the AHP-TOPSIS method is limited only to the analysis of watershed parameters. Therefore, once the analysis of alternatives (Watershed) and the hierarchical order of the same it is necessary to know what are the internal areas of each alternative in which it is feasible to carry out projects of location and construction of dams. For this it is necessary to perform the analysis of each of the rivers belonging to the watersheds Fig. 7.

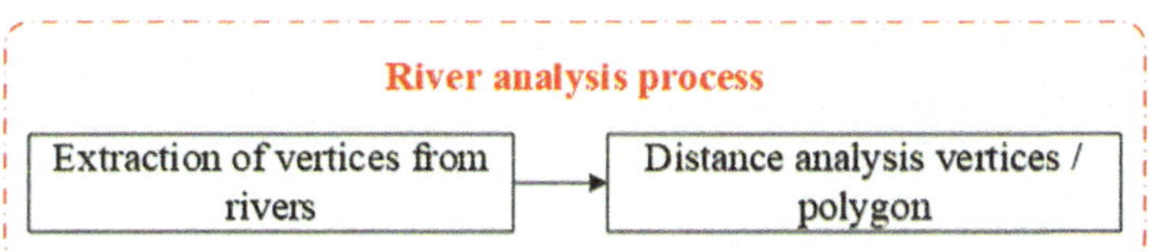

Fig. 7 River analysis process

Extraction of vertices: the drainage network of a watershed is an independent line for each section, characterized by a set of vertices to which a unique identifier is assigned. Through these vertices you can perform distance analysis to nearby polygons, as well as evaluate the possible environmental situations to occur in the analysis area (floods, earthquakes). The fundamental objective of this process is to convert the vector layer of the rivers into a vector layer of vertices, that is, to convert each of the lines type objects of the river layer into vertices type objects in order to perform a better analysis with the locations of each point on the river (Fig. 8) [30].

Distance analysis: the process consists of performing a distance analysis between the vertices resulting from the previous process with respect to each of the areas that the expert decides to protect and as long as it complies with the environmental, social and economic care and protection laws established by the country. Once the distances have been analyzed, the vertices that are within the established distance will be considered as inadequate vertices so they will be eliminated finally obtaining a layer of potential vertices. Finally, these vertices will be ordered hierarchically according to the hierarchical order obtained by AHP-TOPSIS with respect to the watersheds to which they belong [31].

In the *Phase "Output"* the list of potential locations ordered in a hierarchical way for the construction of hydraulic structures is obtained.

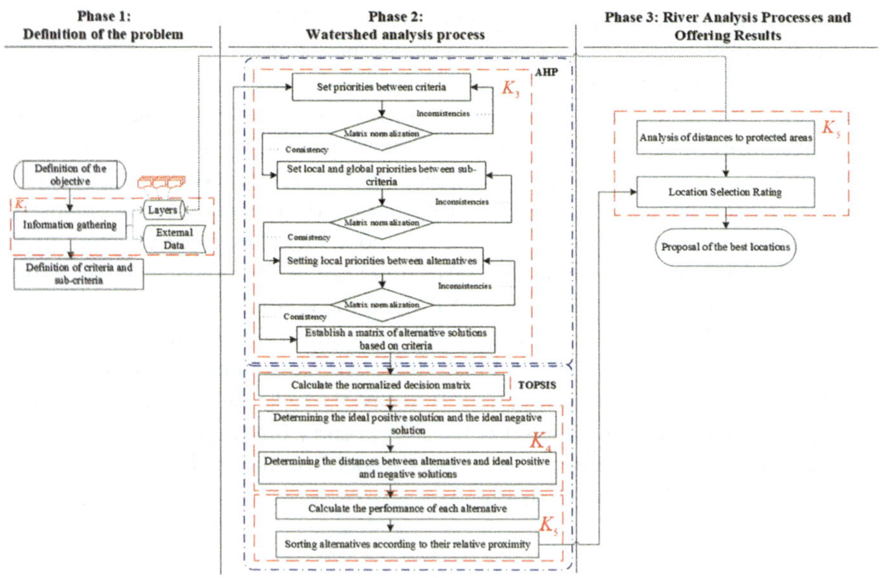

Fig. 8 Process flow diagram of the M-SALD model

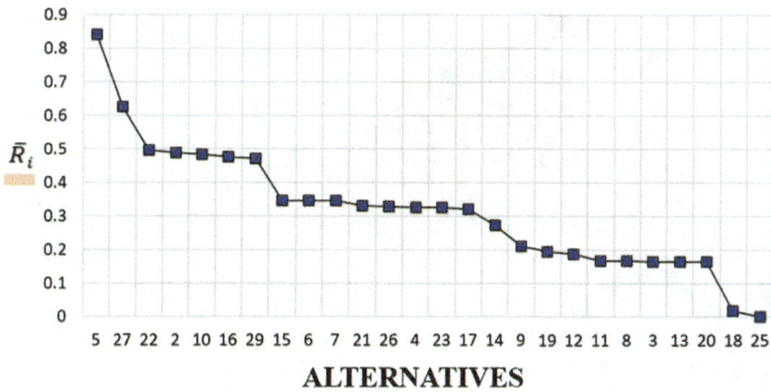

Fig. 9 The hierarchical order of the evaluated alternatives

3 Result

The case study selected to evaluate the hybrid model proposed above focuses on the selection of areas suitable for the construction of dams in the municipality of Manicaragua belonging to the province of Villa Clara, Cuba. The municipality of Manicaragua has a surface area of 1064.4 km^2, of which 309 km^2 constitute a mountain area with limitations for habitat development. Its relief is characterized by a fluvial plain and slightly dissected pre-mountain heights. It presents a group of areas considered as flood risk, highlighting the area within the Jibacoa Valley, which occupies an area of 12 km^2, where only the incidence of heavy rains causes the isolation of the area, thus affecting 176 homes and have to evacuate for such reasons a total of 486 people.

For the evaluation, 4 fundamental criteria were taken into account (Evaluation variables, Fig. 2), 25 parameters distributed among the 4 criteria and 29 alternatives (hydrological watersheds) of which two were false positives. For the selection of the criteria and parameters, the opinion of two hydraulic engineers specializing in the construction of dams was taken into account. Once the evaluation of the hybrid system was completed in the QGIS 3.4.14 tool, the ranking of the alternatives was determined based on their performance.

In this step, the relative proximity (yield, $\overline{R_i}$) for each of the alternatives was calculated using Eq. 10. The obtained value of $\overline{R_i}$ for each of the alternatives ranged from 0.84 to 0. The best alternatives were sorted according to $\overline{R_i}$ in descending order. For the selection of the appropriate areas of the rivers were taken into account, which should be located at a safe distance from those areas that are of the highest priority according to the provisions laid down in the legislation in force in the country (this distance it is not possible to set it as a fixed value, because it depends on the place of the analysis, which in the case of M-SALD the expert is the one who must provide the value of this statistic for the analysis). From this analysis, the areas located in the established distance range were removed as a possible alternative. Later they were

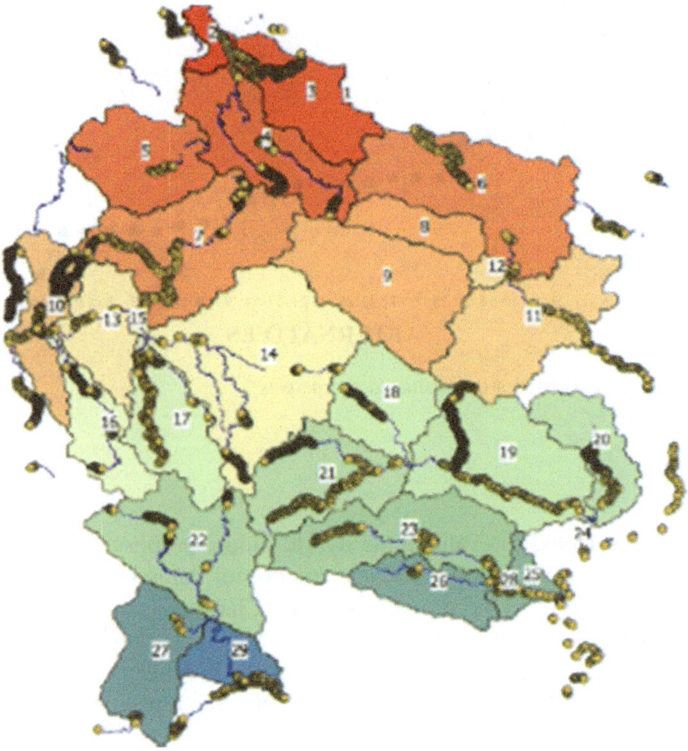

Fig. 10 Final vector layer of the M-SALD model

organized in a hierarchical way according to the watershed to which they belong. For the analyzed case study, the following solution was obtained (Figs. 9 and 10).

As a model output, a vector layer is obtained with the areas that hydrologically represent a power for the location of dams (Fig. 11). A list is also obtained with the analyzed alternatives organized in a descending way from the best evaluated to the worst evaluated. The evaluation of the feasibility of the model showed that, out of a total of 12,591 areas studied, the M-SALD model allowed to reduce to 23% the areas suitable for the construction of dams.

4 Conclusions

The result obtained with the help of the proposed method allows to select hydrologically potential areas of a region for the location of dams. The analysis of the results of the comparison allows to conclude that the M-SALD model proposed for the multicriterial selection of dam locations responds to the indications of the normative documents and to the degree of complexity of the task. The fuzzy preferences of the

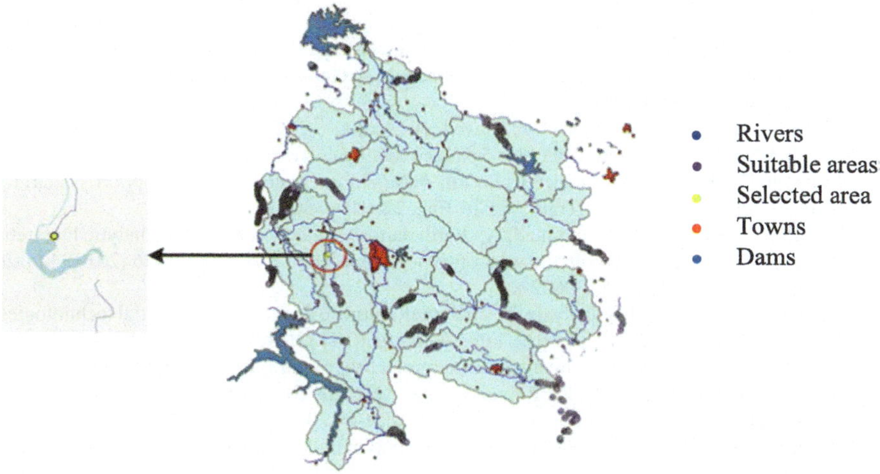

Fig. 11 Analysis of one of the points determined by the model as a possible area for the location of dams

MD in terms of criteria and sub-criteria have been taken into account; Procedures have been used to reduce the number of evaluations by DM and to obtain a solution that is as balanced and consistent as possible. The evaluations of the points obtained through the developed method take into account the quality of the possible subsequent solutions in them. The analysis was carried out according to 4 criteria, two of which are complex due to the number of sub-criteria associated with them. The need for a model for the analysis is fundamentally dictated by the time it takes to carry out this type of selection process, the elimination of manual or perspective selection, the reduction of economic expenses and dependence on external entities to obtain data, dissatisfaction of the sector that should be benefited by the dam, ecological damage to the environment located around the location area. The dam is a facility that will have an impact for many decades, so the use of M-SALD as a support tool is justified.

References

1. Chezgi, J., Pourghasemi, H.R., Naghibi, S.A., Moradi, H.R., Zarkesh, M.K.: Assessment of a spatial multi-criteria evaluation to site selection underground dams in the Alborz Province, Iran. Geocarto Int. **31**, 628–646 (2016). https://doi.org/10.1080/10106049.2015.1073366
2. Noori, A., Bonakdari, H., Morovati, K., Gharabaghi, B.: The optimal dam site selection using a group decision-making method through fuzzy TOPSIS model. Environ. Syst. Decis. **38**, 471–488 (2018). https://doi.org/10.1007/s10669-018-9673-x
3. Dai, X.: Dam Site Selection Using an Integrated Method of AHP and GIS for Decision Making Support in Bortala, Northwest China (2016) [Online]. Available at: https://webapps.itc.utwente.nl/librarywww/papers_2016/msc/gem/dai.pdf. Accessed 11 May 2021

4. Jozaghi, A., Alizadeh, B., Hatami, M., Flood, I., Khorrami, M., Khodaei, N., Ghasemi Tousi, E.: A comparative study of the AHP and TOPSIS techniques for dam site selection using GIS: a case study of Sistan and Baluchestan Province, Iran. Geosciences (2018). https://doi.org/10.3390/geosciences8120494
5. Muhammad, S., Muneer, J., Gul, C., Javed, A., Malik, S.: Identification of suitable site for DAM in Northern Pakistan using Geo-Spatial Techniques. researchgate.net (2017)
6. Minatour, Y., Khazaei, J., Ataei, M., Javadi, A.A.: An integrated decision support system for dam site selection. Sci. Iran. Trans. A Civ. Eng. **22**, 319–330 (2015)
7. Yasser, M., Jahangir, K., Mohmmad, A.: Earth dam site selection using the analytic hierarchy process (AHP): a case study in the west of Iran. Arab. J. Geosci. **6**, 3417–3426 (2013). https://doi.org/10.1007/s12517-012-0602-x
8. Iftikhar, S.: Site suitability analysis for small multipurpose dams using geospatial technologies. J. Remote Sens. GIS **5**, 1–13 (2016). https://doi.org/10.4172/2469-4134.1000162
9. Peng, Z.: Location of dam using analytic hierarchy progress. In: Proceedings of the Advances in Materials, Machinery, Electrical Engineering (AMMEE 2017), pp. 441–444. Atlantis Press, Paris, France (2017). https://doi.org/10.2991/ammee-17.2017.82
10. Peng, J., Peng, F.L.: A GIS-based evaluation method of underground space resources for urban spatial planning: Part 2 application. Tunn. Undergr. Space Technol. **77**, 142–165 (2018). https://doi.org/10.1016/J.TUST.2018.03.013
11. Şen, Z., Al-Suba'I, K.: Hydrological considerations for dam siting in arid regions: a Saudi Arabian study. Hydrol. Sci. J. **47**, 173–186 (2002). https://doi.org/10.1080/02626660209492922
12. Barkhordari, J.: The pre-selection of suitable sites for small underground dams in arid areas using GIS (a case study in Yazd_Ardakan watershed). Int. Geoinformatics Res. Dev. J. **6**, 18–27 (2015)
13. Ibrahim, G.R.F., Rasul, A., Hamid, A.A., Ali, Z.F., Dewana, A.A.: Suitable site selection for rainwater harvesting and storage case study using Dohuk governorate. Water (2019).https://doi.org/10.3390/w11040864
14. Forzieri, G., Gardenti, M., Caparrini, F., Castelli, F.: A methodology for the pre-selection of suitable sites for surface and underground small dams in arid areas: a case study in the region of Kidal, Mali. Phys. Chem. Earth Parts A/B/C **33**, 74–85 (2008). https://doi.org/10.1016/J.PCE.2007.04.014
15. Kharazi, P., Yazdani, M.R., Khazealpour, P.: Suitable identification of underground dam locations, using decision-making methods in a semi-arid region of Iranian Semnan Plain. Groundw. Sustain. Dev. (2019). https://doi.org/10.1016/J.GSD.2019.100240
16. Petheram, C., Gallant, J.C., Read, A.: An automated and rapid method for identifying dam wall locations and estimating reservoir yield over large areas. Environ. Model. Softw. **92**, 189–201 (2017). https://doi.org/10.1016/j.envsoft.2017.02.016
17. Abushand, E., Alataw, S.: Dam site selection using remote sensing techniques and geographical information system to control flood events in Tabuk City. Hydrol. Curr. Res. **6**, 1–13 (2015). https://doi.org/10.4172/2157-7587.1000189
18. Elhag, A.R., Eljack, I.H.: DAM site selection using GIS techniques and remote sensing to minimize flash floods in East Nile Locality (Soba Valley)—Khartoum State. Int. J. Eng. Sci. Invent. **5**, 50–61 (2016)
19. Baban, S.M.J.: Use of remote sensing and geographical information systems in developing lake management strategies. Hydrobiologia **395**(396), 211–226 (1999). https://doi.org/10.1023/A:1017057820780
20. Baban, S.M.J., Wan Yusof, K.: Mapping land use/cover distribution on a mountainous tropical island using remote sensing and GIS. Int. J. Remote Sens. **22**, 1909–1918 (2001). https://doi.org/10.1080/01431160119220
21. Baban, S.M.J., Flannagan, J.: Developing and implementing GIS-assisted constraints criteria for planning landfill sites in the UK. Plan. Pract. Res. **13**, 139–151 (1998). https://doi.org/10.1080/02697459816157

22. Baban, S.M.J., Wan-Yusof, K.: Modelling optimum sites for locating reservoirs in tropical environments. Water Resour. Manag. **17**, 1–17 (2003). https://doi.org/10.1023/A:1023066705226
23. Singh, L.K., Jha, M.K., Chowdary, V.M.: Multi-criteria analysis and GIS modeling for identifying prospective water harvesting and artificial recharge sites for sustainable water supply. J. Clean. Prod. **142**, 1436–1456 (2017). https://doi.org/10.1016/J.JCLEPRO.2016.11.163
24. Singh, J.P., Singh, D., Litoria, P.K.: Selection of suitable sites for water harvesting structures in Soankhad watershed, Punjab using remote sensing and geographical information system (RS&GIS) approach—a case study. J. Indian Soc. Remote Sens. **37**, 21–35 (2009). https://doi.org/10.1007/s12524-009-0009-7
25. Singh, V.P., Chow, V.T.: Handbook of Applied Hydrology. McGraw-Hill Education (2017)
26. Cuban Environment Agency (AMA): Integrating Management of Watersheds and Coastal Area in Caribbean Small Island Developing States, Ciudad de la Habana (2001)
27. Xiao, Q., Shan, M., Xiao, X., Rao, C.: Evaluation model of industrial operation quality under multi-source heterogeneous data information. Int. J. Fuzzy Syst. **22**, 522–547 (2020). https://doi.org/10.1007/s40815-019-00776-x
28. Saeli, M., Micale, R., Seabra, M.P., Labrincha, J.A., La Scalia, G.: Selection of novel geopolymeric mortars for sustainable construction applications using fuzzy TOPSIS approach. Sustainability **12**, 59–87 (2020). https://doi.org/10.3390/su12155987
29. Катаржина, Х.: Выбор технологий с помощью метода TOPSIS. Foresight STI Gov. **14**, 85–96 (2020). https://doi.org/10.17323/2500-2597.2020.1.85.96
30. Vazquez, S.R., Mokrova, N.V.: Comparison of applicability of different computational geometry algorithms for the detection of vertices in river layers in GIS systems. In: 2020 International Multi-conference on Industrial Engineering and Modern Technologies (FarEastCon), pp 1–4. IEEE (2020). https://doi.org/10.1109/FarEastCon50210.2020.9271660
31. Vázquez, S.R., Makrova, N.V.: Vector-spatial analysis of GIS application layers for placing strategic points in dam design. Строительство и техногенная безопасност ь. 43–51 (2021). https://doi.org/10.37279/2413-1873-2021-20-43-51

Chapter 23
Trends in Photogrammetry and Its Integration with Artificial Intelligence

Nayma Martín Amaro, Iliana Pérez Pupo, Dagoberto F. Pérez Montesinos, and Pedro Y. Piñero Pérez

Abstract In this work, important elements about photogrammetry are expressed, such as its concept, the main objective, its principles, stages, techniques that have been used over the years and application examples. Photogrammetry is divided into categories according to the type of frames, it can also be applied in regions where classical methods cannot be used, such as: in impassable regions, such as: jungles, swamps, deserts, territories hit by some epidemic or occupied by enemy forces, etc., due to its intrinsic characteristic, that objects can be measured without the need to be near them.

Keywords Artificial intelligence · Photogrammetry · Photos

1 Introduction

The task of carrying out photogrammetric surveys using aerial photos is carried out systematically in all countries of the world, given its enormous importance in many branches of the economy.

The application of automated means in geodetic and cartographic work dates from the beginnings of computer manufacturing. Particularly, work with digitized aerial images has increased in recent decades due to the technological development achieved.

N. M. Amaro (✉)
Centro de Capacitación, Empresa de Telecomunicaciones ETECSA, Sucursal Sancti Spiritus, Sancti Spiritus, Cuba

I. P. Pupo · P. Y. Piñero Pérez
Grupo de Investigaciones en Gestión de Proyectos, Universidad de las Ciencias Informáticas, CP 54830 La Habana, Cuba
e-mail: ppp@uci.cu

D. F. Pérez Montesinos
Departamento de Informática, Grupo Empresarial CIMEX, Sucursal Sancti Spiritus, Sancti Spiritus, Cuba
e-mail: dagofpm@cimex.com.cu

P. Y. Piñero Pérez et al. (eds.), *Artificial Intelligence in Project Management and Making Decisions*, Studies in Computational Intelligence 1035,
https://doi.org/10.1007/978-3-030-97269-1_23

In Cuba, we mainly depend on systems acquired abroad for geodetic and cartographic purposes, which, in addition to accentuating technological dependence, ultimately hinder our development, due to the impossibility of researching and developing new treatment methods, that deviate from the acquired technology.

Based on this problematic situation, the objective of this work is proposed: to build a theoretical referential framework through a tertiary review that allows analyzing the evolution and trends in the development of photogrammetry, artificial intelligence and its applications. The work is organized as follows: the second Methodology section explains the experimental design used, the Development section shows the results of the analysis and finally the conclusions are presented.

2 Methodology

The study that is presented constitutes an exploratory investigation [1] in it that develops the following design of experiments:

Step 1. Design of a systematic review protocol that allows building a theoretical framework associated with the topic.
Step 2. In-depth analysis of the bibliography consulted by each of the established categories.
Step 3. Application of descriptive statistics and meta-analysis techniques for the presentation of the research results.
Step 4. Identify open lines of investigation.

In line with the silver, a systematic review protocol [2] is designed and the historical evolution of photogrammetry is analyzed in different stages: publications between 1980 and 2000, publications between 2001 and 2017 and from 2018 to 2021.

Systematic review protocol applied:

1. Definition of the object of study, field of action and objective of the research.
   a. Research object: photogrammetry and artificial intelligence.
   b. Course objective: Build a theoretical framework associated with photogrammetry, trends, and applications using artificial intelligence tools and BIM methodologies.
   c. Field Research: structures, methods and applications of photogrammetry and its combination with artificial intelligence tools.
2. Definition of a bibliographic manager.
3. Define fonts of academic information for the development of the review: "Semantic Scholar", "Google Scholar" and "Scopus" and other meta-search engines based on open science.
4. Definition of the following set of key phrases for carrying out the searches: "Photogrammetry" AND ("Artificial Intelligence" OR Usability), Photogrammetry + Artificial Intelligence.

5. Define the goals of the bibliometric analysis in the form of research questions and inclusion–exclusion criteria:

   a. How has it been the trend of publications per year?
   b. Who are the main authors?
   c. What are the affiliations and countries of the main authors?
   d. How I know do they distribute the publications considering the types of documents in: articles, books, theses and reports of conferences or congresses?
   e. Exclusion of works published in spaces with little level of arbitration.

6. Sort out and filter posts in the following set of categories:

   - Classics: refers to pioneering publications in photogrammetry, where the fundamental principles that set guidelines in the theory are exposed.
   - Extensions to theories: it refers to publications that extend the theory proposed in the publications understood as classical, do not set guidelines that significantly change the previously proposed methods, although they do develop contributions to knowledge.
   - Application results: publications that focus on the use of existing theory in concrete practical settings.
   - Reviews tertiary: refers to articles reviewing trends and developments in the subject in question.

7. Synthesize the main trends:

   As a second step in the methodology used, a detailed analysis of each of the works is carried out and they are characterized with respect to the following elements: Areas of application of the proposal.

Then I know they apply descriptive statistics techniques and meta-analysis for the presentation of research results [2]. Finally, open lines are analyzed and proposed in the object of study analyzed.

3 Development

From the study carried out, the largest number of scientific publications on the subject of Photogrammetry was identified in the years 2001–2017 [3–7]. Although we can highlight that in 2020 there is a significant increase in publications on the subject. In addition, 2017 is shown as the year where the largest number of publications on Photogrammetry linked to Artificial Intelligence [8–11] is evidenced.

On the other hand, most of the articles are written by varied authors (Table 1). We can see in the review made several uses of Photogrammetry [10, 12–16], since there are studies of its treatment in different disciplines, such as the development of projects associated with geology, forestry, agriculture, design and construction, cadastre, environmental studies, exploration, military intelligence, medicine, criminology,

Table 1 Authors who have published articles on photogrammetry

Authors	AerialInsights; Úrsula Martínez-Iranzo; Enric Medina-Ripoll; Gonzalo Utrilla Redondo; Cristina García Bermell; Ignacio Bermejo Bosc *, Juan López Pascual; Rubén Sancho Gómez-Zurdo; Omar Mireles-Loera; Llajaruna, GM; Inmaculada Picon-Cabrera, Pablo Rodríguez-Gonzálvez, Isabella Toschi, Fabio Remondino, Diego González-Aguilera; Blachut, T., Burkhardt, R; Garcia, SA; R. Hernández-Sampieri and CPM Torres; Lizeth Daniela Molina Martinez; JH Littell, J. Corcoran, and V. Pillai; Leandro Gómez Mora; Zhang, Bingcai; Honarmand, Mehdi; Shahriari, Hadi; Barbero-García, Inés; Pierdicca, Roberto; Paolanti, Marina; Felicetti, Andrea; Lerma, Jose Luis

traffic studies, oceanography, meteorological observations, architecture and archeology. Generally, these last 6 topics are covered in books called Non-topographic applications of photogrammetry [17–19].

Consulting the journals that publish scientific works [6, 10, 15, 16, 21–27], it can be seen that the works have been published in different countries (Table 2), and therefore, in different languages.

Table 2 Magazines and countries that publish scientific articles related to Photogrammetry and Artificial Intelligence (own elaboration)

Journals	Countries
Editorial People and Education, Cuba	Cuba
Institute of Biomechanics (IBV). Polytechnic University of Valencia. Valen-Inc, Spain	Spain
University Polytechnic of Catalonia. Polytechnic Digital Initiative	
Reports of the construction	
Santander Technology Units	
EGA. Architectural Graphic Expression Magazine	
Unibersitat Politécnica de Valéncia, Faculty of Fine Arts of Sant Carles	
Ox-ford University Press	USA
ISPRS, Falls Church	
University de Los Andes Department of Civil and Environmental Engineering	Colombia
Freedom: Santa Elena Peninsula State University	Ecuador
REV. Archeology	Argentina
McGraw-HillInteramericana México ˆ eD. F DF	México
Rev Mex MedForensic	
Academic repository of the University of Chile	Chile
National UniversityMayor of San Marcos. Lime. Peru	Perú
Advances in Intelligent Systems and Computing	Singapore
Geo-spatial Information science	France
Taylor & Francis	
International Journal of Digital Earth	

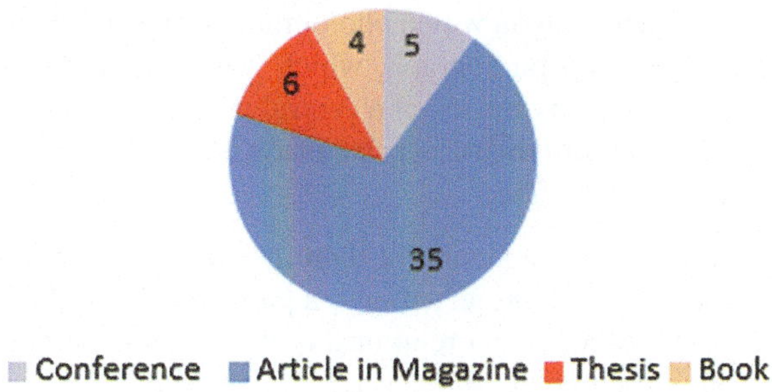

Fig. 1 Bibliography by type of publication

Regarding the h-index, there are varied coefficients, especially the one of the journal Oxford University Press YGeo-spatial Information Science.

A better characterization [28] of the bibliography consulted [29–31] for this work is shown below:

- The majority of the researches consulted are articles in journals, it is identified that in this thematic area there are few doctoral theses, but if there are some degree theses, see Fig. 1.
- The majority of the consulted publications are indexed in Scopus, Scielo and Web Of Science see Fig. 2.
- The sources of key information and the main publishers for the researches consulted were IEEE, Elsevier and Springer, respectively.

In the consulted bibliography [32–35] a set of elements that characterize or form part of the subject area is identified.

Photogrammetry is a discipline that creates 3D models from 2D images, in order to obtain geometric characteristics of the objects they represent, through the use of mathematical relationships established in projective geometry, and the stereoscopic vision that the human being. Since the images of the objects are obtained by photographic means, the measurement is carried out at a distance, without there being physical contact with the object.

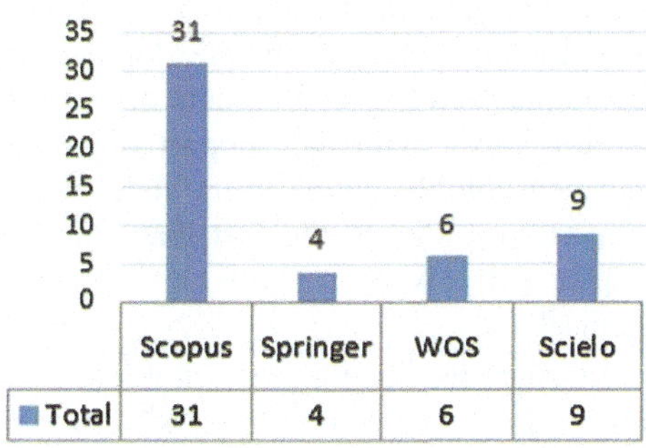

Fig. 2 Indexing of the publications in magazines and conferences analyzed

The principle on which photogrammetry is based consists in projecting the image registered in a photograph, which has been projected onto a negative or onto the sensor matrix by means of a central projection.

In photogrammetry, in principle, it is assumed that the central projection is perfect, which implies that:

- There is no deflection of light rays passing through the camera lens.
- The image is projected onto a perfectly flat surface.
- The mathematical relationship that relates the object and its image is known as the principle of collinearity.

Let's start by exposing the most important elements of photogrammetry such as:

The main objective of photogrammetry is the preparation of topographic maps, through the use of aerial or terrestrial photographs and the quantitative and qualitative analysis of the image [6, 36].

3.1 Principles Detected in the Literature on Photogrammetry

The object of photogrammetry is to go from the central projection, or perspective that constitutes the photogram, to the orthogonal projection that is the topographic plane.

Since its inception, photogrammetry has become the indispensable tool in the production of the cartographic base of all the countries of the world; in fact, most of the topographic mapping of our planet has been carried out through this discipline.

Although photogrammetry had its beginning in the lifting of architectural facades and building plans, through the use of terrestrial photographs, aerial photographs were soon used for the lifting of the base cartography, which gave it the tremendous boom that has maintained to this day. As a consequence of the use of aerial photography, the discipline of photointerpretation, which shares its basic foundations with aerial photogrammetry, was detached from photogrammetry [37].

Starting in the eighties, the accelerated development of computing led to the establishment of remote sensing as a logical consequence of the evolution of photointerpretation, as well as to the development of computerized digital image processing techniques and the development of vision by computer.

Nowadays, with the support of computers, photogrammetry has become an indispensable discipline in the field of cartography, while the number of its users is increasing due to the fact that high-cost photogrammetric equipment is being displaced by expensive programs minor, or by programs developed by the users themselves.

Etymologically, the word photogrammetry is derived from the Greek words φωτος photos, which means light; γραμα, gramma, which means what is drawn or written, and μετρον, metron, which means to measure [37]. Using those words together photogrammetry means measuring graphically by means of light.

In our country, the GEOCUBA company is in charge of studies in the area of photogrammetry. The development of automation in this branch has a variety of specialists trained in cartography and geodesy issues. According to the studies carried out, currently in this center foreign programs are used in the area of aerophotogeodesy, which are: PHOTOMOD, ERDAS, DIAP and ENVI; The latter is used mainly for PDI (Digital Image Processing), classification and obtaining orthophotos with information from aerial and satellite sensors.

From the foregoing, it can be deduced that, in our country, there is a non-negligible dependence on foreign products, which results in investment expenses in the purchase of these inputs, which on the other hand have functionalities that can be obtained through national products.

3.2 Stages of Photogrammetry

The transition from the central projection to the orthogonal projection can be done either by graphic photogrammetry, practically in disuse today, or by stereo photogrammetry, which is currently used in the vast majority of photogrammetric works.

Graphical photogrammetry, using the principles of projective geometry, marked the beginning of this discipline, since for the time it was the only way in which photographs could be restored. This modality is based on the intersection of lines that start from two different stations, that is, from the points where the photographs were taken, towards a common point.

Currently, thank you due to the computational capacity offered by computers, the use of this form of restitution has become, for some special cases, an alternative that can compete with stereo photogrammetry.

Stereo photogrammetry is based on stereoscopic vision to recreate in the mind of the observer a stereoscopic model from a couple of photographs, each one taken from a different position, to be observed separately by the respective eye. In this way, each eye transmits a slightly different image to the brain from the other, just as they do when observing three-dimensional objects. The brain will then interpret these differences as differences in depth, and will form a stereoscopic pattern in the mind of the observer.

3.3 Main Photogrammetry Techniques

In the literature consulted [38–40] identifies the following techniques.

Photogrammetry analog, which emerged in the thirties based on restitution devices and is responsible for the realization of most of the world cartography. In it, a pair of photographs is placed in an optical or mechanical restitution device.

The operator manually performs the interior and exterior orientation to create the stereoscopic model, properly scaled and leveled. The elevation of the planimetric and altimetric information of the model is also carried out manually, by tracking with the floating mark placed on the details of the surface of the model.

This Information is plotted on material that is not very deformable (rigid bases, plastics) placed on the plotting table, related to the model by mechanical or electrical means.

Photogrammetry analytical, which appeared in 1957 as a natural development of the interrelation between analogue restoring devices and the emergence of computing. In it, the information gathering is analog and the geometric modeling is mathematical. Through the use of a monocomparator or a stereocomparator integrated in the renderer, the x, y coordinates of the pertinent points of the photographs are measured, which coordinates are processed by the system's computer programs. East performs the processing of the interior and exterior orientation in an analytical way and processes the information of the model carried out by the operator, to bring it to its correct orthogonal position, and finally store it in a CAD-type database.

Photogrammetry Digital, currently on the rise, arises as a consequence of the great development of computing, which made it possible to carry out all photogrammetric processes through the use of computers. With digital photogrammetry, the possibilities of image exploitation grow, while technologies are simplified, thus allowing the automatic generation of terrain elevation models, ortho-images and stereortho-images, generation and visualization of three-dimensional models. To carry out the digital restitution, the digital images are entered into the computer, and by displaying them on the screen, the operator enters the necessary points to carry out the orientation process in a mathematical way. Restitution can be an iterative process with the operator or be done automatically by image correlation. The output in digital photogrammetry can be in raster or vector format.

3.4 Role of Automation in Aerial Photography Work

Photogrammetry Digital arises as a consequence of the great development of computation, which made it possible to carry out all photogrammetric processes using computers and special software. From the digital era, the possibilities of exploiting images by extracting a lot of information increased, allowing the generation of digital models of elevation of the terrain, ortho-images, stereo-images, three-dimensional visualization of the terrain, obtaining cross-sectional profiles, etc. The simplicity of the digital format makes it very easy to process images to restore them; The digital images are entered into the computer and by means of the visualization on the screen the operator locates the points to carry out the orientation process in a mathematical way.

This allows the operator performs the restitution manually (supervised) or the computer does it automatically (unsupervised). When we operate with digital images, the output can be in two formats, Raster or Vector.

3.5 Advantages and Disadvantages of Photogrammetry

Photogrammetry It is a discipline based on the 3D reconstruction of reality from two-dimensional images; That is why its advantages and disadvantages are closely linked to the forms of registration (generally aerial photographs), and to the restitution methods and equipment [35, 41–43].

Advantages

- Reduction cost: It is related to the size of the area to be restored. Depending on the scale of the map that is made and the characteristics of the area to be surveyed, as the area increases, the competitiveness of the method increases.
- Reduction Fieldwork: Fieldwork is a costly component of all surveying work, the cost of which increases with accessibility and adverse weather conditions. The reduced number of control points required in photogrammetry, reduces the stay in the field.
- Compilation speed: The time required to make a photogrammetric map is less compared to that required by the topographic survey and its subsequent cabinet work.
- Given the short time necessary for the photogrammetric survey with which a faithful reproduction of the terrain is obtained, in a given period, it provides us with very valuable data in cases of sudden changes, such as: during or after natural disasters.
- Flexibility: The photogrammetric method can be carried out on a varied range of scales, depending on the scale of the photographs and the type of compiler used, also depending on the availability of economic and technical resources. For this reason, supply maps or substitutes with different production times, costs and precision.
- Multi-temporal record. It is very useful for verifying photogrammetric maps. Aerial photos provide an accurate record of terrain features on the date they were taken, allowing comparisons between photos from other dates to assess changes in the terrain. Aerial photos can also be used for other uses than the original project, since that also of metric information, aerial photographs provide quantitative and qualitative information.
- Photogrammetry. It can be applied in regions where the classical methods cannot be used, such as: in impassable regions, such as: swamps, deserts, virgin forests, territories plagued by an epidemic or occupied by enemy forces, etc., due to the intrinsic characteristic of photogrammetry, that objects can be measured without the need to be close to them.
- Aero photogrammetry. It also provides a series of advantages, such as the photograph itself, which is a document that allows any control to be carried out at any given time. Critical legal, geological, historical and geographic data can also be obtained from it.

Table 3 Advantages and disadvantages of digital photogrammetry

Advantage	Digital information output
	It is possible to automate processes
	Image treatment requires soft-ware
	Ease display and find the homologous points
	Images of great dimensional stability
Disadvantages	It is a recent technique and still in development
	Requires a high volume of storage lie (currently processes are faster)

Disadvantages

- Impossibility of seeing the surface of the land when there is dense vegetation cover. In this case, it is impossible to locate the floating mark on the ground, so an average height of the vegetation with respect to the ground must be assumed. However, as the vegetation cover tends to soften the topographic features of the terrain, there will always be errors in the location of the contour lines, although the elevation can be verified in the gaps that exist in the vegetation.
- Location of contour lines on flat surfaces. Determining the trajectory of a contour line on flat terrain has a high degree of difficulty, due to the imprecision in the placement of the floating mark. Consequently, bounded points are placed in the restitution or it is complemented with field work.
- The site must be inspected to determine those elements that are not satisfactorily visible, or whose exact nature cannot be determined on the stereo model.
- Forever a field check is necessary.
- The Application of photogrammetry requires a considerable investment of equipment and specialized personnel, which is why its cost is high.
- To carry out new surveys it is necessary to obtain new photographs.

3.6 Advantages and Disadvantages of Digital Photogrammetry

According to in the literature, we can list a series of advantages and disadvantages of digital photogrammetry [4], see Table 3.

3.7 Classification of Photogrammetry Techniques and Applications

Photogrammetry is divides into categories according to the type of frames or the sensor used:

1. Photogrammetry land
2. Photogrammetry aerial
3. Photogrammetry space.

If the vehicle is in space, three types can be established:

1. Photogrammetry space
2. Photogrammetry satellite
3. Photogrammetry otherworldly.

The first use of photogrammetry consisted in making topographic maps and plans [9, 19, 44–47]. In fact, the base maps of the cartography of any country are obtained through it. Currently, in addition to the realization of these base maps, many other types of maps of a special nature are made, which can present a great variety of scales, and are used in the project and design of works such as highways, roads, roads. of railways, bridges, pipelines, oil pipelines, gas pipelines, transmission lines, hydroelectric dams, urban studies, etc.

What's more of these maps, mainly oriented to the development of civil engineering works, we can mention maps made for cadastral use, geological maps, soil maps, forest maps, etc.

Inside Of the disciplines that benefit from non-topographic photogrammetry we can mention architecture, in the raising of monuments and sites; archeology, in applications similar to those used in architecture; biostereometry, in the study of forms of living beings; shipbuilding, automotive and heavy machinery also make use of this discipline.

An important amount of cartographic information produced through the use of photogrammetry is used as a spatial reference in digital databases. These are integrated with other data obtained by different means, generally of a qualitative and descriptive nature, to form geographic information systems (GIS).

From the analysis carried out [7, 36, 47, 48] it is identified that the area with the highest number of applications in the period 2000–2017 was that associated with the sectors of medicine and computer science. However, the trend in the last five years (2017–2021) has focused on the development of projects associated with medicine, engineering, construction and architecture, see Fig. 3.

4 Conclusions

In this work, a review of the state of the art of Photogrammetry has been carried out from the following perspectives: the areas of application and their evolution, principles, stages, techniques, objective, classification, as well as their advantages and disadvantages.

Photogrammetry has been become an innovative trend applicable to countless areas. In recent years, there has been a growth in the application of photogrammetry, also linked to artificial intelligence and augmented reality, in many areas, such as

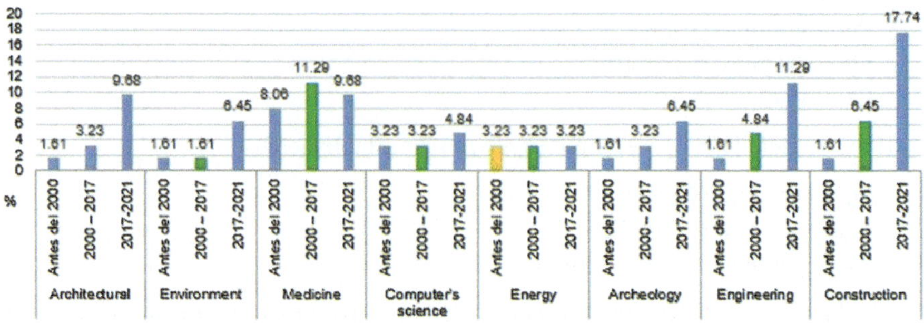

Fig. 3 Areas of application of photogrammetry according to the publications analyzed

scientific, educational, medical and social. The foregoing allows us to speak of a technological revolution that provides an expansion in the fields of application of photogrammetry, as well as its use in many impact projects.

It can be determined that there are still limitations for the use of Photogrammetry, such as: high cost for small jobs, use of highly qualified and experienced personnel.

The use of Photogrammetry in combination with artificial intelligence tools is identified as an open line for research, which we value could be more economical, efficient and effective, to provide solutions in various sectors of engineering, medicine and projects. of construction.

References

1. Hernández-Sampieri, R., Torres, C.P.: Research Methodology, vol. 4. McGraw-Hill Interamericana México (2018)
2. Iglhaut, J., Cabo, C., Puliti, S., Piermattei, L., O'Connor, J., Rosette, J.: Structure from motion photogrammetry in forestry: a review. Curr. Forestry Rep. **5**(3), 155–168 (2019). https://doi.org/10.1007/s40725-019-00094-3
3. Rodríguez, L.P., Gómez, S.A., Buitrago, Y.: Comparative analysis between topographic surveys with total station as a direct method and the use of photogrammetry with drones as an indirect method (2021) [Online]. Available at: http://reposito-rio.uts.edu.co:8080/xmlui/handle/123456789/6740. Accessed 22 May 2021
4. Luhmann, T.: A historical review on panorama photogrammetry. Int. Arch. Photogramm. Remote Sens. Spat. Inf. Sci. **34**(5), 8 (2004)
5. Buill, F., Jordana, J.J., Andrés, M.A.: Architectural Photogrammetry. University Polytechnic of Catalonia, Iniciativa Digital Politecnica (2008) [Online]. Available at: https://upcommons.upc.edu/handle/2099.3/36829?locale-attribute=en. Accessed 21 May 2021
6. Thompson, M., et al.: Manual of Photogrammetry. American Society of Photogrammetry (1966)
7. Cunliffe, A.M., Tanski, G., Radosavljevic, B., Palmer, W.F., Sachs, T., Lantuit, H., Myers-Smith, I.H.: Rapid retreat of permafrost coastline observed with aerial drone photogrammetry. Cryosphere **13**(5), 1513–1528 (2019)
8. Condorelli, F.: Processing historical photographs and film footage with Photogrammetry and Artificial Intelligence for Cultural Heritage documentation and virtual reconstruction. ELCVIA Electron. Lett. Comput. Vis. Image Anal. **19**(2) (2020). https://doi.org/10.5565/rev/elcvia.1323

9. Mireles-Loera, O.: Artificial intelligence and its application on the Istanbul Protocol. Rev. Mex. Med. Forense **6**(1), 12–22 (2021). https://doi.org/10.25009/revmedforense.v6i1.2882
10. Hélicéo: Geomatic Innovation & Technology (2019) [Online]. Available at: http://www.heliceo.com/es/industrias/foto-grametry-by-drones. Accessed 12 May 2021
11. Barbero-García, I., Pierdicca, R., Paolanti, M., Felicetti, A., Lerma, J.L.: Combining machine learning and close-range photogrammetry for infant's head 3D measurement: a smartphone-based solution. Measurement **182** (2021). https://doi.org/10.1016/j.measurement.2021.109686
12. Gudiño, J.L., Aguilar, F.J., Hernández-Rivera, R.: Application of digital photogrammetry for the registration of dinosaur tracks from the Cerro del Pueblo Formation (late Campanian), Coahuila, Mexico. Bol. Soc. Geológica Mex. **70**(2), 307–324 (2018). https://doi.org/10.18268/bsgm2018v70n2a3
13. Gómez-Zurdo, R.S., Martín, D.G., González-Rodrigo, B., Sacristán, M.M., Marín, M.: Application of drone photogrammetry to deformational control of structures and terrain. Inf. Constr. **73**(561) (2021). https://doi.org/10.3989/ic.77867
14. Varela, I., Alexis, B.: Application of conventional topography and photogrammetry as support in forensic road surveys (2021). Accessed July 28 [Online]. Available at: http://repository.udistrital.edu.co/handle/11349/26159. Accessed 21 May 2021
15. Llajaruna, G.M.: Automation of ecchymosis dating in Peruvian medical expertise using artificial neural networks and image processing. Bachelor's thesis, Faculty of Computer and System Engineering. National University of San Marcos, Lime, Peru (2011)
16. Picon-Cabrera, I., Rodríguez-Gonzálvez, P., Toschi, I., Remondino, F., González-Aguilera, D.: Building reconstruction and urban analysis of historic centers with airborne photogrammetry. Inf. Constr. **73**(562) (2021). https://doi.org/10.3989/ic.79082
17. Lehmann, G.: Photogrammetry. Editorial Reverté, Barcelona (1975). ISBN: 978-84-7146-154-4 [Online]. Available at: https://www.reverte.com/libro/fotogrametria_82628/. Accessed 2 May 2021
18. Wolf, P.R., Dewitt, B.A., Wilkinson, B.E.: Elements of Photogrammetry with Applications in GIS (4th edn.). McGraw-Hill Education (2014). ISBN: 978-0-07-176112-3
19. Ruiz, A.M.: The application of photogrammetry (SFM) and new technologies for the improvement of documentation, dissemination and dissemination of archaeological heritage. University of Granada (2020) [Online]. Available at: http://purl.org/dc/dcmitype/Text. Accessed 5 May 2021
20. Kacprzyk, J., Zadrożny, S.: Linguistic database summaries using fuzzy logic, towards a human-consistent data mining tool. Studia i Materiały Polskiego Stowarzyszenia Zarządzania Wiedzą **20**, 10 (2009)
21. Dong, W., Gong, Q., Yuan, K.: Application of artificial intelligence-based UAV photogrammetry technology in electric power surveying and mapping engineering. In: Big Data Analytics for Cyber-Physical System in Smart City, Singapore, pp. 1138–1144 (2021). https://doi.org/10.1007/978-981-33-4572-0_163
22. Barbero-García, I., Pierdicca, R., Paolanti, M., Felicetti, A., Lerma, J.L.: Combining machine learning and close-range photogrammetry for infant's head 3D measurement: a smartphone-based solution. Measurement **182**, 109686 (2021). https://doi.org/10.1016/j.measurement.2021.109686
23. Oliver, S., Seyedzadeh, S., Rahimian, F., Dawood, N., Rodriguez, S.: Cost-effective as-built BIM modeling using 3D point-clouds and photogrammetry. Curr. Trends Civ. Struct. Eng. CTCSE **4**(5), 000599 (2020). https://doi.org/10.33552/CTCSE.2020.04.000599
24. Gómez, M.L., Furió, V.D.: Tranzyxion Virtual Prosthesis Through Photogrammetryand Augmented Reality. Unibersitat Politécnica de Valéncia, Faculty of Fine Arts of Sant Carles (2019) [Online]. Available at: https://riunet.upv.es/bitstream/handle/10251/160056/G%C3%B3mez%20-%20TRANZYXION.%20Pr%C3%B3tesis%20virtuales%20mediante%20fotogrametr%C3%ADa%20y%20realidad%20aumentada.pdf?sequence=1&isAllowed=y. Accessed 21 June 2021
25. Littell, J.H., Corcoran, J., Pillai, V.: Systematic Reviews and Meta-analysis. Oxford University Press (2008)

26. Structure from Motion (SfM): a low-cost photogrammetric technique for the characterization and monitoring of rock masses. In: 10th Simp. Nac. Geotechnical Engineer, A Coruña, Spain, vol. 1 (2016)
27. Molina, L.D.: Work Monitoring Through Digital Elevation Models Generated by Photogrammetry with Drones. Universidad de Los Andes, Department of Civil and Environmental Engineering (2019)
28. Hernández-Sampieri, R., Torres, C.P.: Research Methodology. McGraw-Hill Interam. Mexico DF, vol. 4 (2018)
29. Heipke, C., Rottensteiner, F.: Deep learning for geometric and semantic tasks in photogrammetry and remote sensing. Geo-Spat. Inf. Sci. **23**(1), 10–19 (2020). https://doi.org/10.1080/10095020.2020.1718003
30. Montes, B.A., Henríquez, C.H., Rodríguez, T.R., Zelaya, K.L.: Estimation of yield in the cultivation of sugarcane (Saccharum officinarum) from photogrammetry with unmanned aerial vehicles (UAV), pp. 67–80 (2021). https://doi.org/10.15517/rac.v45i1.45695
31. Tello-Cifuentes, L., et al.: Evaluation of damage in flexible pavement using terrestrial photogrammetry and neural networks. TecnoLógicas **24**(50), 59–71 (2021). https://doi.org/10.22430/22565337.1686
32. García, A.: Automated photogrammetry and video game engines for the dissemination of architectural heritage. Las Reales Atarazanas del Grao de València, Final Degree Project, Polytechnic University of Valencia (2021)
33. Geissler, H.J.: Bens culture is digital photogrammetry, Laguna digital/Photogrammetry in cultural heritage, Laguna. Braz. J. Dev. **7**(5) (2021). https://doi.org/10.34117/bjdv.v7i5.30029
34. Alonso, I.L., Gorostidi, J.Z., Garate, A.M., Fabra, J.A.: Portable photogrammetry in process with optical markers applied to high range dimensional metrology. DYNA **96**(1), 35–39 (2021). https://doi.org/10.6036/9401
35. Lanza, D., Vidal, L.: Photogrammetry. Notes for a Practical Introduction (2020). https://eprints.ucm.es/id/eprint/60840/. Accessed 28 Jul 2021
36. Cruz, R.O., Lorite, L.: Aerofotogeodesia. Editorial Pueblo y Educación, Cuba (1987)
37. Blachut, T., Burkhardt, R:. Historical Development of Photogrammetric Methods and Instruments. ISPRS Falls Church (1988)
38. Yongjun, Z., Zuxun, Z., Jianya, G.: Generalized photogrammetry of space-borne, airborne and terrestrial multi-source remote sensing datasets. Acta Geod. Cartogr. Sin. **50**(1) (2021). https://doi.org/10.11947/j.AGCS.2021.20200245
39. Honarmand, M., Shahriari, H.: Geological mapping using drone-based photogrammetry: an application for exploration of vein-type Cu mineralization. Minerals **11**(6) (2021). https://doi.org/10.3390/min11060585
40. Fernández, G., Vázquez L., Palacio, J.L., Peralta, A., García, A.: Geomorphometry and calculation of water erosion in different lithologies through digital photogrammetry with drones: Invest. Geogr. **96** (2018). https://doi.org/10.14350/rig.59548
41. Mancini, F., Pirotti, F.: Innovations in photogrammetry and remote sensing: modern sensors. New processing strategies and frontiers in applications. Sensors **21**(7) (2021). https://doi.org/10.3390/s21072420
42. Zhang, B.: Intelligent photogrammetry for digital elevation model production. Automatic Target Recognition XXXI, vol. 11729, p. 1172902 (2021).https://doi.org/10.1117/12.2595802
43. Rodríguez, P.: Automated digital photogrammetry in front of systems based on active 3D sensors. EGA Archit. Graphic Expr. Mag. **17**(20), 100–111 (2012). https://doi.org/10.4995/ega.2012.1408
44. Mora, G., Leandro, J.: Tranzyxion, virtual prostheses using photogrammetry and augmented reality. Project/Final Degree Project/Degree, Polytechnic University of Valencia (2021)
45. Durgan, S.D., Zhang, C., Duecaster, A., Fourney, F., Su, H.: Unmanned aircraft system photogrammetry for mapping diverse vegetation species in a heterogeneous coastal wetland. Wetlands **40**(6), 2621–2633 (2020). https://doi.org/10.1007/s13157-020-01373-7
46. Qin, R., Gruen, A.: The role of machine intelligence in photogrammetric 3D modeling—an overview and perspectives. Int. J. Digit. Earth **14**(1), 15–31 (2021). https://doi.org/10.1080/17538947.2020.1805037

47. Martínez, U., Medina, E., Utrilla, G., García, C., Bermejo, I., López, J.: Application of Artificial Intelligence to biomechanical analysis. Inst. Biomecánica IBV Univ. Politècnica València, Building 9C Camino Vera Sn, 46022 Valencia, España (2021)
48. Rubén, D.G., Gómez, S., González, B., Marchamalo, M., Martínez, R.: Application of drone photogrammetry to the control of deformations of terrain and structures. Inf. Constr. **73**(561) (2021). https://doi.org/10.3989/ic.77867

The manufacturer's authorised representative in the EU is Springer Nature Customer Service Centre GmbH, Europaplatz 3, 69115 Heidelberg, Germany. If you have any concerns regarding our products, please contact ProductSafety@springernature.com

Printed and bound by CPI Group (UK) Ltd, Croydon, CR0 4YY
22/09/2026
02220954-0003